Progress in Optical Science and Photonics

Volume 19

The purpose of the series Progress in Optical Science and Photonics is to provide a forum to disseminate the latest research findings in various areas of Optics and its applications. The intended audience are physicists, electrical and electronic engineers, applied mathematicians, biomedical engineers, and advanced graduate students.

More information about this series at https://link.springer.com/bookseries/10091

Euth Ortiz Ortega · Hamed Hosseinian ·
Ingrid Berenice Aguilar Meza ·
María José Rosales López ·
Andrea Rodríguez Vera · Samira Hosseini

Material Characterization Techniques and Applications

Springer

Euth Ortiz Ortega
School of Engineering and Sciences
Tecnologico de Monterrey
Monterrey, Mexico

Ingrid Berenice Aguilar Meza
School of Engineering and Sciences
Tecnologico de Monterrey
Monterrey, Mexico

Andrea Rodríguez Vera
School of Engineering and Sciences
Tecnologico de Monterrey
Monterrey, Mexico

Hamed Hosseinian
School of Engineering and Sciences
Tecnologico de Monterrey
Monterrey, Mexico

María José Rosales López
School of Engineering and Sciences
Tecnologico de Monterrey
Monterrey, Mexico

Samira Hosseini
School of Engineering and Sciences
Tecnologico de Monterrey
Monterrey, Mexico

ISSN 2363-5096 ISSN 2363-510X (electronic)
Progress in Optical Science and Photonics
ISBN 978-981-16-9568-1 ISBN 978-981-16-9569-8 (eBook)
https://doi.org/10.1007/978-981-16-9569-8

This Springer imprint is published by the registered company Springer Nature Singapore Pte Ltd.
The registered company address is: 152 Beach Road, #21-01/04 Gateway East, Singapore 189721, Singapore

To the true spirit of science,

And to scientists who serve beyond the notions of "I" and "mine".

Preface

Newly developed materials require a thorough characterization of bulk and surface properties for any application they may serve. The materials characterization domain involves a wide variety of characterization techniques each promoting a different level of understanding of the materials, their features, and functions. The present book offers detailed information about the commonly applied characterization techniques in material science, their brief history and origins, their mechanism of operation, the advantages and disadvantages, their applications, and a troubleshooting section for each technique that addresses the challenges researchers may face when working with these techniques, in addition to the possible causes and the recommended solutions to counteract the errors. This book lends its focus to a profound understanding of the nature of materials including analyses of morphology, topography, mass, as well as the characterization of elemental, structural, thermal, wettability, electrochemical, and chromatography properties of samples. Additionally, the book features the benefits of using characterization techniques simultaneously by covering some of the widely applied coupled analytical methods.

Monterrey, Mexico

Euth Ortiz Ortega
Hamed Hosseinian
Ingrid Berenice Aguilar Meza
María José Rosales López
Andrea Rodríguez Vera
Samira Hosseini

Acknowledgments The authors would like to acknowledge the financial and technical support of Writing Lab, Institute for the Future of Education (IFE), Tecnologico de Monterrey, Mexico. Additionally, the authors would like to express their ample gratitude for the relentless work and support of Michelle Alejandra Espinosa-Hernandez and Katia Donjuan in editing the final draft of the book.

Contents

Chapter 1
Characterization Techniques for Morphology Analysis

Euth Ortiz Ortega, Hamed Hosseinian, María José Rosales López, Andrea Rodríguez Vera, and Samira Hosseini

Abbreviations

3D-KSCs	Three-dimensional macroporous carbon
AC	Anno domini
AChE	Acetylcholinesterase
ATCl	Acetylthiocholine
AuNPs	Gold nanoparticles
CCD	Charge-coupled device
CS	Chitosan
Cu-BTC MOFs	Copper–trimesic acid metal–organic frameworks
DFM	Dark-field microscopy
EMCCD	Electron multiplying charge-coupled device
EVs	Extracellular vesicles
FITC	Fluorescein isothiocyanate
FRAP	Fluorescence recovery after photobleaching
GFP	Green fluorescent protein
GOD	Glucose oxidase
GRA	Graphene
HCl	Hydrochloric acid
iELISA	Imaging enzyme-linked immunosorbent assay
IR	Infrared
LOD	Limit of detection
LSPR	Localized surface plasmon resonance
MPBA	Mercaptophenylboronic acid
NA	Numerical aperture
ND	Neutral density
NeSy	Netherton syndrome
NS	Not specific
OPPs	Organophosphorus pesticides
PANI	Polyaniline

E. O. Ortega et al., *Material Characterization Techniques and Applications*,
Progress in Optical Science and Photonics 19,
https://doi.org/10.1007/978-981-16-9569-8_1

PCD	Primary ciliary dyskinesia
PCV2	Porcine circovirus type 2
PFP	Platelet-free plasma
PPIs	Protein-protein interactions
rGO	Reduced graphene oxide
SED	Secondary electron detector
SEM	Scanning electron microscope
SPD	Single-particle detection
STED	Stimulated emission depletion
STORM	Stochastic optical reconstruction microscopy
TEM	Transmission electron microscopy
TI	Trichorrhexis invaginata
TiO_2	Titanium dioxide
TIRFM	Total internal reflection fluorescence microscopy
UV	Ultraviolet

1.1 Optical Microscopy for Material Characterization

1.1.1 History of Optical Microscopy

Humans are curiosity-driven individuals who have always been passionate about the venture into the unknown world surrounding them. The history of microscopy began in the fifteenth century when science enthusiasts felt the necessity to observe object details that could not be seen with a naked eye, thus exploring the possibility of bringing objects closer to human sight. The word microscope derives from the Greek words "mikró" (small) and "vlépo̱" (to view), and it refers to an instrument that can be used to observe small objects. Figure 1.1 displays a timeline depicting the significant developments that have led to the evolution of microscopy techniques.

In 100 AC, Lucius Annaeus Seneca described magnification with water drops, explaining that letters seemed larger when viewing them through a glass of water. A thousand years later, Ibn al-Haytham (known in the West as Alhazen) worked on developing the principles of optics and defined the eye anatomy emphasizing the lens's role in focusing images [1, 2]. In 1624, the term microscope was coined in Italy by the first established Accademia Dei Lincei, to which Galileo Galilei, the creator of the first microscope, was affiliated. In 1631, Isaac Beeckman drew the first image of a microscope [3]. Later in 1646, Athanasius published microscopy images in the book Ars Magna Lucis et Umbrae. However, the images were of little use since they presented very poor proportions [2]. In 1665, the British scientist Robert Hooke designed a two-lens microscope to study multiple microorganisms. As a result of his finding, Hooke published the famous book Micrographia and coined the term "Cell". Around the same time, the Dutch merchant and biology enthusiast

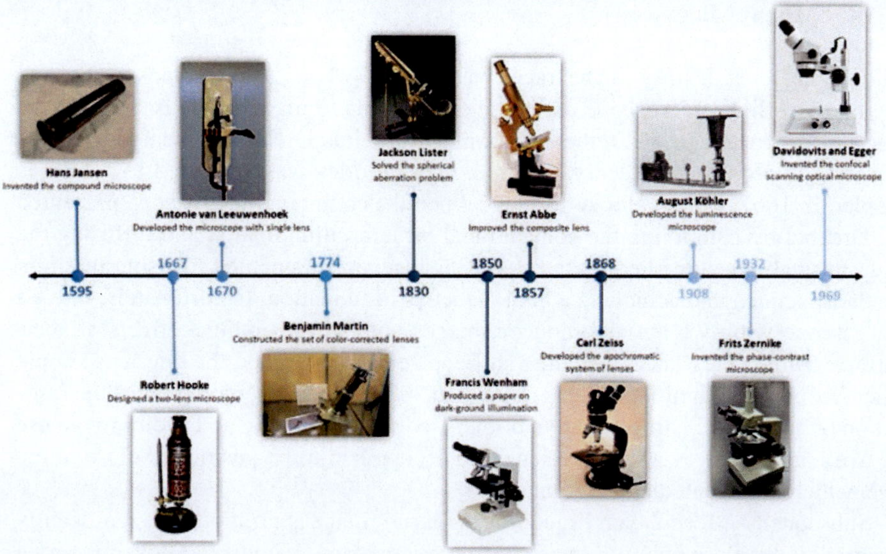

Fig. 1.1 History of optical microscopy from 1595 to 1969. Microscopes have been evolving from robust devices to sophisticated and accurate systems

Antonie van Leeuwenhoek developed a single-lens microscope to observe insects and other specimens. Throughout his life, Leeuwenhoek built over 500 microscopes and contributed significantly to microbiology, first describing bacteria and other microorganisms (Table 1.1) [4].

Although several opticians dedicated their work to constructing telescopes at the end of the sixteenth century [2], three Dutch spectacles manufacturers named Hans Jansen, Zacharias Jansen, and Hans Lippershehy took the credit for the invention of the compound microscope in 1595 [5]. In 1733, Chester Moor Hall coupled a convex glass lens with a concave glass lens to correct the chromatic aberration of telescopes, a lens failure that caused all colors to focus on the same spot thus degrading the quality of the displayed image. Later, in 1759, John Dolland developed an achromatic (non-color-distorting) lens; nevertheless, the instrument exhibited optical distortion due to the spherical aberration which Joseph Jackson Lister solved later in 1830 [1]. Fifteen years later, Benjamin Martin created a set of color-corrected lenses. In 1791, the Dutch Francois Beeldsnijder introduced two different glasses with different refractive indices to minimize chromatic aberration, which he later called achromatic microscope [4]. In the mid-nineteenth century, Joseph Jackson Lister published a work illustrating the complete design of microscope objectives, providing different assemblies to focus light waves [1]. In 1857, Carl Zeiss, Ernst Abbe, and Otto Schott improved the composite lens of the microscope, thereby developing a high-performance stand-type microscope [5]. Carl Zeiss continued his work, and in 1868, he created an apochromatic system of lenses characterized by superior color correction than achromatic lenses. He published a study on lens theory in 1873 [4].

1.1.1.1 Light Microscopy

Light microscopy history can be traced back to 1595 when Zacharias Jansen invented the two-lens light microscope using the second lens to magnify the image that the first has produced. In 1620, the astronomer Cornelius Drebbel presented a microscope of two converging lenses, even though the idea was attributed to Johannes Kepler. In 1667, Robert Hooke further refined the compound microscope presented by Drebbel by combining the conventional oil lamp illumination and diffusing the light through a water-filled glass flask. This mechanism enabled focusing the light on the specimen and achieving a homogeneous illumination. Unfortunately, using a microscope with two lenses produced images with severe additive effects of aberrations. Three years later, Antonie van Leeuwenhoek invented the one-lens simple microscope and magnified objects more than 200×, thus overcoming twice the resolution by the best existing compound microscopes of that time and avoiding the use of two lenses. However, the one-lens microscope had to be positioned close to the eye, which represented a disadvantage [6].

Subsequently, Joseph von Fraunhofer reduced image aberrations by significantly improving the glass quality used in optical microscopy. Moreover, Giovanni Amici developed water and oil immersion objectives [5] in the mid-nineteenth century to augment the resolution of the microscopy image by using a solution with a greater refractive index than air. In 1873, the most significant advance in microscope optics was achieved when Carl Zeiss and Ernst Abbe added a condenser to the equipment to vastly reduce chromatic and spherical aberration, allowing improved resolution and higher resolution magnification [7]. Finally, in 1893, August Koehler established the Koehler illumination to standardize microscope illumination and, to date, is considered to be a generally accepted illumination procedure [6].

1.1.1.2 Dark-Field Microscopy

Although microscopes and telescopes were invented in the late sixteenth century, such devices were commonly considered novelties instead of scientific instruments due to low-quality produced images and aberrations. Illumination played a vital role in resolving such optical difficulties; thus, additional components were implemented. From Descartes's (1637) to those recommended by David Brewster (1831), condensers served to illuminate the studied sample uniformly. Furthermore, Humphry Davy's arc-light and Albert T. Thompson's right-angled lamp were necessary discoveries that improved the dark-field microscope. Among a number of scientists who have contributed to the development of such instruments, Joseph Jackson Lister, J. B. Reade, Francis Wenham, and Edmund Wheeler have pioneered the use of the dark-field microscopy (DFM) in their research [8]. In the early 1950s, one of the most common devices of the dark-ground illumination technique was the prism illuminator, first implemented in the well-known Amici Lenticular Illuminator. At the same time, Shadbolt and Wenham jointly developed the Wenham–Shadbolt Illuminator, used for many decades to come [9].

1.1.1.3 Fluorescence Microscopy

The expression of color in light was first observed in 1845, as Sir Fredrick W. Herschel noticed that a colorless and transparent solution exhibited a blue light when illuminated under specific incidences of sunlight. In 1852, Sir George Stokes named the phenomenon fluorescence which also made the mineral fluorspar emit red light when displayed under ultraviolet (UV) light. Almost fifty years later, physicist Siegfried Czapski and the microscopist August Köhler visited the world-famous optician Carl Zeiss and developed the modern light microscope. Simultaneously, Czapski and Moritz von Rohr developed an ultraviolet microscope, which led to the subsequent discovery of grid lighting. Additionally, a suggestion by Köhler led to the development of parfocal lenses to keep the sample in focus when an objective is changed [10]. In 1904, Köhler observed that structures under the microscope exhibited a luminous phenomenon when the specimen was irradiated with UV light, and subsequently, he developed the first luminescence microscope in 1908 [11].

Once the technique became popular, fluorescent microscopy became widely investigated and used in science. Since their first employment in the 1930s, fluorochromes, responsible for molecules' fluorescence, have been extensively used to dye tissue constituents and pathogens. The high specificity of these applications accelerated the evolution of fluorescence microscopy [12, 13]. An essential step in the evolution of fluorescence was the Green Fluorescent Protein (GFP) discovery. Osamu Shimomura isolated the GFP from the jellyfish *Aequorea victoria*, first describing the protein in 1961 and later receiving the Nobel Prize in Chemistry. Shimomura shared the Nobel Prize with Douglas Prasher and Martin Chalfie, who developed the extracted GFP and its expression using it as a marker for other proteins in the 1990s. In order to advance in the field, several fluorescence techniques such as the Fluorescence Recovery After Photobleaching (FRAP) and the Total Internal Reflection Fluorescence Microscopy (TIRFM) were developed and improved in the following years [6]. The contemporary fluorescence microscope executes sophisticated analysis by coupling high-performance optical components with computerized instrument control [12].

1.1.2 Mechanism of Operation of Optical Microscopy

1.1.2.1 Mechanism of Operation of Light Microscopy

One of the first techniques used to examine the morphology of a surface was optical microscopy, also known as light microscopy. The equipment usually has a single eyepiece fitted with a camera for imaging. Conventional optical microscopy has a resolution limited to a submicron scale and uses the wavelength of visible light (400–700 nm). There exist two types of optical microscopy depending on the type of light interaction:

1. Transmission: a beam of light passes through the sample. Examples are the polarizing and petrographic microscopes, which are usually used to analyze fine powders or thin slices (transparent) (Fig. 1.2a).

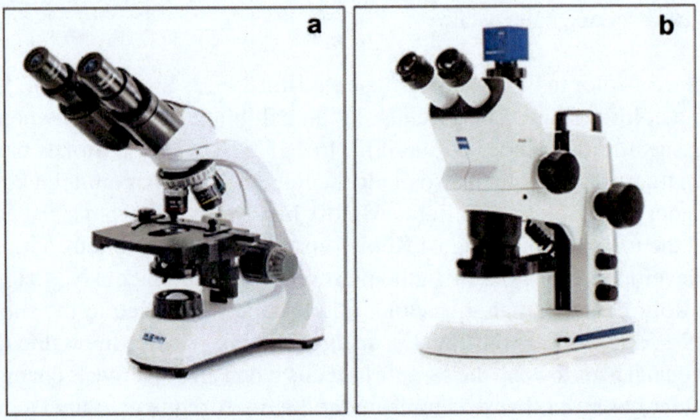

Fig. 1.2 Optical microscopes. **a** Transmitted light microscopy. **b** Reflected light microscopy

2. Reflection: a beam of light reflected by the sample surface. An example is the reflected light microscope, principally used to analyze opaque surface materials (Fig. 1.2b) [14].

An object observed by the human eye is considered a magnification of 1×, and the field of view is the length and width of the scene captured by the human eye. When using a simple lens configuration, a smaller portion of the field of view can be mapped, resulting in the object appearing larger which is referred to as magnification. Magnification is the enlargement of the appearance of an object. In contrast, a negative magnification or demagnification indicates a larger field of view mapped onto the eye. For example, Fig. 1.3a represents an object seen by the naked eye which has a magnification of 1×. In another word, a magnification power of 1× indicates 100% of the perceived size of the object viewed through the lens, as viewed with the naked

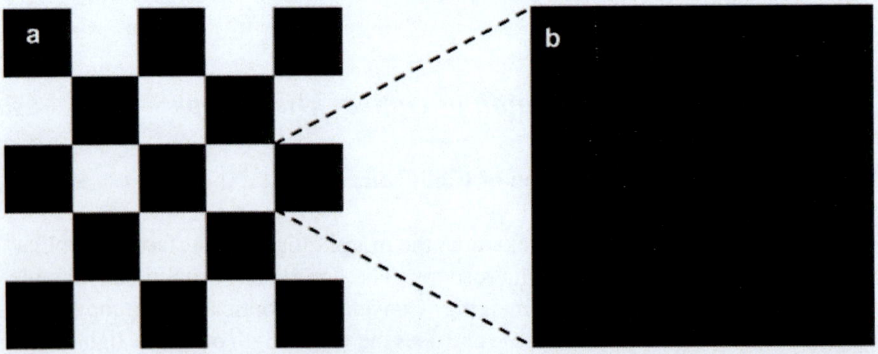

Fig. 1.3 Another way to look at magnification is the reduction of the size of the field of view projected onto an image plane. **a** Magnification 1×. **b** Magnification 5×

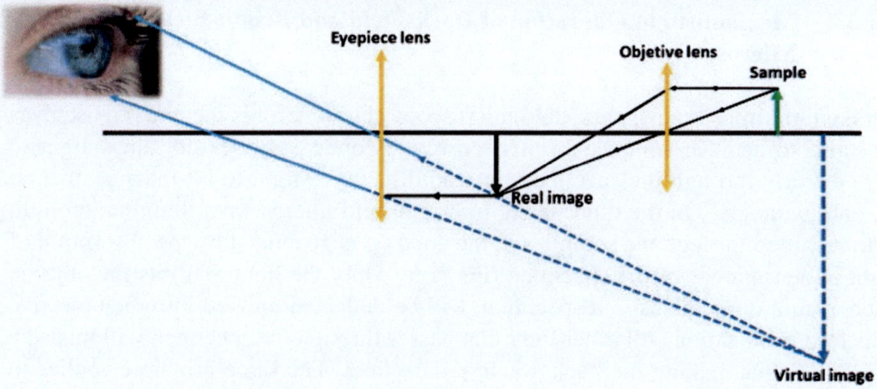

Fig. 1.4 Optical principle of microscope imaging

eye [15]. In Fig. 1.3b, only one-fifth of the original object's width and height is projected onto the eye, thus corresponding to a magnification of 5×.

An ordinary optical microscope includes two lenses: (i) the objective lens, and (ii) the eyepiece lens (Fig. 1.4). The objective lens produces an image that serves as the object for the eyepiece lens, which creates the virtual image observed by the eye. The final image that the viewer sees is an inverted virtual image that is magnified according to the original object's magnification. The modern optical microscope relies on several combined lenses in the eyepiece and the objective lenses to reduce all possible distortions and produce a better-quality image. However, objective lenses are usually considered single lenses with a specific focal length and numerical aperture (NA). As a supplementary kit, it is also possible to replace the eyepiece lens with a camera to retrieve the image and present it on a computer or large screen [16].

Even though the eyepiece can magnify the image, it is not possible to increase the resolution of the microscopy images. As a consequence, the quality and detail of an object's image, known as the spatial resolution of an optical microscope, is mainly determined by the objective lens and is given by the Rayleigh equation (Eq. 1.1), as follows [17]:

$$\Delta r_0 = \frac{0.62\lambda}{n \sin \sin \alpha} \tag{1.1}$$

where Δ is the minimum resolvable distance, λ is the wavelength of the light source, n is the refractive index among the lens and the object, α is the half-angle aperture minus the half inclination of the lens to the objective points, and n is the NA of the objective lens.

1.1.2.2 Mechanism of Operation of Dark-Field and Bright-Field Microscopy

Unlike light microscopy, dark-field microscopy (DFM) avoids the need to use dyes or stains to achieve contrast. In this sense, the condenser does not allow light to pass directly through the specimen, instead directs the light to hit the specimen at an oblique angle. On the other hand, in bright-field microscopy, illumination light is transmitted through the sample and the contrast is generated by the absorption of light in dense areas of the specimen (Fig. 1.5). Only the light that hits the objects, such as microorganisms in a specimen, will be deflected upward into the objective lens for visualization. All other light that passes through the specimen will miss the objective, thus making the background a dark field. The same principle applies to dark-field microscopy [18].

In Fig. 1.5, it is worth noticing that the diameter of the dark-field stop is equal to the diameter of the aperture opening in the bright-field. In the bright-field structure, the light that enters the objective lens must be inside the cone of light; otherwise, the light would be bent too far, and would not enter the objective lens. This fact is crucial for the dark-field system to block the center of cone light with the dark-field stop and let the outside light pass inside via opening the condenser aperture. Subsequently, the light that passes inside the condenser is refracted at an oblique angle and would generally pass outside the objective lens in a hollow cone of light. For instance, a specimen placed on the stage would be located at the apex of the hollow cone of light which will receive the light. Therefore, the light will scatter into different directions, and some may reflect upward into the objective lens. This process results in a bright image of the sample against a dark background [19].

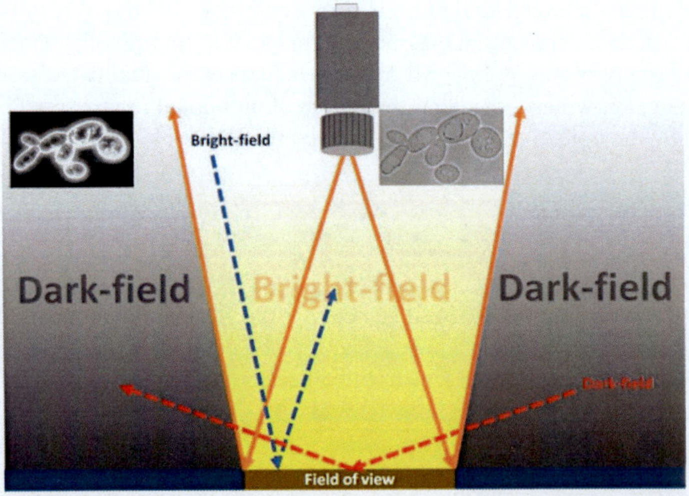

Fig. 1.5 Cells are visualized by different types of light microscopies, such as bright-field and dark-field microscopy. Cells average 8–10 μm wide

1.1.2.3 Mechanism of Operation of Fluorescence Microscopy

Photoluminescence is the ability of fluorescent molecules, named fluorophores or fluorochromes, to absorb and emit distinct light particles called photons. The Jablonski energy diagram illustrates the relationship between a fluorophore's light absorption and emission (Fig. 1.6a). The absorption of photons (Fig. 1.6a, blue arrows) describes the activity when a fluorophore is excited from its ground state (S_0) to higher energy states (S_1 or S_2) in a timescale of femtoseconds. Shortly after, within picoseconds, electrons in the excited state undergo nonradiative vibrational relaxation (i.e., within an electronic energy state) and internal conversion (i.e., among neighboring electronic energy states, Fig. 1.6a, yellow arrows). The electrons return to their ground state at a nanosecond scale, and a photosensitive surface detects the radiated fluorescence signals (Fig. 1.6a, green arrow).

Phosphorescence signals (Fig. 1.6a, red arrow) are emitted when excited-state electrons return to their ground state from an electronic triplet energy state (T_1). The emission of photons occurs at a longer wavelength than the initial excitation wavelength, resulting in the Stokes shift, a phenomenon of difference in the wavelength of the maximum of the first absorption band and the maximum of the fluorescence emission (Fig. 1.6b). Fluorophores may permanently lose their ability to fluoresce due to

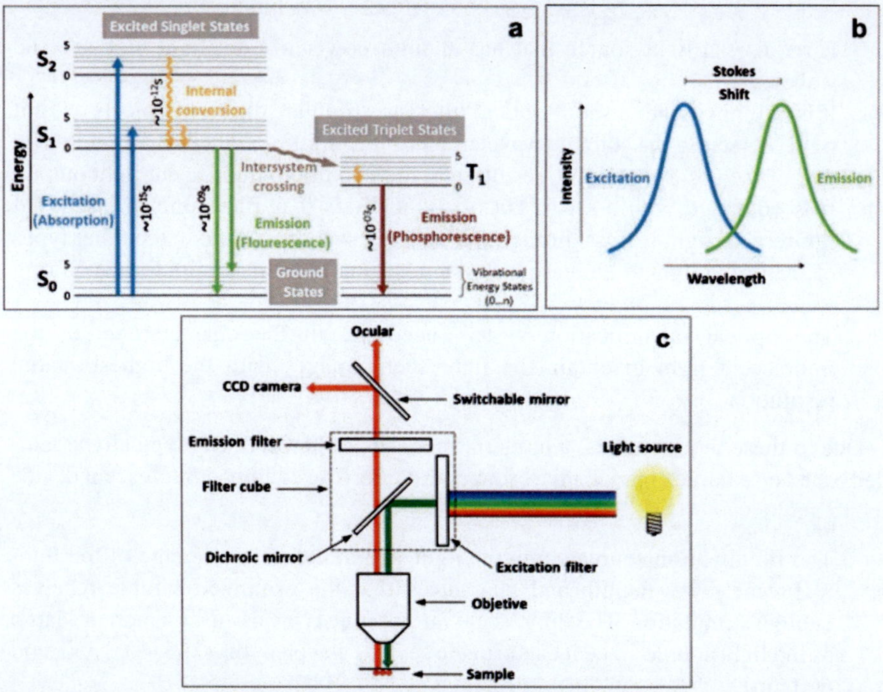

Fig. 1.6 **a** Jablonski energy diagram. Horizontal lines indicate electronic states, whereby thick lines depict the respective vibrational ground states. **b** The Stokes shift, a phenomenon that describes the shift among excitation and emission wavelength. **c** Working mechanism of fluorescence microscopes

photon-induced chemical damage and covalent modification occurring through interaction with other molecules in the long-lived excited triplet state. Photobleaching is the name of the above-described photochemical destruction of a fluorophore phenomenon, and it depends on the molecular structure of the fluorophore and the cellular environment. It can become a limiting factor when performing fluorescence microscopy, especially for live imaging studies.

The fluorescence microscope has many unique image generation and application characteristics in the biomedical imaging research fields, further discussed. The fluorescence microscope [20] uses short-wave light to illuminate the examined specimen and express fluorescence from a specific fluorochrome. The emitted fluorescent light allows the human eye to observe the shape and location of the target objects of the studied sample. One of the most significant differences between fluorescence and conventional optical microscopes is that the fluorescence microscope uses a high-voltage mercury lamp to provide all band light illumination, using a method in which the light is projected to the specimen. Due to the weakly reflected fluorescent light, the microscope must have a large numerical aperture (NA) to allow the objects to be observed. As a result, the NA of the fluorescent microscopes is more sizeable than conventional optical microscopes. The fluorescent microscopes also have a particular filtering system, including the illumination and cut-off filters assembly. Specifically, the fluorescent microscope has the following unique characteristics and requirements:

(1) It requires a light source that has enough power to emit light of a specific wavelength.
(2) It is equipped with a set of optical filters to stimulate different objects to emit light. Selecting the suitable wavelength of the emitting light can coincide with that of the absorption light, resulting in the maximum fluorescent light output.
(3) It is equipped with a set of cut-off light filters that allow only the selected fluorescent light to pass through the imaging system and block the other types of scattering light to increase the system's signal-to-noise ratio, thus acquiring weak fluorescent images.
(4) The optical magnification system needs to fit the characteristic of the fluorescent light to obtain the fluorescent images with the highest spatial resolution.

Due to these requirements, a modern fluorescent microscope is typically assembled based on a duplex optical microscope structure. The equipped fluorescent device mainly includes:

(1) The illumination source emits the light within the specific range of the fluorochrome's wavelength and guarantees that the examined sample receives enough stimulation. The fluorescent microscope typically uses a mercury lamp as the light source since it can provide several wavelengths (365, 405, 550, and 600 nm).
(2) An exciting light path serves as a fluorescent lighting device that contains a group of condenser lenses and light focusing adjustment devices. The light

path is installed with adjustable light and field apertures, neutral density (ND) filters, light splitter, and processing lens are installed in the light path.

(3) Light filters are a set of optical filtering components. Each has a certain wavelength width that selectively allows emitted light to pass through the optical system of the microscope [21].

In order to observe the results through a fluorescence microscope, the studied sample must have been marked with a fluorophore previously. An excitation filter allows only lower wavelengths to pass through and removes all non-specific wavelengths since the mercury vapor arc lamp generates several unused wavelengths. The filtered light is reflected by a dichroic filter and falls on the sample (i.e., fluorophore-labeled). The fluorochrome absorbs the lower wavelength with higher energy and emits a longer wavelength with the lower energy that passes through an emission filter. The emission filter suppresses any residual excitation light and passes the desired longer emission wavelengths to the detector; hence, the microscope forms glowing images of the fluorochrome-labeled microorganisms against a dark background (Fig. 1.6c).

1.1.3 Advantages and Disadvantages of Optical Microscopy

Among the high-quality available techniques to explore and study the specimens' hidden-to-the-eye details, optical microscopy outstands for being a specific, rapid, non-invasive, and simple technique [22–24]. The optical microscope is present in the most basic laboratory infrastructure due to its low cost and high functionality. It allows physical features such as color, shape, and size to be monitored and determined in samples. In addition, the portability of optical microscopes allows bringing the equipment to several places as field tests, industries, and laboratories, while not leaving behind essential accessories such as fine tweezers, open face filter samplers, illuminators, and others. One of the advantages that particularly stand out is the efficiency of the apparatus. For instance, 50 μm specimens can be analyzed with satisfactory accuracy using a low magnification (7×). On the other hand, specimens with a size around 1 or 2 μm can be examined using magnifications near to 1000×, and characteristics as the shape, color, and size can be determined [22, 24, 25].

When using high-resolution techniques such as transmission electron microscopy (TEM) and scanning electron microscopy (SEM), samples are often mandatory to be conductive materials compatible with a vacuum environment. Nevertheless, specimen manipulation is easier in optical microscopy since the vacuum is avoided and the sample is no longer needed to dry or coat with conductive substances [23]. To analyze organic and biological units, fluorescence microscopy allows the study of near to 50 molecules per cubic micron of a specimen. In addition, supported by digital imaging, it is possible to acquire quantitative data efficiently and quickly in real time. Inorganic materials can also be monitored and explored [26]. Furthermore, the development of advanced optical microscopy techniques such as the stimulated emission depletion (STED) microscopy and the stochastic optical reconstruction microscopy (STORM) succeeded in overcoming the original problem attributed to the diffraction resolution limit. Researchers who use such techniques can now observe specimens beyond the surface, reduce the resolution in the order of nanometers (30–60 nm), create multicolor imaging, and study molecular interactions [27].

Although all samples can be examined in detail by optical microscopy, the resolution and magnification limit the operation in this microscopy technique. One of the principal disadvantages of optical microscopes is the spatial resolution attributed to the employed wavelength of the light source (theoretical resolution limit of 200–300 nm) [26–28]. Such limitation is produced by the interaction between the light and the glass lenses. Consequently, the above results in a low-resolution distortion or a blurred image acquired from the specimen. Therefore, the capability to observe detailed and comprehensive images is limited during analysis [22, 23, 25]. The optical microscope can reach a maximum magnification of 1500×, which is attractive for several practical applications (counting particles, single-molecule detection, diagnosing diseases, analyzing live cells, among others). However, it is considerably low compared to electron microscopy (greater than 16,000×). In the particular case of fluorescence microscopy, the fluorescent probes present a limited contrast in some cases; hence, not all molecules can remain in a dark state when illuminated [24, 26, 29]. Furthermore, the incorporation of fluorescent probes or dyes and prolonged time under light can potentially impact the properties of the specimen, sometimes leading to photobleaching or decreased fluorescence intensity. For this reason, selecting the optimal fluorescent dye is critical [25]. Table 1.1 summarizes the advantages and disadvantages of optical microscopy.

Table 1.1 Advantages and disadvantages of optical microscopy

Advantages	Disadvantages
Rapid, simple, relatively low cost, non-invasive, specific, and flexible technique [22–26]	Obtained image can become blurry since the diffraction of the light source (theoretical resolution of 200–300 nm) may limit the spatial resolution [22–24, 26–28]
Provides overall data of physical features such as color, homogeneity, shape, degree of aggregation, and size [22, 25]	Low magnification (range 500×–1500×) [24, 26, 29]
Usage of multiple probes for a sample [25]	Inadequate to produce conclusive observations and comprehensive information of a specimen when compared with the details provided by other techniques [25]
Data on membrane components' structure and dynamics can be obtained [25]	
STED microscopy and STORM allow observing beyond the surface of a specimen [27]	Photobleaching and loss of fluorescence intensity due to prolonged exposure to light of a specific wavelength [25]
Multicolor imaging can be reached with Fluorescence Microscopy [27]	The properties of the specimen can be affected by the incorporation of probes and dyes [25]
Obscured subcellular images can be processed [27]	

STED stimulated emission depletion; *STORM* stochastic optical reconstruction microscopy

1.1.4 Applications of Optical Microscopy

1.1.4.1 Applications of Light Microscopy

Optical microscopy allows imaging of the sample under many light conditions [30]. Single-molecule detection, thin-film analysis, Langmuir–Blodgett films examination, biological sciences research, protein localization, and chromosome mapping are only some examples of applications of such techniques [23]. For biological samples, details such as shape and color can be distinguished. For instance, liver (Fig. 1.7a) and *Chlorella* sp. cells (Fig. 1.7b) are imaged by a normal optical microscopy setup as shown below. Additionally, physical processes including cell division can be recorded in real time (Fig. 1.7c).

Optical microscopy is also valuable for biosensing applications and for the development of diagnostic methods. Taking Netherton Syndrome (NeSy) as an example, it is an autosomal recessive congenital keratinizing disorder characterized by hair shaft deficiency, atopic diathesis, and the triad of congenital ichthyosis. The clinical diagnosis commonly relates to the appearance of Trichorrhexis Invaginata (TI). Nevertheless, TI frequently improves with age. Utsumi et al. developed a method to diagnose NeSy using light and polarized microscopy. The authors discovered that

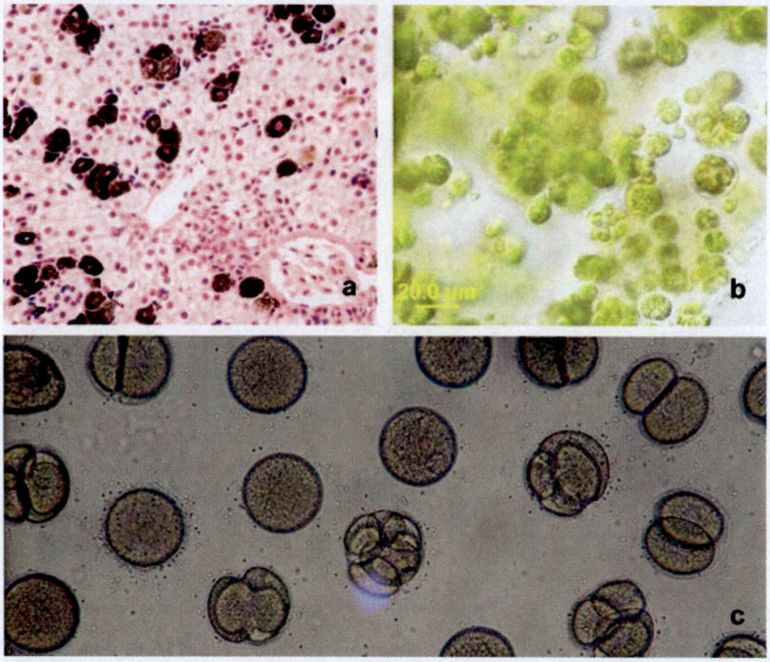

Fig. 1.7 Images obtained by light microscopy. **a** Liver cells. **b** *Chlorella* sp. cells. **c** Isolated cells from sea urchin

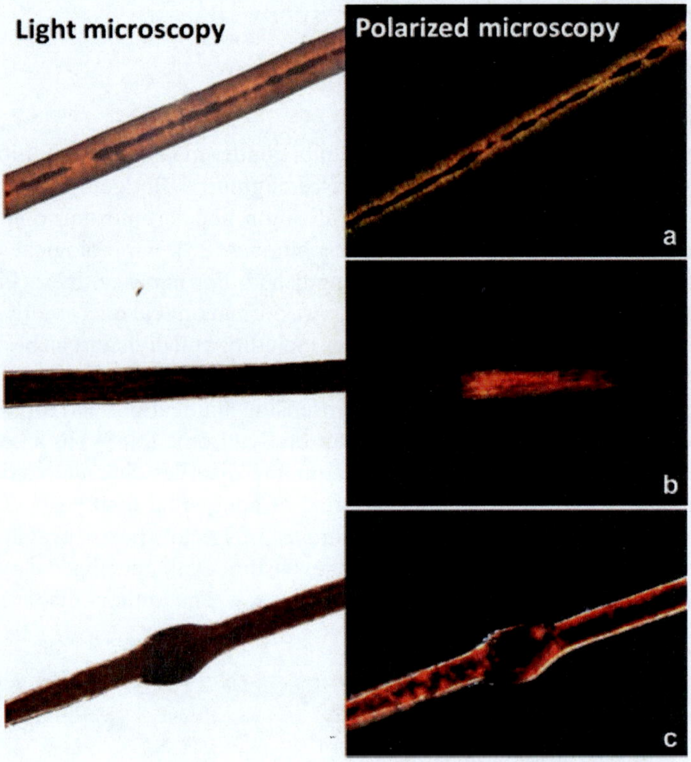

Fig. 1.8 Light and polarized microscopy in the diagnosis of Netherton Syndrome (NeSy). **a** Control, **b** hair strand from diagnosed NeSy patient and without Trichorrhexis Invaginata (TI), **c** hair strand from diagnosed NeSy patient and with TI. Reproduced (or adapted) with permission [31], ©2020, Elsevier

the hair of NeSy patients forms a band-like pattern under polarized light. This observation provided an easy and frequent method for diagnosing NeSy. The hair of an individual without NeSy presents a uniform light pattern (Fig. 1.8a). Meanwhile, a patient with NeSy presented a band-like pattern, mostly purple to dark green, in the hair (Fig. 1.8b). The hair of the same patient also presented a band-like pattern when having TI (Fig. 1.8c) [31].

1.1.4.2 Applications of Dark-Field Microscopy

Dark-field microscopy (DFM) is particularly helpful for unstained specimens [26]. In this technique, an inclined light beam illuminates the sample, and its features can be seen in a dark background as the microscope's detectors collect only the scattered light (Fig. 1.9) [32].

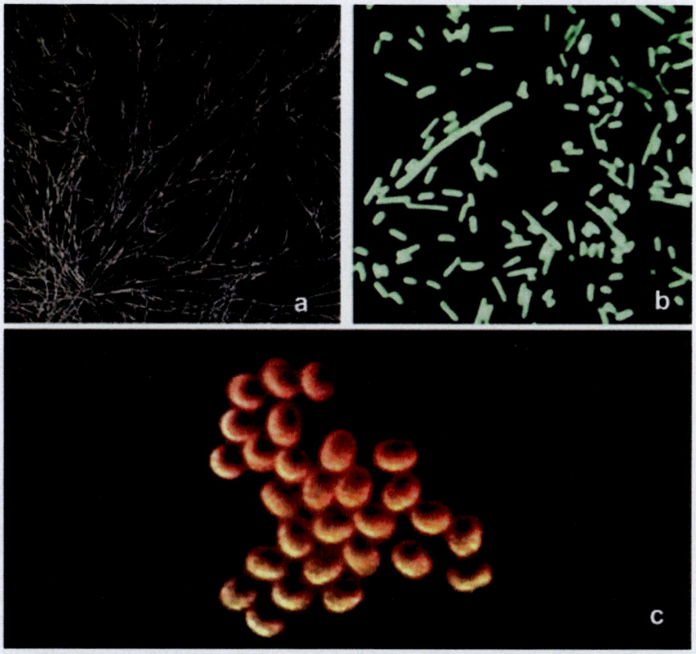

Fig. 1.9 Different samples are observed by dark-field microscopy. **a** Fibrous caffeine crystals. **b** *Shigella dysenteriae*. **c** Pollen grains. The images are obtained from open source without an access limit

Conventional assays to diagnose diseases can include methods based on DFM. Qi et al. developed a method to measure galactose concentration levels in the human body to diagnose Galactosemia at early stages. The quantification of galactose was based on the single-particle detection (SPD) sensing method, coupled with DFM and the localized surface plasmon resonance (LSPR) sensitivity. For this proposal, gold nanoparticles (AuNPs) were functionalized with mercaptophenylboronic acid (MPBA) as a signal probe since the galactose's cis-diol groups could react with MPBA. The complex Galactose-MPBA-AuNPs generated a plasmon oscillation resulting in a color change from green to yellow observed by DFM (Fig. 1.10a). Figure 1.10b exhibits the scattering color change of MPBA-AuNPs as a result of galactose presence at different times. The results demonstrated the MPBA-AuNPs reaction in a solution where galactose is present. Some green spots gradually turned yellow indicating the capability of AuNPs to absorb galactose. The limit of detection (LOD) could be calculated by aggregating different galactose concentrations and recording the DFM images. MPBA-AuNPs were added to Tris–hydrochloric acid (HCl) buffer solution to obtain a control image where no scattering color changes (the green color was maintained) and no particle aggregation occurred. Subsequently, galactose was gradually incorporated (from 1 to 75 nM) resulting in yellow particles in DFM images. The fraction of yellow particles increased with time (Fig. 1.10c).

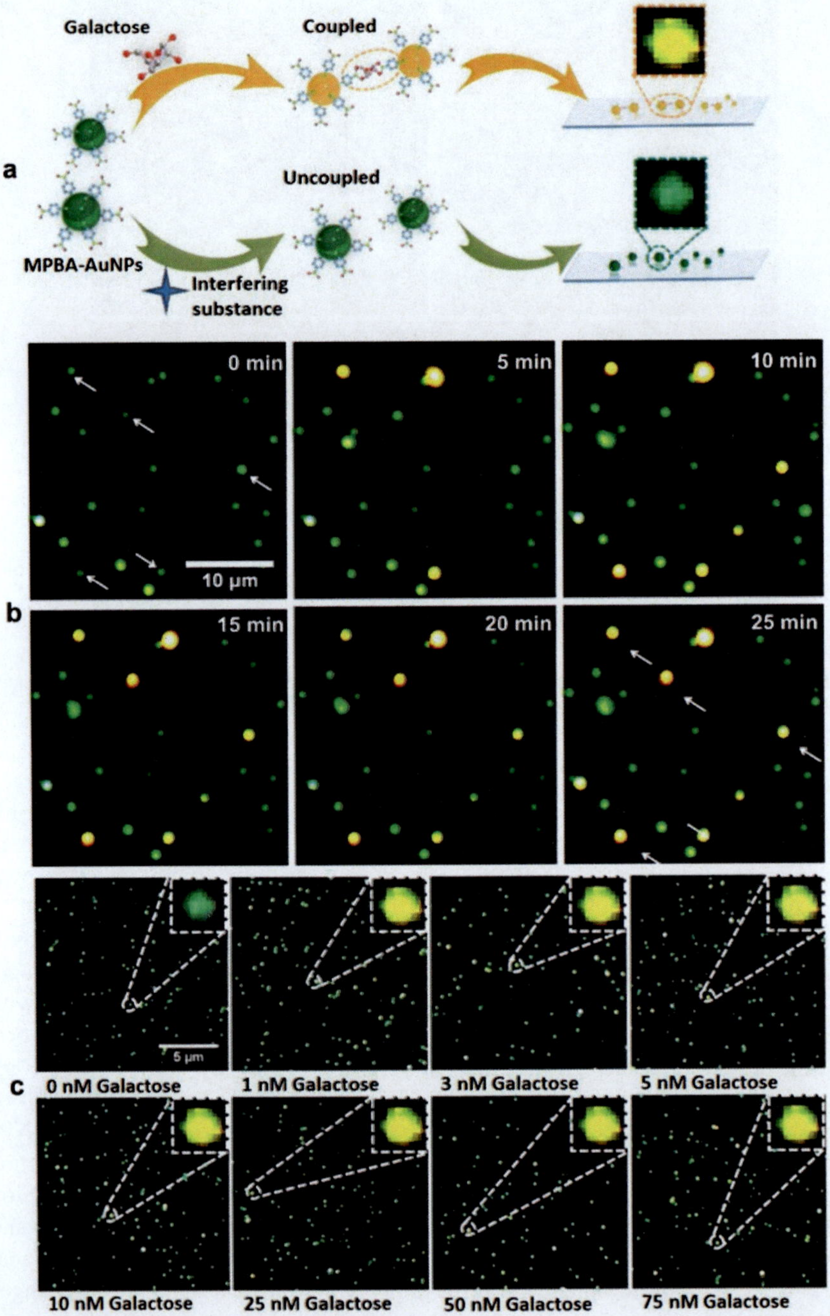

Fig. 1.10 Quantification of galactose using dark-field microscopy (DFM). **a** The schematic diagram of a single nanoparticle undergoing DFM imaging. DFM images of the complex MPBA-AuNPs in the presence of galactose at different **b** times and **c** concentrations. Reproduced (or adapted) with permission [33], ©2020, Elsevier

The data analysis demonstrated a lower LOD (0.83 nM) than that presented by electrochemical sensing, fluorescence, and capillary electrophoresis methods [33].

1.1.4.3 Applications of Fluorescence Microscopy

Fluorescent molecular probes can analyze biological systems with a satisfactory spatial and temporal resolution, advantageous for biochemical, biological, and medical research [34]. For instance, the observation and analysis of protein organization and cellular structural changes are achieved by staining those biochemical structures with different fluorophores (Fig. 1.11). Other applications include detecting intracellular protein–protein interactions (PPIs) [35], visibility of components and structures on polymer samples [36], chromophore-target molecule interaction, molecular organization [37], and live-cell imaging [38]. Additionally, a greater variety of rapid and sensitive detection systems can be developed based on the chosen fluorescence technique [39].

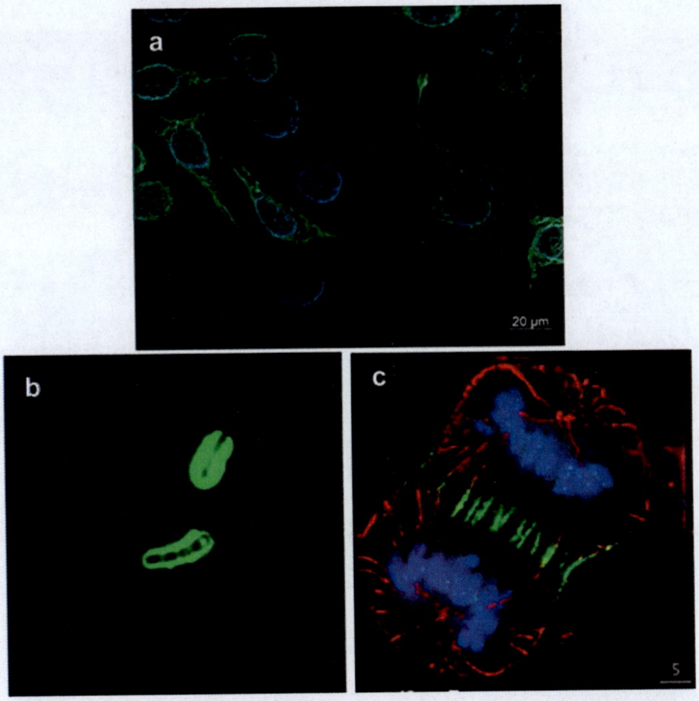

Fig. 1.11 Images obtained by fluorescence microscopy. **a** Human-derived cells. **b** *Bacillus anthracis* antibody subjected to a direct immunofluorescence test. **c** Division of a human cancer cell

Fluorescence microscopy can be implemented in the quantitative detection of Porcine Circovirus Type 2 (PCV2) Cap protein using Imaging Enzyme-Linked Immunosorbent Assay (iELISA). Figure 1.12a shows a step-by-step sequence of the iELISA system principle that results in a technique with a total duration of 1-h, excellent protein selectivity (LOD of 7 ng/mL), and anti-interference capability. The protein of interest, PCV2 Cap, was detected by a secondary antibody labeled with fluorescein isothiocyanate (FITC). A TIRFM system adapted with a micro-objective (100×, NA = 1.49), and an electron multiplying charge-coupled device (EMCCD) camera was used to obtain the fluorescent images of the PCV2 Cap protein. Additionally, the time exposure of the sample to the EMCCD camera was 100 ms, and the device was set to obtain a gain of 300 compared to input fluorescence signals.

The effectiveness of the detection method of PCV2 Cap protein via single-molecule iELISA was tested by adding several protein concentrations and recording the corresponding fluorescent images (Fig. 1.12b). The thresholding method provided a number of fluorescent signals after data analysis (Fig. 1.12c). These values extended with the PCV2 Cap protein concentration from 150 ± 17 to 494 ± 36 for a concentration of 2.8 and 56 ng mL^{-1}, respectively. A t-test provided a LOD of 250 pM; hence, the authors concluded that single-molecule iELISA presents exceptionally high sensitivity to PCV2 Cap protein [40].

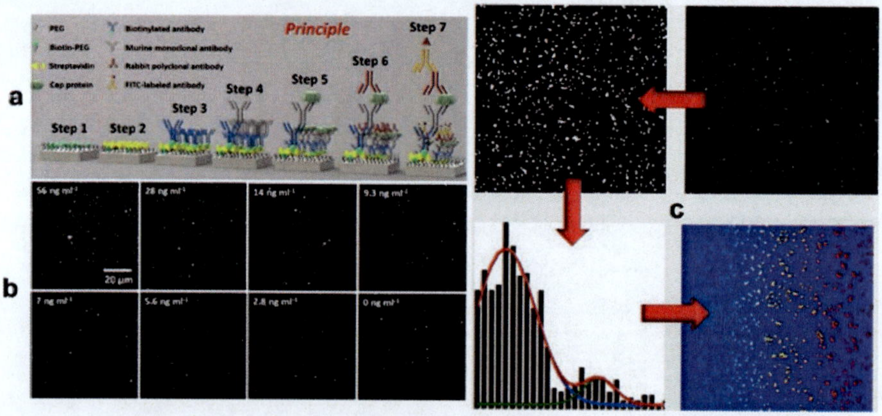

Fig. 1.12 **a** Scheme of a step-by-step detection using single-molecule iELISA. **b** The fluorescent images at distinct concentrations of PCV2 Cap protein. **c** The thresholding method. Reproduced (or adapted) with permission [40], ©2019, Elsevier

1.2 Scanning Electron Microscopy for Material Characterization

1.2.1 History of Scanning Electron Microscopy

The development of electron optics led to the emergence of electron microscopy (Fig. 1.13). In 1926, Hans Busch analyzed the trajectory of charged particles when subjected to axially symmetric electric and magnetic fields and settled the ground of geometrical electron optics by demonstrating that fields act as particle lenses [41]. Around the same time, the French physicist Louis de Broglie studied the corpuscular theory of light, which states that light is composed of small particles called corpuscles for which he introduced the concept of corpuscle waves. However, the establishment of the electron optics field occurred when wavelengths were associated with charged particles [42]. The idea of an electron microscope was, therefore, formed as a continuation of these two discoveries in electron optics.

In Berlin 1931, Ernst Ruska's research group eagerly worked on the development of electron microscopy until evidence showed that wavelengths would limit the resolution even using electrons. Subsequently, de Broglie revealed that electron wavelengths were smaller than light wavelengths. In other words, the de Broglie equation implied that electron microscopes could render a better resolution than light instruments, and the Ruska team decided to continue the project.

In 1932, Max Knoll and Ruska attempted to evaluate the resolution limit of the electron microscope by assuming that the resolution limit equation for light microscopes was applicable for material waves. The electron's wavelength replaced the light wavelength at an accelerating voltage of 75 kV in the equation. Knoll and Ruska calculated a theoretical limit of 0.22 nm for this wavelength. Although such value was experimentally replicated until 40 years later, the previous calculations

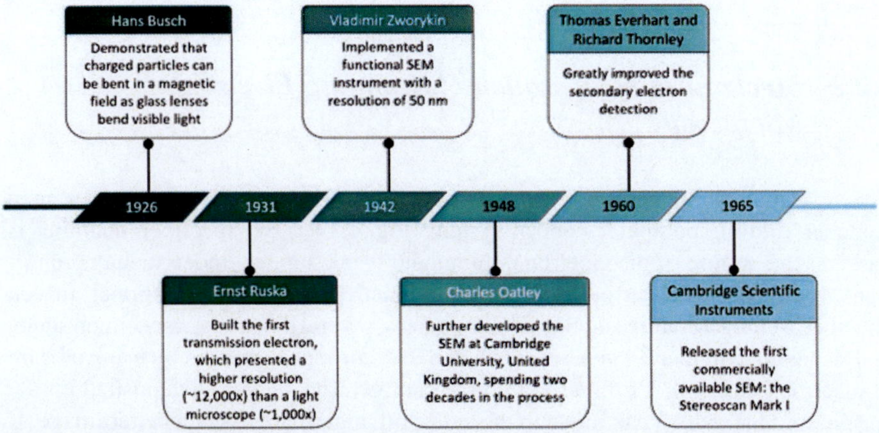

Fig. 1.13 History of SEM from 1926 to 1965

demonstrated that it was theoretically possible to overcome the resolution of light microscopy at high magnifications. Conversely, investigators faced many technical limitations when taking these parameters into practice. For example, in 1935, Knoll obtained a resolution limit of approximately 100 mm due to the absence of magnification lenses to produce a fine probe. A colossal advance took place in 1938 when Manfred von Ardenne clearly expressed the used-to-date fundamental theoretical principles regarding the scanning microscope. Researchers oriented the development of scanning microscopy toward studying sample surfaces since it was challenging to compete with the resolution that the transmission electron microscopy (TEM) accomplished for thin samples [43].

Vladimir Zworykin first described and developed SEM in 1942. He implemented an electron multiplier tube as a secondary-electron emission current preamplifier, and although the achieved resolution was 50 nm, he demonstrated that secondary electrons provided topographic contrast by diverting the collector positively relative to the specimen. Even with this result, the TEM community refused to find meaningful functions for equipment with a lower resolution than TEM. Fortunately, the SEM's ability to acquire three-dimensional information from the surfaces of the specimens was rapidly recognized [44].

Charles Oatley built a SEM in 1948 while continuing Vladimir Zworykin's work. Following this development, signal processing was proposed as a means to improve micrographs, and later nonlinear signal amplification was introduced to correct cylindrical imperfections in the SEM. In 1960, Thomas Everhart and Richard Thornley significantly enhanced secondary-electron detection using an electron-collecting detector based upon a positively biased grid, a scintillator to turn the electrons into light, and a light pipe [44]. In 1963, Roger Fabian W. Pease and William C. Nixon integrated all refinements to create an instrument with an Everhart–Thornley detector (ETD) and three magnetic lenses. They called the instrument SEM V. The 1965 Mark I Stereoscan by Cambridge Scientific Instruments was the first SEM commercially available [45]. Modern SEM is not remotely different from this first instrument.

1.2.2 Mechanism of Operation of Scanning Electron Microscopy

The principal components of the SEM are the electron column, the specimen chamber, and the computer control system (Fig. 1.14). The specimen chamber is based at the bottom of the electron column and is maintained under vacuum conditions during operation. Moreover, this component consists of an optional airlock chamber, a charge-coupled device (CCD) camera, several detectors, a specimen stage, and a specimen holder. In general terms, the SEM produces images by using a beam of electrons that collides directly with the sample. The sample is deposited on the specimen stage, which can move to the sides, tilt, and rotate to get a better image. If it is required to maintain vacuum conditions inside the specimen chamber for long

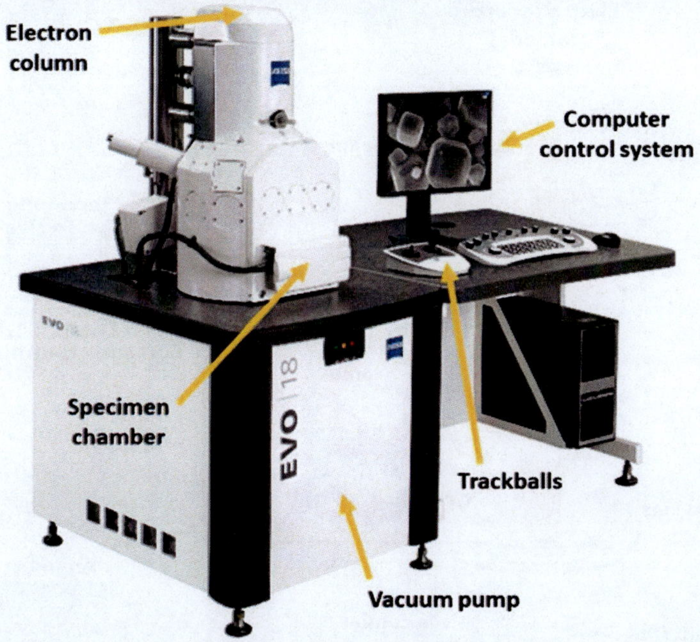

Fig. 1.14 Principal components of the SEM

periods, it is recommended that specimen holders carry multiple samples simultaneously. Researchers have developed different specimen stages over history, such as highly stable stages to achieve high-resolution microscopy, hot stages for experiments at elevated temperatures, and cold stages for specimens requiring staying under low temperatures.

The modern SEM includes an infrared (IR) camera inside the specimen chamber to allow displaying the images on the computer monitor in real time. Although the other components of the SEM are indispensable, the one that does most of the work is the electron column. Inside the column, the electron gun or cathode generates a small electron beam. The electron beam can only travel effectively through the space inside the chamber if it is in a vacuum. It only takes around three minutes to obtain enough vacuum in the specimen chamber. The diameter of the electron beam is determined by the type of electron gun used. The achieved spatial resolution depends on the probe size, for instance, high spatial resolution is reached with a small probe size [46, 47].

The different electron guns used are the tungsten hairpin, the lanthanum hexaboride (LaB6) emitter, the Schottky field emission, and the cold field emission gun. The oldest and most used nowadays is the tungsten hairpin gun, consisting of a 100 μm diameter filament bent to a V-shape and heated to 2700 K. On the other hand, a block of LaB6 produces a 5–10 times brighter source. However, it requires a much lower vacuum compared to other gun types [1].

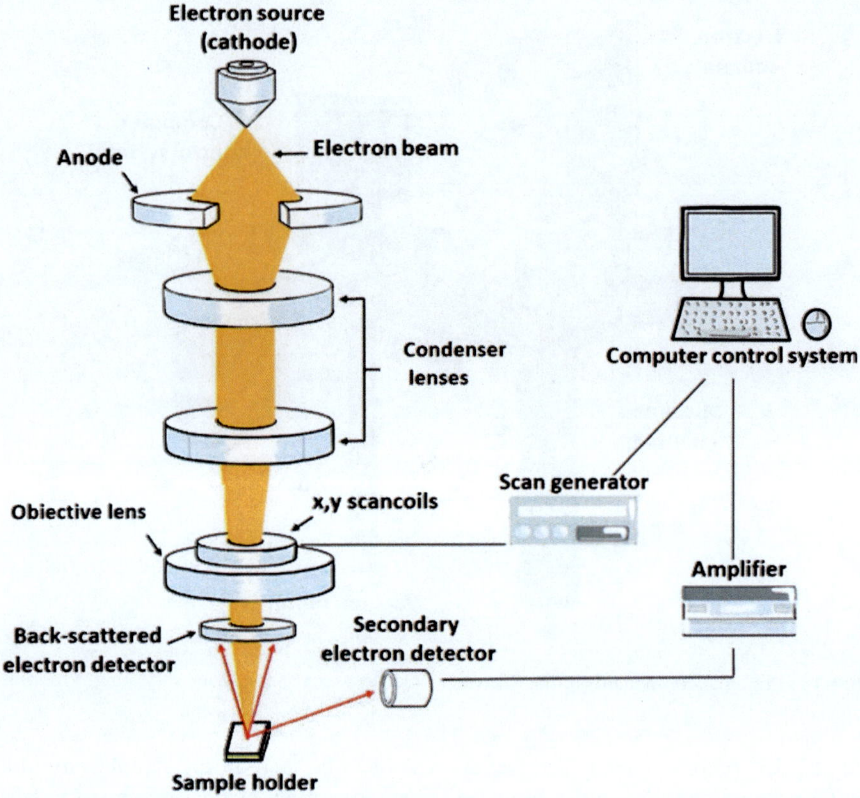

Fig. 1.15 Principle of operation of the SEM

The principle of operation of SEM consists of two electrodes connected to a high voltage source. The negatively charged cathode is located at the top, and the positively charged anode is at the bottom of the column (Fig. 1.15). The strong electric field generated by the electrodes accelerates the movement of the electron beam. Subsequently, electromagnetic lenses focus the beam on the specimen surface. Condenser lenses are placed just below the anode to reduce the diameter of the electron beam by approximately 0.4–5 nm, depending on the extent of demagnification required. Consequently, the objective lens controls the movement of the probe crossover along the column's optic axis, hence focusing on the image [1, 46, 47].

The electron beam impacts the sample with a force that depends on the accelerating voltage used, the type of specimen examined, the nature of the analysis, and the required information (typically varied from 2 to 30 kV) [46].

Finally, the specimen surface releases secondary electrons due to the electron beam impact, and this is primarily why the SEM can provide images of three-dimensional objects. Since the secondary electrons do not have to be focused, the secondary detector only collects them. The higher the atomic number, the greater the

Fig. 1.16 Image of pollen grains obtained by SEM

backscattering and the lesser the electron transmission, resulting in higher quality and more detailed images. Figure 1.16 shows an image obtained by SEM [1, 47, 48].

Knowing the processes a sample goes through is also important. Once the beam of electrons hits the sample surface, an electron from the innermost shell of the specimen is released, generating an empty place. The empty place generates an unstable energetic state; hence, an electron from the next layer fills the space. The energy difference caused by the movement of electrons in the shell is released as a unique X-ray photon since its energy is different for each element. Finally, an electron from the beam fills the remaining space in the outermost shell. A positively charged grid can be found at the beginning of the secondary electron detector (SED) to increase the number of detected electrons. In this regard, the grid attracts more secondary electrons and generates an image of the sample surface [47]. The electrons that penetrate the surface of materials at depths of less than 10 nm (100 Å) can escape and be detected as secondary electrons. At such shallow depths, the electron beam moves parallel to the incidence direction due to minimal elastic scattering. Several electrons will release from the surface at a higher tilt [48]. An electron microscope converts electron intensity to a visible and photographable light intensity [46–49]. The secondary electron detector (SED) uses the energy released in electronic transitions within the atom to generate an image through the SEM's computers and specialized software systems [46, 49].

SEM may present spherical aberration due to electrons traveling in trajectories apart from the optical axis (Fig. 1.17). Astigmatism in the SEM is defined as the inability to bring light rays of electrons to a specific point. It can be caused due to asymmetry in the windings, machining deviations, inhomogeneities within the iron pieces, and accumulation of contaminants in the electron-probe lens. Cleaning the SEM avoids the accumulation of carbon and other compounds that may deviate from the electrons beam [48].

The SEM produces the images in a sequence similar to the knitting process. The scan generator directs the electron beam to every corner of the specimen and stops

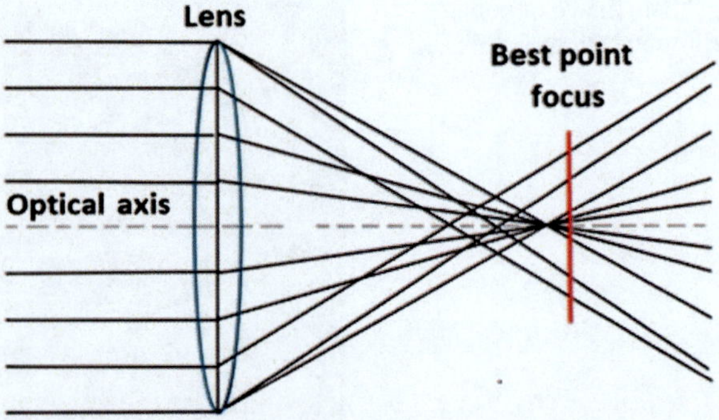

Fig. 1.17 Diagram of the spherical aberration phenomenon

there briefly. For instance, if the scanner starts in the upper left corner of the sample, the corresponding image will be shown in the upper left corner of the monitor. Afterward, the electron beam moves to the right and stops shortly again, and the process is repeated until the completion of the first row. The scanner moves down to repeat the process until it reaches the opposite corner of the specimen, which following the example, it would be the lower-left corner. Depending on the number of electrons collected by the SED, a white, gray, or black dot will be displayed at the equivalent corner of the monitor. Numerous electrons show a bright dot, a few electrons show a gray dot, and no electrons show a black dot in the monitor. Within seconds, a gray-color-scale image of the specimen appears on the monitor [47].

1.2.3 Advantages and Disadvantages of Scanning Electron Microscopy

Among the main advantages of the SEM, the sub-nanometer resolution ($\leq$1 to $\leq$10 nm), detailed morphological analysis, elemental composition, and topography studies of the surface can be named. Additionally, the modern SEM allows simultaneous imaging of several perspectives of a material's surface, which makes this characterization technique suitable for many applications, including nanotechnology, material science, biotechnology, and biomedical fields [44, 50–52].

Adversely, the high electron beam energy may damage sensitive surfaces, for instance, samples containing soft polymers, biological samples, and hydrogel structures [50, 51]. Furthermore, the electron beam broadening effect limits the high resolution of this technique [51]. The quality of images produced by the SEM also depends on the operational conditions including low-resolution imaging while capturing large sample areas with fast acquisition time [52], projection distortion (foreshortening) when the specimen tilts around an axis, weak contrast due to a small beam size with low beam current, and electron scattering caused by inadequate environmental vacuum conditions. Using a high beam current enhances the visibility of low-contrast objects even though the downside could be the deterioration of the resolution. Also,

Table 1.2 Advantages and disadvantages of SEM

Advantages	Disadvantages
Sub-nanometer resolution (≤1 to ≤10 nm) [44, 50, 52]	Limited resolution due to the electron beam broadening effect [51]
Suitable for a vast spectrum of applications [44, 50, 51]	Possible damage to the samples due to the high electron beam energy [50, 51]
Provides information related to the compositional microstructure and morphology of the surface [44, 50]	Low resolution while capturing large areas with a fast acquisition time [52]
Simultaneous imaging of several perspectives of the surface [44, 52]	The complete computer control system of SEM is not adequate for categorizing and registering a large quantity of produced data [53]
Relatively simple operation [53]	Selection of small beam size with low beam current compromises the visibility of features producing a weak contrast in the image [44]
The computational control of the SEM serves as a multi-user facility and generates a substantial quantity of images per user and application [53]	Subject to projection distortion (foreshortening) when the specimen tilts around an axis [44]
High beam current enhances the visibility of low-contrast objects [44]	A high beam current may deteriorate the resolution [44]
The divergence angle of the beam can be minimized to increase the depth of field [44]	Risk of electron scattering produced by inadequate environmental vacuum conditions [44]

SEM scanning electron microscope

the divergence angle of the beam can be minimized to reach a more extensive sample's depth of field to allow the image to have a higher sharpness [44].

The computer control of the SEM allows a simple operation that only requires simple training. It serves as a multi-user facility and generates a substantial quantity of images per user and application. Nevertheless, the computer system is not enough for categorizing and registering a large quantity of produced data. For this reason, neural network techniques have been additionally implemented for adequate recognition, categorization, and labeling [53]. Table 1.2 summarizes the benefits and limitations of using the SEM as a material characterization technique.

1.2.4 Applications of Scanning Electron Microscopy

The scientific community primarily uses SEM for evaluating and studying morphology (size and shape) and topography (texture, smoothness, or roughness) of a wide variety of specimens (Fig. 1.18). The elastic and inelastic interactions between the beam of electrons and the surface of the sample produce signals that enable the production of images with high-detail definition (spatial resolution of

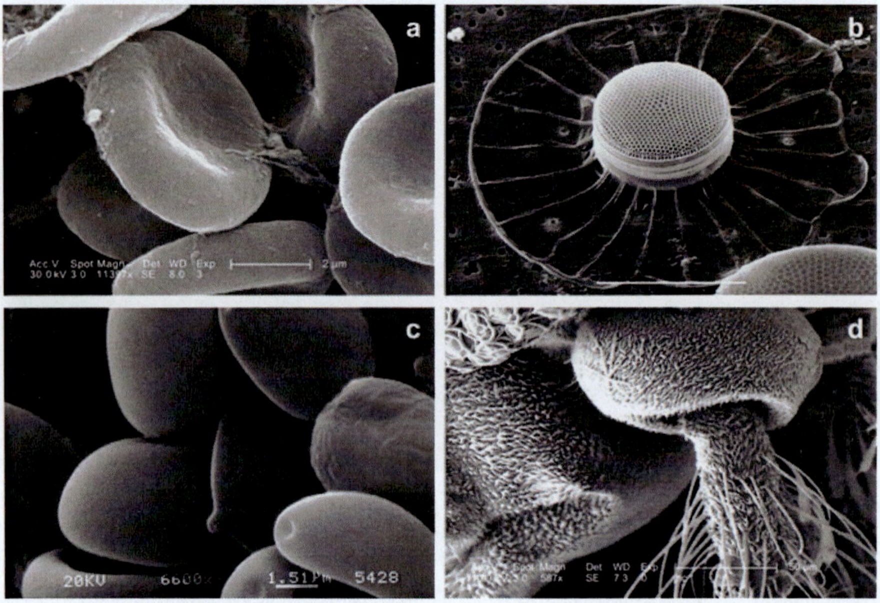

Fig. 1.18 Images obtained by the SEM: **a** Erythrocytes. **b** Diatom algae. **c** *Psilocybe aztecorum* spores. **d** *Anopheles dirus* mosquito's antennae

1 nm). The quality of the image mainly relies on the energy of the incident beam of electrons, the incident angle, and the composition of the solid sample. The SEM is incompatible with wet specimens, making the drying methods fundamental when examining biological samples since organisms, organs, tissues, biomolecules, and cells must be hydrated to remain in their native condition. SEM can characterize the sample's thickness, morphology distribution, lateral dimensions, and structural defects studies. SEM is widely applied in nanoscience because it is a non-destructive tool and provides clear high-quality images. For instance, the size of the nanomaterials, which a SEM can determine, defines their properties in photocatalysis and electrocatalysis applications [54, 55].

The presence of organophosphorus pesticides (OPPs) in the human body causes damage to the nervous system due to the accumulation of acetylcholine. Cui et al. developed an electrochemical biosensor to detect OPPs based upon the oxidation reaction of acetylthiocholine (ATCl) catalyzed by the enzyme acetylcholinesterase (AChE). The authors used the specific surface area, non-toxicity, biocompatibility, and excellent thermal/chemical stability properties of chitosan (CS), titanium dioxide (TiO_2) sol–gel, and reduced graphene oxide (rGO) to create an enzymatic electrode on glassy carbon (GC) with high catalytic activity (Fig. 1.19a). The authors chose dichlorvos as the organophosphorus pesticide model to evaluate the electrochemical biosensor performance. In this case, the SEM was used to analyze the surface morphology of the enzymatic electrode after each fabrication step. The rGO showed a typical blown surface morphology (Fig. 1.19b). Homogeneous interconnected

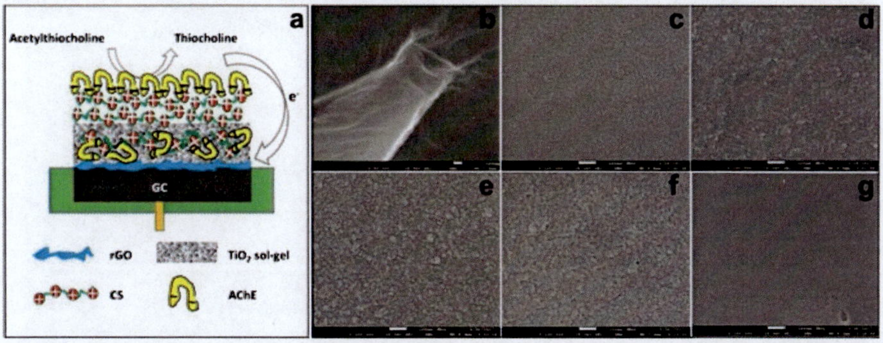

Fig. 1.19 **a** Scheme of electrochemical AChE biosensor and SEM images of: **b** rGO/GC complex, **c** TiO_2/rGO/GC, **d** $TiO_2^{99.0}$-$CS^{0.005}$/rGO/GC, **e** $TiO_2^{90.0}$-$CS^{0.005}$/rGO/GC, **f** $CS^{@}TiO_2^{m}$-CS^{m}/rGO/GC, and **g** AChE/$CS^{@}TiO_2^{m}$-CS^{m}/rGO/GC electrodes. Reproduced (or adapted) with permission [56], ©2018, Elsevier

nanoparticles and pores showed in the mesoporous TiO_2 sol–gel film (Fig. 1.19c). Meanwhile, the incorporation of CS in the TiO_2 sol–gel matrix caused the electrode surface to get rougher (Fig. 1.19d, e). Moreover, nanoparticles agglomerated when CS was electrodeposited on TiO_2-CS composite. Nevertheless, the electrode surface remained rough and porous (Fig. 1.19f). Finally, the pores filled with AChE enzymes resulted in a smooth electrode surface (Fig. 1.19g) [56]. This is an example of how the SEM can help in the development of new technologies such as biosensors.

Diabetes is a worldwide health problem characterized by abnormally high glucose levels in the blood. Such a condition can be caused due to an alteration in the pancreatic beta-cell behavior or insulin insufficiency. An electrochemical biosensor is an excellent tool for accurately detecting and supervising blood glucose concentration while offering high performance, low cost, and ease of fabrication. Feng et al. developed a biosensor by immobilizing glucose oxidase (GOD) with CS on a conductive nanocomposite made from graphene (GRA) and polyaniline (PANI). The SEM observed the biosensor surface in different construction stages to perform a morphology analysis. The produced images showed that GRA presented a flake-like structure, exhibiting excellent properties for a physical isolation film (Fig. 1.20a). PANI formed homogeneous coral-shaped rods with a length of approximately 1.5 μm and a diameter of nearly 300 nm (Fig. 1.20b). The PANI was previously deposited on the electrode surface, which physically absorbed GRA, resulting in enwrapped-PANI GRA nanocomposite (GRA-PANI) with a larger diameter and shorter length than PANI (Fig. 1.20C). Finally, the immobilized GOD presented a 3D net framework where the GRA-PANI nanocomposite wrapped the immobilized GOD (Fig. 1.20D). The biosensor had a high sensitivity (22.1 $\mu A\ mM^{-1}\ cm^{-2}$) and a limit of detection of 2.769 μM. Moreover, the selectivity was excellent since molecules of similar structure, such as D-galactose, did not cause interference. The authors attributed the aforementioned characteristics to the high conductivity and the physical isolation

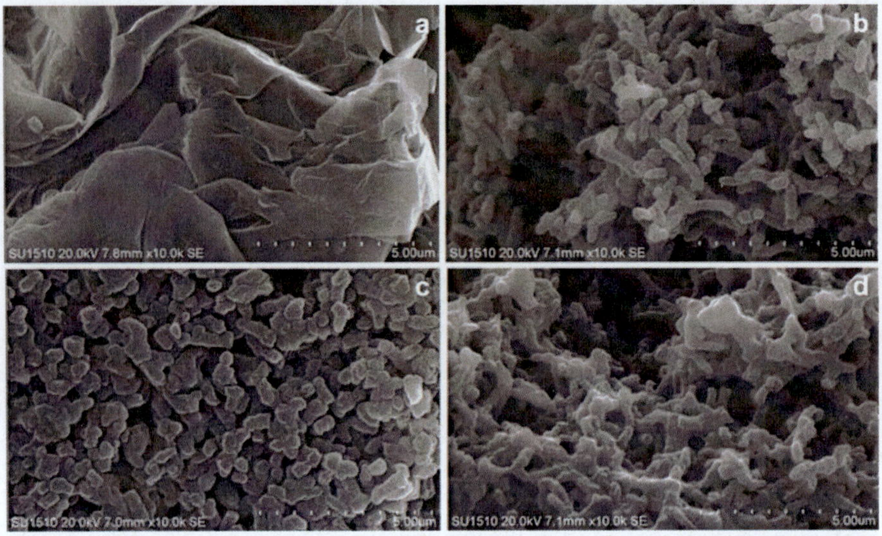

Fig. 1.20 Images produced by the SEM. **a** GRA. **b** PANI. **c** GRA-PANI nanocomposite. **d** GRA-PANI/CS-GOD complex. Reproduced (or adapted) with permission [57], ©2015, Elsevier

provided by the GRA-PANI nanocomposite [57]. In this study, SEM helped identify the characteristics and differences of each nanocomposite.

The medical field is interested in the early diagnosis of diabetes as it prevents more severe health difficulties. As mentioned, one of the most common tools to detect glucose concentration is enzyme-based glucose biosensors. Nevertheless, the enzyme electrodes usually have low selectivity, repeatability, and vulnerability to ambient conditions when used as a biosensor. The ratiometric electrochemical sensors produce separable signals from two different electroactive materials: The first material generates an internal reference signal to overcome device, sample, or environmental interferences, and the second material monitors the analyte. Therefore, this electrochemical principle allows enhancing the biosensor's stability and reproducibility. Song et al. developed a high-performance, selective, and reproducible electrode for a ratiometric electrochemical biosensor. The electrode was made of three-dimensional macroporous carbon (3D-KSCs) as secondary material. As the name suggests, Glucose Oxidase (GOD) is responsible for glucose oxidation, while this chemical reaction consumes oxygen molecules, which, in turn, increases the copper-trimesic acid (Cu-BTC) in blood. In the study, deposited copper–trimesic acid metal–organic frameworks (Cu-BTC MOFs) on 3D-KSCs catalyzed the glucose oxidation to obtain the reference signal. Subsequently, gold nanoparticles (AuNPs) electrodeposited on Cu-BTC MOFs/3D-KSCs improved the electrical conductivity and allowed immobilizing GOD via Au–S covalent bond (Fig. 1.21a). The morphology characterization performed by the SEM exhibited that 3D-KSCs had flat walls and uniform macropores with a diameter close to 25 μm. Therefore, it created a large surface area to load activity materials and enhanced the transfer of

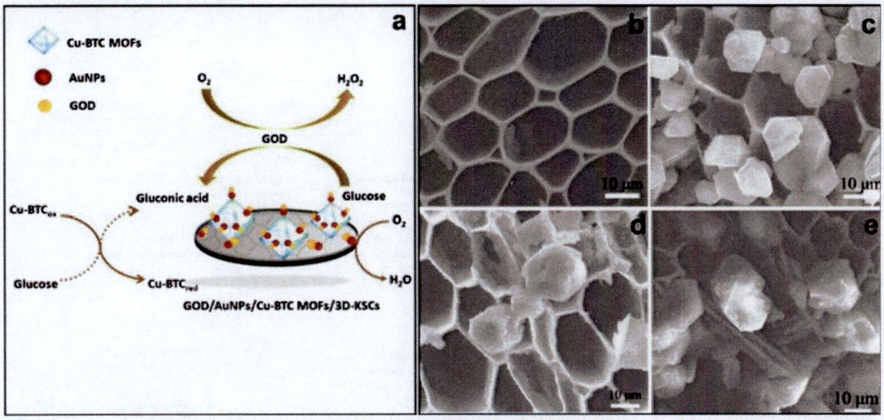

Fig. 1.21 Results from a biosensor developed for early diagnosis of diabetes. **a** Principle of the ratiometric electrochemical glucose biosensor and obtained images by the SEM of **b** 3D-KSCs, **c** Cu-BTC MOFs/3D-KSCs, **d** AuNPs/Cu-BTC MOFs/3D-KSCs, and **e** GOD/AuNPs/Cu-BTC MOFs/3D-KSCs. Reproduced (or adapted) with permission [58], ©2018, Elsevier

mass (Fig. 1.21b). The Cu-BTC MOFs generated octahedron nanostructures on the 3D-KSCs and decreased the macropore's diameter to 10 μm (Fig. 1.21). Furthermore, the Cu-BTC MOFs surface became rough with the AuNPs electrodeposition (Fig. 1.21d), and no significant changes were found after the immobilization of GOD (Fig. 1.21e). Finally, the biosensor performance proved to be suitable for glucose level detection, demonstrating linearity in the range from 44.9 μM to 4.0 mM and from 4.0 to 19 mM and obtaining 14.77 μM as the detection limit [58].

1.3 Transmission Electron Microscopy for Material Characterization

1.3.1 History of Transmission Electron Microscopy

In the first decades of the 1900s, scientific and technological advances gave rise to improved microscopy resolution and the new technology surpassed the light microscope at the time. The following timeline (Fig. 1.22) illustrates the chronology of such events.

During 1923 and 1925, Louis de Broglie proposed that particles had specific frequencies and wavelengths and stated that electrons possessed a notably smaller wavelength than visible light [59–61]. In 1926, Hans Busch revealed that a rotationally symmetric magnetic field produced a focusing effect that the lens formula could describe. One year later, in 1927, George Paget Thomson, Alexander Reid, Clinton

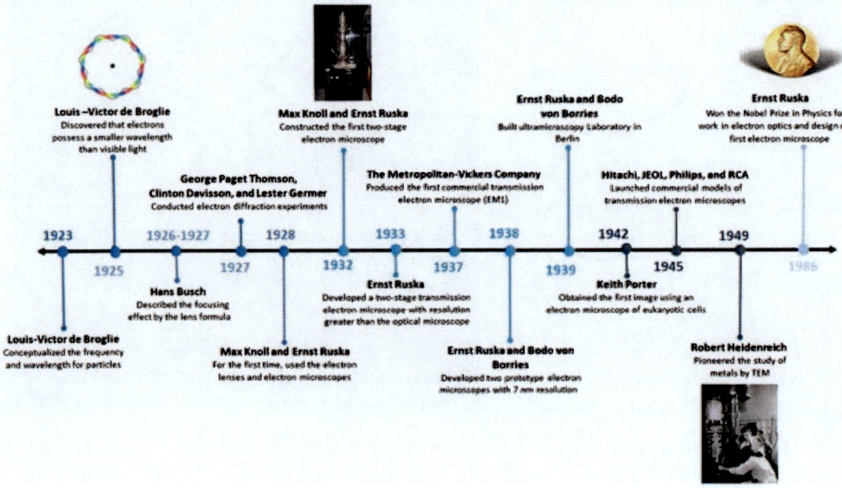

Fig. 1.22 History of TEM from 1923 to 1986

J. Davisson, and Lester H. Germer conducted experiments related to electron diffraction, which contributed to revealing characteristics of electrons. This discovery was particularly useful to Max Knoll and Ernst Ruska, who applied these notions to develop electron lenses and constructed the first two-stage transmission electron microscope (TEM) in 1932 [60–63]. In the following year, Ruska's microscope (containing three magnetic lenses, condenser, objective, and projector) achieved a resolution more significant than the optical microscope [60, 64, 65]. In 1986, Ruska was awarded a Nobel Prize in Physics for his fundamental research in electron optics and the design of the first electron microscope [61].

In 1937, the Metropolitan-Vickers Company produced the first commercially available TEM in Manchester. Nonetheless, the resolution of the equipment did not surpass that of light microscopes. That same year, Ruska and Bono von Borries worked with Carl Zeiss to build an Ultramicroscopy Laboratory in Berlin. As a result of this collaboration, in 1938, the researchers involved developed two prototype electron microscopes with a resolution of 7 nm. After the end of the Second World War, companies such as Hitachi, JEOL, Phillips, and RCA widely introduced transmission electron microscopes to the international market [60, 61, 64]. The transmission electron microscope was greatly accepted, and this characterization technique acquired relevance in the biology and material science fields. Approximately ten years after the development of the first TEM, in 1942, Keith Porter took the first image of eukaryotic cells using the equipment while Robert Heidenreich of Bell Labs pioneered the study of metals by TEM in the late 1940s [60, 61, 64, 66, 67].

1.3.2 *Mechanism of Operation of Transmission Electron Microscopy*

Electron microscopy is an outstanding tool for examining specimens such as cells, tissues, molecular assemblies, organisms, individual proteins, and organelles in macro- or microscopic resolution. Nevertheless, the technique requires high vacuum conditions to make high-quality imaging, which represents a time-consuming and challenging step when processing the specimen. The law of Ernst Abbé states that the wavelength of an electron depends on its velocity. Therefore, to obtain a better resolution, it is necessary to achieve a high electron speed resulting in a small wavelength. For TEM, the acceleration voltage sets the speed, and the practical resolution is close to 0.5 nm at 100 kV (Fig. 1.23). The diffraction effect caused by the electron beam and a small wavelength allows imaging particles in the nanometer scale. As a consequence, the TEM produces images with high magnification and improved resolution [55, 68–70].

Fig. 1.23 Resolution limit of different characterization techniques for morphological analysis

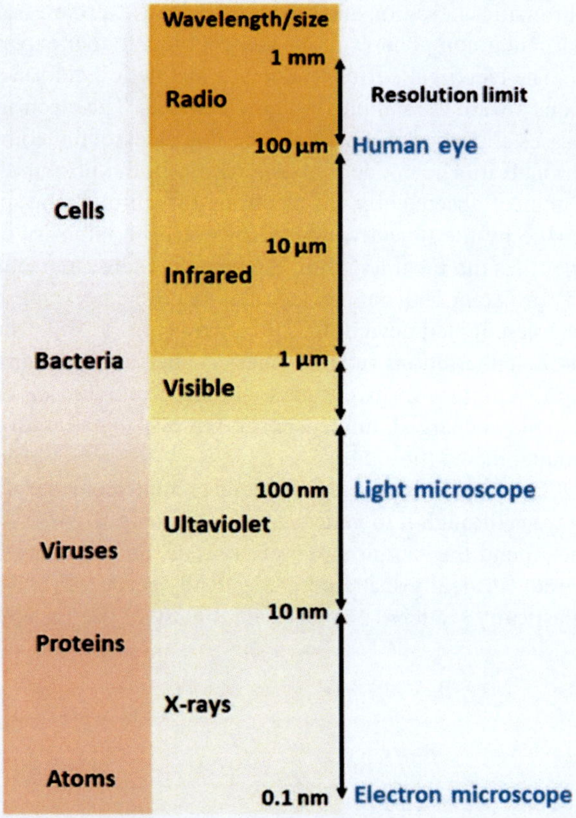

The operation mechanism of TEM is comparable to that of an optical microscope. The most main differences are the evolution of the illumination source from light to electrons, the development of electromagnetic/electrostatic lenses instead of glass lenses, and the operation of TEM in a high vacuum. The last aspect is crucial when operating the TEM because the image resolution is affected by the scattering of electrons colliding with a residual gas molecule. The high vacuum system consists of three main zones. The electron source reaches a 10^{-7} to 10^{-10} millibar in the first zone to avoid damaging the heated filament. The column remains at a constant vacuum of 10^{-5} to 10^{-7} millibar in the second zone. Finally, the third zone reaches a high vacuum of 10^{-0} to 10^{-2} millibar [68–70].

One of the main TEM components is the electron source, called a cathode, where cold field emission or thermionic emission produces electrons. In thermionic emission, a thin tip made by tungsten filament, LaB6 crystal, or a ZrO/W Schottky is heated using an electrical current to generate the electron beam inside the TEM's column. On the other hand, the cold field emission is caused when a thin tungsten filament tip is heated at room temperature and produces the electron beam. This process enables us to obtain an image at atomic resolution due to the electrons' low chromatic aberration and high efficiency. Nevertheless, the specific high vacuum and particular components make the technique rather expensive [68–70].

The electrical current induces a magnetic field to accelerate and direct the electrons toward the sample through the lenses. The electromagnetic lens system encompasses all lenses of the TEM, and it is an essential component since electrons travel through it. The condenser lens regulates the brightness and leads the beam to the sample. Subsequently, the electrons pass through the specimen to create the primary image by the objective lens. However, the intensity of the applied magnetic field modifies the focal length in the objective lens, impacting the magnification. Lastly, the projector lens defines the final magnification and screens the final image on the charge-coupled device (CCD) camera (Fig. 1.24a). The electromagnetic lenses can present aberrations such as spherical aberration, astigmatism, chromatic, and distortion. Astigmatism is the most common problem in TEM, and it may occur if the sample is charged, the lenses are not distributed uniformly, or the lenses system is contaminated [68–71].

The sample thickness is a crucial factor to consider since the electrons must be able to pass through it to generate an image (Fig. 1.24b). The interaction of the electron beam and the sample produces one of the following types of electrons: backscattered, elastically scattered, inelastically scattered, and unscattered (Fig. 1.24c). The elastically scattered electrons are mainly of use for biological analysis [68–71].

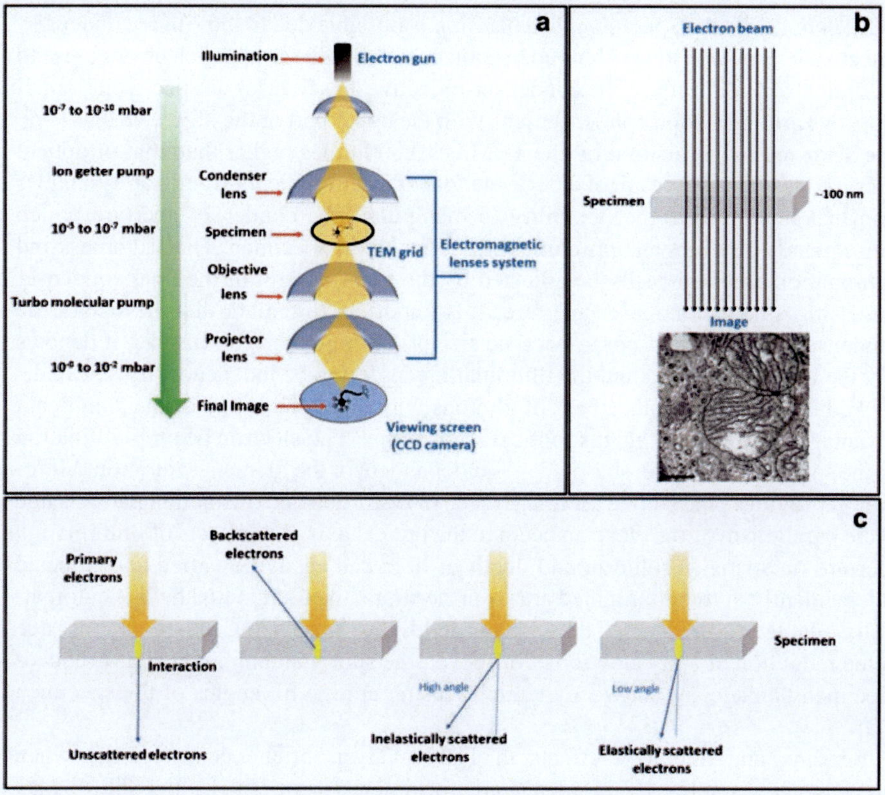

Fig. 1.24 Mechanism of operation of the TEM. **a** Operational principle of the TEM, **b** general scheme of the imaging process of the TEM, and **c** electrons produced by different interaction types between electron beam and sample

1.3.3 Advantages and Disadvantages of Transmission Electron Microscopy

The TEM operates on the same basic principles as the light microscope. The differences take place in electrons replacing photons, electromagnetic lenses supplanting glass lenses, and screens substituting eyepieces as the equipment for observing the produced images. The TEM invention has dramatically impacted the scientific processes since this microscopy technique creates higher resolution images, provides crystallographic and atomic data, creates 2D images easy to interpret, allows visualization of nano-scaled structures, and renders the element and compound structure.

Such high-quality specimen visualization is possible due to the single condenser-objective lens, which lowers lens aberration coefficients by keeping electrons close to the optical axis. Moreover, the condenser-objective lens focuses on specimens' small spots by forming a solid condenser lens with the upper part of the objective lens [72]. The achieved magnification by the TEM is 10,000 times higher than that of optical microscopes. This is a result of the wavelength of electrons being significantly smaller than that of light and the flexibility to manipulate the condenser aperture which consequently provides an optimum illumination to the specimen. The brightness and illumination area can easily be adjusted by the focus control of the final condenser lens using a fixed condenser aperture. It is important to remark that the size of the condenser aperture does not impact the size of the source image. Instead, it denotes that the illuminated area and the illuminating angle can be independently selected.

Additionally, the replacement of photons with electrons is one of the significant advantages of the TEM against optical microscopy. The electron beam illumination on the specimen results in sharp images independent of the filament saturation. Moreover, it provides phase-contrast images free of distortion and diffraction patterns due to the parallelism of the electron beam to the optical axis. The effects of illuminating aperture on spatial resolution and depth of field can easily be varied and selected independently of the illuminated area. Furthermore, the illuminated field is automatically adapted to the size of the viewing field, thus optimizing image contrast due to the reduction of stray electrons. Moreover, the spot scanning in the TEM reduces specimen damage and allows dynamic focusing at high tilt angles of the specimen [72].

Bending and thickness effects in TEM analysis often occur together when analyzing a sample. If the analyzed specimen is too thick, the TEM may produce incorrect conclusions about particle composition, structure, and shape [72]. Contrarily, if the sample is too thin, it could bend elastically and physically rotate the lattice planes. Even a 0.1 angle rotation can significantly affect the image. The TEM also may present inconsistencies representing the sample's thickness since there are cases where the thicker areas of the specimen are brighter than the thinner areas, which is contradictory to the result that should be displayed [73].

A particularly disadvantageous feature for researchers in the biology field is that TEM cannot be used for observing living compounds [74] since samples need to withstand the vacuum chamber and have electron transparency, which involves a thickness usually less than 100 nm so that electrons can pass through the sample. Delicate samples are subjected to previous chemical preparation to prevent damage produced by the electron beam. Types of preparation include dehydration, sputter coating of non-conductive materials, cryofixation, sectioning, and staining, which add to the total time and cost of the analysis. These treatments, however, may permanently damage the specimen [75]. The TEM's operation and analysis require special training, infrastructure, and maintenance. For instance, the equipment requires constant calibration including maintaining voltage, currents to the electromagnetic coils, cooling water, and isolation from vibration and electromagnetic fields. The price of the TEM setups manufactured by Zeiss, Jeol, Philips, and Hitachi are considerably high, representing a heavy initial investment [76]. Examples of prices for new

Table 1.3 Advantages and disadvantages of utilizing the TEM

Advantages	Disadvantages
Low lens aberration coefficients [72]	The specimens could bend elastically affecting the quality of the image [73]
Sharp images due to the illumination system [72]	Living specimens cannot be observed due to the use of vacuum [74]
The image contrast is optimized automatically [72]	The electron beam may damage the samples [75]
Depth of field can be easily selected independent of the illuminated area [72]	Operation of the TEM and the analysis of images require special training [76]
	The machinery and accessories are expensive [77]
The compound structure of the specimen is provided [72]	The TEM may present inconsistencies due to the sample's thickness [73]
The magnification can reach 10,000 times compared to the optical microscopes [72]	Requires frequent calibration and isolation from vibration and electromagnetic fields [76]

TEM transmission electron microscope

TEM models include $95,000USD for Joel's 1200EXII, $95,000USD for Philips' EM10, and $100,000USD for Hitachi's 7000. The focused ion beam tool may cost between 30 and 80% of the total cost of a TEM, which is a considerable expense for a specimen preparation tool [77]. The advantages and disadvantages of TEM are summarized in Table 1.3.

1.3.4 Applications of Transmission Electron Microscopy

The TEM provides topographical, morphological, compositional, and crystalline information of a specimen, making it an ideal technique for nanotechnology, biology, material research, forensic analysis, and other areas. The images produced by TEM allow researchers to view samples on a molecular level and analyze their structure and texture, which is valuable information in the study of crystals and metals for industrial applications. Furthermore, technology-based companies use the technique to identify defects, fractures, and damages in micro-sized objects and devices. Different businesses can identify such problems in order to manufacture a more durable and efficient product. The TEM has also opened its way into the medical and biological research field [78]. The technique is most frequently used in cell biology to produce a contrasting image of plastic-embedded cells previously stained in thin sections. This microscopy technology is relevant to analyze cellular components, including the cytoskeleton, membrane systems, organelles, and cilia, as well as specialized structures, such as microvilli and the synaptonemal complex (Fig. 1.25). There is practically no other alternative to visualize the complexity of cells and observe cellular structures using another technique [79].

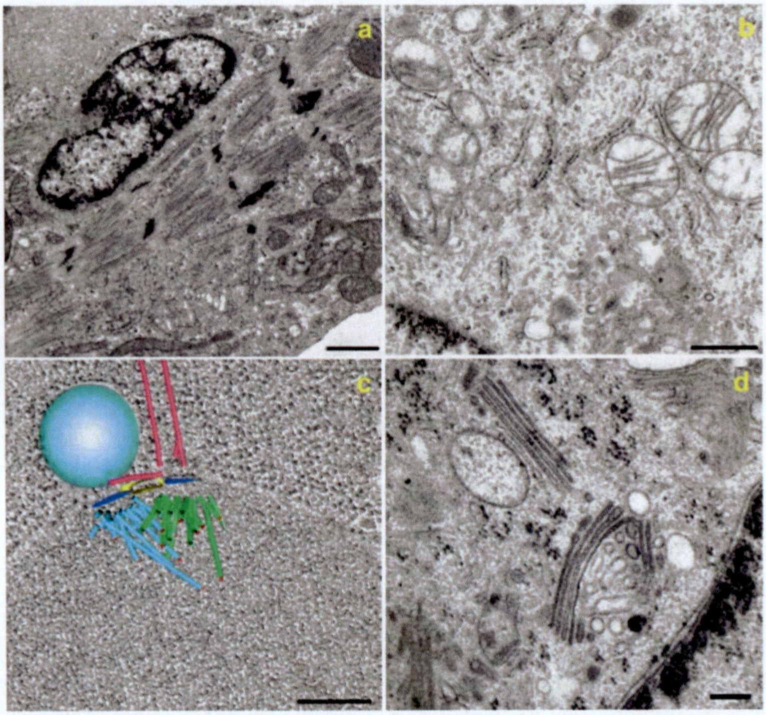

Fig. 1.25 Cell structure as visualized by the TEM. **a** Actin-myosin cytoskeleton (Bar 1 μm), **b** cytoplasmic organelles in a mouse macrophage (Bar 700 nm), **c** Golgi membranes in a cultured 3T3 cell (Bar 200 nm), and **d** three-dimensional tomographic model of a forming mitotic spindle from budding yeast (Bar 200 nm). Reproduced (or adapted) with the permission [79], © 2013, Molecular Biology of the Cell

The imaging and quantification of extracellular vesicles (EVs) in biofluids is another field for applying the TEM. EVs characterization is relevant to investigators to understand the diverse physiopathological roles and characteristics of the samples. Nevertheless, their small size poses a challenge for analyzing EVs. TEM has made significant contributions to the discovery and research of EVs. TEM combined with a receptor-specific gold labeling provides a detailed description of the size, morphology, distribution, and phenotypes of the EVs populations present in platelet-free plasma (PFP) and other samples of body fluids [80].

TEM is also a powerful tool for structural analysis of soft matter and supramolecular chemistry. This technology allows imaging a wide range of biological systems, including cells, proteins, materials, EVs, and aggregating materials. These samples are examined using the drying, freeze-drying, negative or positive staining technique, embedding, and are commonly analyzed by quick-freeze deep TEM and cryo-TEM. TEM is a suitable method for nanomaterials characterization and could avoid over-interpretations. Coupling TEM results with other analytical methods leads to further advancement in the structural analysis of biological samples and soft materials [81].

Primary ciliary dyskinesia (PCD) is a heterogeneous genetic anomaly characterized by motile cilia dysfunction. The diagnosis of PCD is confirmed by identifying ciliary ultrastructure defects or by detecting biallelic pathogenic mutations in the PCD gene. The assessment of ciliary ultrastructure by TEM has been central to its diagnosis. TEM provides improved spatial information and resolution compared to the single micrograph. Overall, electron microscopy has rendered new insight into the ciliary ultrastructure and ciliary function at a molecular and cellular level [82].

Researchers have attempted to use liposomes as a tool for disease diagnosis. However, alike EVs, their small size and distribution make their characterization rather difficult. TEM is one of the most frequently used methods to characterize the ultrastructure of liposomes. A heavy metal salt surrounds the liposomal particles and becomes adsorbed on a carbon film grid during the dry process. The liposomes appear bright compared to the dark background of the heavy metal in a TEM image, hence the technique is widely used for the evaluation of liposomes and nanomaterials as they provide high-contrast images [83].

1.4 Troubleshooting of Morphology Analysis Techniques

Whether newly designed or existing, morphological analysis of samples is one of the primary characterization methods any researcher would need within a laboratory setup. Table 1.4 summarizes the troubleshooting of the characterization techniques involved in the morphology analysis of a specimen. The first section is dedicated to optical microscopy and describes the commonly reported problems faced by users. The initial section also offers the potential causes for these problems and subsequently recommends possible solutions to address these difficulties. The following two sections review the same set of elements for SEM and TEM along with possible tips to benefit from these powerful analytical methods.

Table 1.4 Troubleshooting for characterization techniques used in morphology analysis

Technique	Problem	Cause	Solution
OM	Equipment runs out of power [84, 85]	Incorrect plugging or unsealed cables [85]	Verify plugging, turn on the switch, and substitute unsealed cables [85]
	Inadequate focus [84–86]	Absence, incorrect fitting or placement of the objective lens, or the use of an incorrect objective lens for the desired task [85, 86]	Place the corresponding lens and verify correct fitting on the objective head [85, 86]
	Shadows, reduced resolution, and non-uniform illumination on images [85–87]	Dust and dirt along the light path and optical surfaces [86]	Non-intensively clean the device with a dust blower and clean the internal components with the help of specialized personnel [86]
			Avoid the use of compressed air and cleaning agents that include diluted ammonia, xylene, or organic solvents [86]
	Unequal distribution of light through the eyepieces [85–87]	The iris diaphragm, condenser, or light source are not aligned [85–87]	The condenser, diaphragm, and light source must be centered in that order. Additionally, check that the condenser is focused, and the diaphragm is completely open [85]
	Lack of light level adjustment [85]	Contrast, brightness, and/or exposure specifications are not correct [85]	Use a bright light for higher magnification, simplify image capture, reduce thick or heavily stained samples, wash away a layer of the sample, or overexpose the sample to light [85, 87]
			Use a low light to add contrast and prevent deterioration of live samples [85, 87]
	Bias in the obtained data [85, 86]	Variances in refractive index, light scattering, and wavelength of light [85, 86]	Provide calibration and maintenance periodically utilizing ideal standard samples as a reference [85, 86]

(continued)

Table 1.4 (continued)

Technique	Problem	Cause	Solution
SEM	The sample is too close and almost hitting the lens column [88, 89]	NS	Use the ball controller to move the sample stage or remove the sample and measure the proper height [88, 89]
	The sample is not aligned [88, 89]	NS	Correct the wobbler, the stigmatism and make sure to work on the sample in the optimum focus position [88, 89]
	The image does not appear on the screen [88, 89]	NS	Reduce the magnification to the lowest value using the magnification knob on the control panel [88, 89]
	The sample cannot be inserted into the chamber [88, 89]	NS	Check the position of the holder in the stage so that the arrows are placed toward the user [88, 89]
	The system cannot reach an ideal vacuum [88, 89]	NS	Wait for the sample to degas. If the vacuum is still too high after some time, remove the sample and check that all pieces in the chamber are in their designated place [88, 89]
	The sample fell off the stage within the chamber [88, 89]	NS	Turn off the beam from the computer connected to SEM and the software and contact staff members for assistance [88, 89]
	The microscope control is frozen [88, 89]	NS	If the user clicked the transfer image button too soon, reboot the computer and ask for staff assistance [88, 89]
	The image is too noisy [88, 89]	NS	Try a slower scan rate, increase the beam intensity on the sample or try a higher current [88, 89]
	Chamber does not open after venting [88, 89]	NS	Wait for vacuum status to become ideal and vent again [88, 89]
	Image drifting [88, 89]	NS	Pause the process and wait 10 min before trying to image again [88, 89]

(continued)

Table 1.4 (continued)

Technique	Problem	Cause	Solution
	The image shifts whenever the focus is adjusted [88, 89]	NS	Adjust the lens alignment by trained staff [88, 89]
TEM	The condenser image does not expand and contract concentrically [90, 91]	The condenser aperture is misaligned	Realign and center the filament image [90, 91]
		The condenser lens is not stigmatized	Adjust stigmatism of lens using "C-2 Stigma" wheels and a desaturated filament image. Adjust the system for the greatest clarity of shadow detail [90, 91]
	The image cannot be stigmatized with objective stigmata [90, 91]	The objective aperture is not centered	Center objective aperture [90, 91]
		Objective aperture is contaminated	If no position of the objective aperture reduces astigmatism, the hole may be contaminated. Check and update the "aperture map" on the user message board [90, 91]
		The condenser aperture is misaligned, or the condenser is not stigmatized	Check centering and alignment [90, 91]
	Image is distorted or cannot be focused [90, 91]	The specimen holder is contaminated	The holder should be cleaned or replaced. Vigorous washing should be avoided during staining as that may damage the integrity of the sections [90, 91]
		Grid or specimen is contaminated	Distortion is evident near the grid bar or contamination. Avoid distorted areas and use cleaner grids in the future [90, 91]
		The device is poorly stigmatized	See stigmation section [90, 91]
		The microscope is grossly out of alignment [90, 91]	Ensure appropriate alignment of the microscope by a trained user

(continued)

Table 1.4 (continued)

Technique	Problem	Cause	Solution
	The beam is too dim [90, 91]	The filament is grossly undersaturated	Filament saturation and voltage should be checked [90, 91]
		Gun is misaligned	Check gun alignment, which may also be adjusted with changes in accelerating voltage [90, 91]
		The condenser aperture is too small	Replace the condenser to one with a larger aperture [90, 91]
		Gun bias is too low	Adjust gun bias upward and recenter the filament [90, 91]
	No beam [90, 91]	NS	Make sure the specimen is completely inserted into the microscope column [90, 91]
		NS	Turn down the magnification to check the specimen position to be at the center of the specimen traverse [90, 91]
		Objective, intermediate, or condenser aperture obscure the beam	Recenter the aperture or remove the aperture for rechecking the details [90, 91]
		Magnification is too high	Reduce the magnification [90, 91]
	Image drifting [90, 91]	Specimen is unstable	Continue irradiating the specimen until the movement ends, or cut thicker parts of specimen [90, 91]
		The microscope specimen holder is grossly contaminated [90, 91]	Holder should be cleaned or replaced [90, 91]

NS not specific; *OM* optical microscopy; *SEM* scanning electron microscopy; *TEM* transmission electron microscopy

References

1. W.J. Croft, *Under the Microscope: A Brief History of Microscopy*, vol. 5 (World Scientific Publishing, 2006)
2. S. Bradbury, *The Evolution of the Microscope*, 1st edn. (Pergamon, 1967)

3. Invention of the microscope. Sci. Am. **13**(41), 321–321 (1858). https://doi.org/10.1038/scientificamerican06191858-321a
4. R.R. Shannon, B.J. Ford, Microscope, in *Encyclopedia Britannica* (2020). https://www.britannica.com/technology/microscope
5. T. Araki, The history of optical microscope. Mech. Eng. Rev. **4**(1), 16-00242 (2017). https://doi.org/10.1299/mer.16-00242
6. W. Ockenga, *A Brief History of Light Microscopy: From the Medieval Reading Stone to Super-Resolution* (2015) [Online]. Available: https://www.leica-microsystems.com/science-lab/a-brief-history-of-light-microscopy-from-the-medieval-reading-stone-to-super-resolution/
7. S. Bradbury, B. Bracegirdle, *Introduction to Light Microscopy* (Springer Singapore, Oxford, 1998)
8. S.H. Gage, Modern dark-field microscopy and the history of its development. Trans. Am. Microsc. Soc. **39**(2), 95–141 (1920)
9. J. James, H.J. Tanke, *Biomedical Light Microscopy* (Springer, The Netherlands, Dordrecht, 1991)
10. A. Köhler, W. Loos, Das Phasenkontrastverfahren und seine Anwendungen in der Mikroskopie. Naturwissenschaften **29**(4), 49–61 (1941). https://doi.org/10.1007/BF01476460
11. S. Bradbury, P.J. Evennett, *Contrast Techniques in Light Microscopy* (Garland Science, 1996)
12. F.H. Kasten, The origins of modern fluorescence microscopy and fluorescent probes, in *Cell Structure and Function by Microspectrofluorometry* (Elsevier, 1989), pp. 3–50
13. M. Renz, Fluorescence microscopy-A historical and technical perspective. Cytom. Part A **83**(9), 767–779 (2013). https://doi.org/10.1002/cyto.a.22295
14. S. Ebnesajjad, Surface and material characterization techniques, in *Handbook of Adhesives and Surface Preparation* (Elsevier, 2011), pp. 31–48
15. P.C. Goodwin, A primer on the fundamental principles of light microscopy: optimizing magnification, resolution, and contrast. Mol. Reprod. Dev. **82**(7–8), 502–507 (2015). https://doi.org/10.1002/mrd.22385
16. M. Walls, Optical and electron microscopy, in *Physical Methods, Instruments and Measurements* (EOLSS Publications, 2009), p. 27
17. L.C. Martin, *The Theory of the Microscope Blackie* (Blackie, London, 1966)
18. M.T. Madigan, K.S. Bender, D.H. Buckley, W.M. Sattley, D.A. Stahl, *Brock Biology of Microorganisms*, 15th edn. (Pearson Education, NJ, 2017)
19. J.E. Hayden, Adventures on the dark side: an introduction to Darkfield microscopy. Biotechniques **32**(4), 756–761 (2002). https://doi.org/10.2144/02324bi01
20. B. Herman, Fluorescence microscopy. Curr. Protoc. Cell Biol. **00**(1) (1998). https://doi.org/10.1002/0471143030.cb0402s13
21. X. Chen, B. Zheng, H. Liu, Optical and digital microscopic imaging techniques and applications in pathology. Anal. Cell. Pathol. **34**(1–2), 5–18 (2011). https://doi.org/10.3233/ACP-2011-0006
22. R.W. Welker, Size analysis and identification of particles, in *Developments in Surface Contamination and Cleaning* (Elsevier, 2012), pp. 179–213
23. S. Okazaki, T. Nagamura, Near-field scanning optical microscopy. J. Jpn. Soc. Precis. Eng. **57**(7), 1155–1158 (1991). https://doi.org/10.2493/jjspe.57.1155
24. A. Diaspro (ed.) *Optical Fluorescence Microscopy* (Springer, Berlin, Heidelberg, 2011)
25. A.-L. Robson et al., Advantages and limitations of current imaging techniques for characterizing liposome morphology. Front. Pharmacol. **9**, 80 (2018). https://doi.org/10.3389/fphar.2018.00080
26. G.F. Vander Voort, Optical microscopy, in *Encyclopedia of Condensed Matter Physics* (Elsevier, 2005), pp. 175–182
27. B. Huang, Super-resolution optical microscopy: multiple choices. Curr. Opin. Chem. Biol. **14**(1), 10–14 (2010). https://doi.org/10.1016/j.cbpa.2009.10.013
28. I.I. Smolyaninov, Optical microscopy beyond the diffraction limit. HFSP J. **2**(3), 129–131 (2008). https://doi.org/10.2976/1.2912559
29. E.G. van Putten, D. Akbulut, J. Bertolotti, W.L. Vos, A. Lagendijk, A.P. Mosk, Scattering lens resolves Sub-100 nm structures with visible light. Phys. Rev. Lett. **106**(19), 193905 (2011). https://doi.org/10.1103/PhysRevLett.106.193905

30. P.R. Lewis, The failure of synthetic polymeric medical devices, in *Durability and Reliability of Medical Polymers* (Elsevier, 2012), pp. 183–224
31. D. Utsumi, M. Yasuda, H. Amano, Y. Suga, M. Seishima, K. Takahashi, Hair abnormality in Netherton syndrome observed under polarized light microscopy. J. Am. Acad. Dermatol. **83**(3), 847–853 (2020). https://doi.org/10.1016/j.jaad.2019.08.024
32. H. Xiao et al., The study on starch granules by using darkfield and polarized light microscopy. J. Food Compos. Anal. **92**, 103576 (2020). https://doi.org/10.1016/j.jfca.2020.103576
33. F. Qi, Y. Han, H. Liu, H. Meng, Z. Li, L. Xiao, Localized surface plasmon resonance coupled single-particle galactose assay with dark-field optical microscopy. Sens. Actuators B Chem. **320**, 128347 (2020). https://doi.org/10.1016/j.snb.2020.128347
34. Z. Zou et al., Photoactivatable fluorescent probes for spatiotemporal-controlled biosensing and imaging. TrAC Trends Anal. Chem. **125**, 115811 (2020). https://doi.org/10.1016/j.trac.2020.115811
35. D. Boassa et al., Split-miniSOG for spatially detecting intracellular protein-protein interactions by correlated light and electron microscopy. Cell Chem. Biol. **26**(10), 1407-1416.e5 (2019). https://doi.org/10.1016/j.chembiol.2019.07.007
36. D.T. Grubb, Optical microscopy, in *Polymer Science: A Comprehensive Reference* (Elsevier, 2012), pp. 465–478
37. L. Chen et al., Advances of super-resolution fluorescence polarization microscopy and its applications in life sciences. Comput. Struct. Biotechnol. J. **18**, 2209–2216 (2020). https://doi.org/10.1016/j.csbj.2020.06.038
38. C.C. Moura, A. Miranda, R.O.C. Oreffo, P.A.A. De Beule, Correlative fluorescence and atomic force microscopy to advance the bio-physical characterisation of co-culture of living cells. Biochem. Biophys. Res. Commun. **529**(2), 392–397 (2020). https://doi.org/10.1016/j.bbrc.2020.06.037
39. S.G. Liu, D. Zhang, Y. He, W. Gao, X. Shi, A split aptamer sensing platform for highly sensitive detection of theophylline based on dual-color fluorescence colocalization and single molecule photobleaching. Biosens. Bioelectron. **166**, 112461 (2020). https://doi.org/10.1016/j.bios.2020.112461
40. C. Wu, Y. Shan, X. Wu, S. Wang, F. Liu, Quantitative protein detection using single molecule imaging enzyme-linked immunosorbent assay (iELISA). Anal. Biochem. **587**, 113466 (2019). https://doi.org/10.1016/j.ab.2019.113466
41. C.W. Oatley, The early history of the scanning electron microscope. J. Appl. Phys. **53**(2), 1–13 (1982). https://doi.org/10.1063/1.331666
42. P. Hawkes, Recent advances in electron optics and electron microscopy. Ann. Fond. Louis Broglie **29**(1), 837–855 (2004)
43. E. Ruska, The development of the electron microscope and of electron microscopy (nobel lecture). Angew. Chem. Int. Ed. Engl. **26**(7), 595–605 (1987). https://doi.org/10.1002/anie.198705953
44. J.I. Goldstein, D.E. Newbury, J.R. Michael, N.W.M. Ritchie, J.H.J. Scott, D.C. Joy, *Scanning Electron Microscopy and X-Ray Microanalysis*, 4th edn. (Springer, New York, NY, 2018)
45. P.J. Breton, From microns to nanometers: early landmarks in the science of scanning electron microscope imaging. Scan. Microsc. **13**(1), 1–6 (1999)
46. A. Ul-Hamid, *A Beginners' Guide to Scanning Electron Microscopy*, vol. 1 (Springer International Publishing, Cham, 2018)
47. S.-V. Berlin, H. Gmbh, *Springer Series in Optical Sciences*
48. T.G. Rochow, E.G. Rochow, *An Introduction to Microscopy by Means of Light, Electrons, X-Rays, or Ultrasound* (Springer, US, Boston, MA, 1995)
49. P. Echlin, Scanning electron microscopy and analysis of moist, wet and liquid specimens. Microsc. Microanal. **15**(2), 1116–1117 (2009). https://doi.org/10.1017/S1431927609097815
50. K. de Haan, Z.S. Ballard, Y. Rivenson, Y. Wu, A. Ozcan, Resolution enhancement in scanning electron microscopy using deep learning. Sci. Rep. **9**(1), 1–7 (2019). https://doi.org/10.1038/s41598-019-48444-2

51. G. Rong, et al., Liquid-phase electrochemical scanning electron microscopy for in situ investigation of lithium dendrite growth and dissolution. Adv. Mater. **29**(13) (2017). https://doi.org/10.1002/adma.201606187
52. S. Tsiper, O. Dicker, I. Kaizerman, Z. Zohar, M. Segev, Y.C. Eldar, Sparsity-based super resolution for SEM images. Nano Lett. **17**(9), 5437–5445 (2017). https://doi.org/10.1021/acs.nanolett.7b02091
53. M.H. Modarres, R. Aversa, S. Cozzini, R. Ciancio, A. Leto, G.P. Brandino, Neural network for nanoscience scanning electron microscope image recognition. Sci. Rep. **7**(1), 13282 (2017). https://doi.org/10.1038/s41598-017-13565-z
54. W. Zhou, R. Apkarian, Z.L. Wang, D. Joy, Fundamentals of Scanning Electron Microscopy (SEM), in *Scanning Microscopy for Nanotechnology*. ed. by W. Zhou, Z.L. Wang (Springer, New York, 2006), pp. 1–40
55. S.K. Sharma, D.S. Verma, L.U. Khan, S. Kumar, S.B. Khan, *Handbook of Materials Characterization* (Springer International Publishing, 2018)
56. H.-F. Cui, W.-W. Wu, M.-M. Li, X. Song, Y. Lv, T.-T. Zhang, A highly stable acetylcholinesterase biosensor based on chitosan-TiO_2-graphene nanocomposites for detection of organophosphate pesticides. Biosens. Bioelectron. **99**, 223–229 (2018). https://doi.org/10.1016/j.bios.2017.07.068
57. X. Feng, H. Cheng, Y. Pan, H. Zheng, Development of glucose biosensors based on nanostructured graphene-conducting polyaniline composite. Biosens. Bioelectron. **70**, 411–417 (2015). https://doi.org/10.1016/j.bios.2015.03.046
58. Y. Song et al., Ratiometric electrochemical glucose biosensor based on GOD/AuNPs/Cu-BTC MOFs/macroporous carbon integrated electrode. Sens. Actuators B Chem. **257**, 792–799 (2018). https://doi.org/10.1016/j.snb.2017.11.004
59. L. De Broglie, Recherches sur la théorie des Quanta. Ann. Phys. (Paris) **10**(3), 22–128 (1925). https://doi.org/10.1051/anphys/192510030022
60. F. Haguenau, P.W. Hawkes, J.L. Hutchison, B. Satiat-Jeunemaître, G.T. Simon, D.B. Williams, Key events in the history of electron microscopy. Microsc. Microanal. **9**(2), 96–138 (2003). https://doi.org/10.1017/S1431927603030113
61. D.B. Williams, C.B. Carter, The transmission electron microscope, in *Transmission Electron Microscopy* (Springer US, Boston, MA, 1996), pp. 3–17
62. G.P. Thomson, The diffraction of cathode rays by thin films of platinum. Nature **120**(3031), 802–802 (1927). https://doi.org/10.1038/120802a0
63. Gaede, Das elektronenmikroskop, DMW Dtsch. Med. Wochenschr. **66**(31), 858–860 (1940). https://doi.org/10.1055/s-0028-1122268
64. P. Harris, Transmission electron microscopy of carbon: a brief history. J. Carbon Res. **4**(1), 4 (2018). https://doi.org/10.3390/c4010004
65. E. Ruska, Über Fortschritte im Bau und in der Leistung des magnetischen Elektronenmikroskops. Z. Phys. **87**(9–10), 580–602 (1934). https://doi.org/10.1007/BF01333326
66. Ł. Mielańczyk, N. Matysiak, O. Klymenko, R. Wojnicz, Transmission electron microscopy of biological samples, in *The Transmission Electron Microscope—Theory and Applications* (InTech, 2015)
67. R.D. Heidenreich, Electron microscope and diffraction study of metal crystal textures by means of thin sections. J. Appl. Phys. **20**(10), 993–1010 (1949). https://doi.org/10.1063/1.1698264
68. A. Kaech, I. Analysis, *An Introduction to Electron Microscopy Instrumentation, Imaging and Preparation*
69. P. Goodhew, General introduction to transmission electron microscopy (TEM), in *Aberration-Corrected Analytical Transmission Electron Microscopy* (Wiley, Chichester, UK, 2011), pp. 1–19
70. E.J. Kirkland, The transmission electron microscope, in *Advanced Computing in Electron Microscopy* (Springer International Publishing, Cham, 2020), pp. 9–36
71. W. Probst, G. Benner, J. Bihr, E. Weimer, An 'omega' energy filtering TEM—principles and applications. Adv. Mater. **5**(4), 297–300 (1993). https://doi.org/10.1002/adma.19930050417

72. G. Benner, W. Probst, Köhler illumination in the TEM: fundamentals and advantages. J. Microsc. **174**(3), 133–142 (1994). https://doi.org/10.1111/j.1365-2818.1994.tb03461.x
73. D.B. Williams, C.B. Carter, Thickness and bending effects, in *Transmission Electron Microscopy: A Textbook for Materials Science* (Springer US, Boston, MA, 2009), pp. 407–417
74. S.B. Surman et al., Comparison of microscope techniques for the examination of biofilms. J. Microbiol. Methods **25**(1), 57–70 (1996). https://doi.org/10.1016/0167-7012(95)00085-2
75. J.P. McCaffrey, J.-M. Baribeau, A transmission electron microscope (TEM) calibration standard sample for all magnification, camera constant, and image/diffraction pattern rotation calibrations. Microsc. Res. Tech. **32**(5), 449–454 (1995). https://doi.org/10.1002/jemt.1070320507
76. L.A. Giannuzzi, F.A. Stevie, A review of focused ion beam milling techniques for TEM specimen preparation. Micron **30**(3), 197–204 (1999). https://doi.org/10.1016/S0968-4328(99)00005-0
77. D.B. Williams, C.B. Carter, *Transmission Electron Microscopy* (Springer, US, Boston, MA, 2009)
78. N.K. Patel, M.A. Chaudhary, A.J. Parmar, *Life Sciences Leaflets*, vol. 4297 (2013), pp. 20–28
79. M. Winey, J.B. Meehl, E.T. O'Toole, T.H. Giddings Jr., Conventional transmission electron microscopy. Mol. Biol. Cell **25**(3), 319–323 (2014). https://doi.org/10.1091/mbc.E12-12-0863
80. R. Linares, S. Tan, C. Gounou, A.R. Brisson, Imaging and quantification of extracellular vesicles by transmission electron microscopy. Methods Mol. Biol. **1545**, 43–54 (2017). https://doi.org/10.1007/978-1-4939-6728-5_4
81. L.E. Franken, E.J. Boekema, M.C.A. Stuart, Transmission electron microscopy as a tool for the characterization of soft materials: application and interpretation. Adv. Sci. **4**(5) (2017). https://doi.org/10.1002/advs.201600476
82. A. Shoemark, Applications of emerging transmission electron microscopy technology in PCD research and diagnosis. Ultrastruct. Pathol. **41**(6), 408–414 (2017). https://doi.org/10.1080/01913123.2017.1365789
83. U. Baxa, Imaging of liposomes by transmission electron microscopy. Methods Mol. Biol. **1682**, 73–88 (2018). https://doi.org/10.1007/978-1-4939-7352-1_8
84. A.F.A. Foad, Comparing the use of virtual and conventional light microscopy in practical sessions: virtual reality in Tabuk University. J. Taibah Univ. Med. Sci. **12**(2), 183–186 (2017). https://doi.org/10.1016/j.jtumed.2016.10.015
85. D. Lawlor, Troubleshooting, in *Introduction to Light Microscopy* (Springer International Publishing, Cham, 2019), pp. 139–154
86. R.C. Deagle, T.-L.E. Wee, C.M. Brown, Reproducibility in light microscopy: maintenance, standards and SOPs. Int. J. Biochem. Cell Biol. **89**, 120–124 (2017). https://doi.org/10.1016/j.biocel.2017.06.008
87. J. Sanderson, *Understanding Light Microscopy* (Wiley, 2019)
88. P. Echlin, *Handbook of Sample Preparation for Scanning Electron Microscopy and X-Ray Microanalysis* (Springer, 2009)
89. L. Reimer, *Scanning Electron Microscopy: Physics of Image Formation and Microanalysis* (Springer, Berlin, Heidelberg, 2013)
90. D. Ye, K.A. Dawson, I. Lynch, A TEM protocol for quality assurance of in vitro cellular barrier models and its application to the assessment of nanoparticle transport mechanisms across barriers. Analyst **140**(1), 83–97 (2015). https://doi.org/10.1039/c4an01276c
91. R. Operation, *Operating Instructions for HITACHI H-7000 TEM AIR* (2018)

Chapter 2
Characterization Techniques for Mass Spectrometry Analysis

Hamed Hosseinian, Euth Ortiz Ortega, María José Rosales López, Andrea Rodríguez Vera, and Samira Hosseini

Abbreviations

AgNWs	Silver nanowires
BBB	Blood–brain barrier
DART-MS	Direct analysis in real-time mass spectrometry
DESI-MS	Desorption electrospray ionization mass spectrometry
EESI-MS	Extractive electrospray ionization mass spectrometry
FT	Fourier transform
FT-ICR	Fourier transform ion cyclotron resonance
GLU	Glutamate
H&E	Hematoxylin and eosin
ICR	Ion cyclotron resonance
LVC-MS	Liquid vortex capture mass spectrometry
m/z	Mass-to-charge ratio
MAA	Methacrylic acid
MALDI	Matrix-assisted laser desorption/ionization
MALDI-TOF	Matrix-assisted laser desorption/ionization time-of-flight
MALDI-TOF-MS	Matrix-assisted laser desorption/ionization time-of-flight mass spectrometry
MMA	Methyl methacrylate
MS	Mass spectrometry
MSC	Mass spectrometric biosensing chip
PMMA-co-MAA	Polymethyl methacrylate-co-methacrylic acid
PKA	Protein kinase A
PMF	Protein mass fingerprinting
QMS	Quadrupole mass analyzer
SCP-LVC-MS	Single-cell printer liquid vortex capture mass spectrometry
TOF	Time-of-flight
TOF-SIMS	Time-of-flight secondary ion mass spectrometry

E. O. Ortega et al., *Material Characterization Techniques and Applications*,
Progress in Optical Science and Photonics 19,
https://doi.org/10.1007/978-981-16-9569-8_2

UCA	Urocanic acid
UV	Ultraviolet

2.1 Mass Spectrometry for Material Characterization

2.1.1 *History of Mass Spectrometry*

In 1912, J. J. Thomson invented the world's first mass spectrometer, originally naming it a parabola spectrograph. The constructed instrument was widely used at the beginning of the twentieth century to discover fundamental aspects of atomic and molecular structure in addition to bringing to light stable isotopes [1]. In this device, a parabolic slit in a metal plate was behind a Faraday cup connected to an electroscope. The slit deflected each positive ion beam, and the intensity was measured by adjusting the magnetic field. Subsequently, a mass spectrum of ion abundance against relative mass was plotted [2]. Following the progress of their advisor, F. W. Aston, and A. J. Dempster developed a more advanced version of this device, achieving isotopes mapping (Fig. 2.1).

Furthermore, Sir J. J. Thomson, following the creation of the first mass spectrometry (MS) device, uncovered the evidence of multiple isotopes of a single stable element. By 1919, Francis William Aston developed the mass spectrograph to map isotopes and discovered ^{20}Ne and ^{22}Ne [1]. In this mass spectrograph, an electrical field produces ion beams within a tube. The ions are then separated according to their mass-to-charge ratio (m/z), using a magnetic field and based upon the principle that lighter isotopes bent more than heavier ones. In 1940, an American physicist and mass spectrometry pioneer named Alfred Nier designed a new mass spectrometer in which the ion source and detector were removed to avoid the influence of

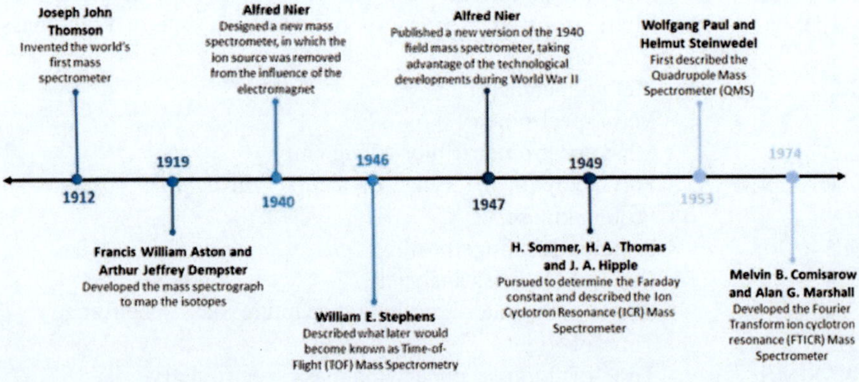

Fig. 2.1 Milestones that shaped the history of MS

the electromagnet. These adjustments resulted in the reduction of power consumption and fabrication costs without any decline in resolution [3]. Nier's improvements allowed more sensitive and precise measurements of isotopes and their ratios [1] and strongly influenced World War II's course. For instance, the Manhattan project employed MS to purify ^{235}U, the fissionable isotope of uranium. Increasing the fuel quality by monitoring the petroleum processing was also an important task given to mass spectrometers as the demand grew for higher-octane fuel destined to fighter aircraft. After the war ended, the petroleum industry continued to use MS for the molecular characterization of oil and the development of oil-based products.

In 1946, William E. Stephens first described the time-of-flight mass spectrometry (TOF-MS). The TOF-MS uses the time-of-flight (TOF) to measure the m/z. This technique separates ions of different masses based on the time the ion takes to travel a fixed distance from the ion source to the ion detector. The movement is achieved by an electrically accelerating method, seeking that ions of the same charge possess the same kinetic energy. In this regard, ions of the same kinetic energy and different masses take distinctive time intervals to travel a definite distance [2]. Taking advantage of the advancements in vacuum technology during World War II, Nier published a modified version of his previous field mass spectrometer in 1947. This instrument represented an important event in the history of MS as it made the knowledge of isotope widely available to the scientific community. Nier also discovered and purified the ^{13}C isotope, which is now employed in tracer experiments to understand metabolic pathways [3].

In 1949, the initial aim of J. A. Hipple, H. Sommer, and H. A. Thomas was to determine the Faraday constant by measuring the cyclotron resonance frequency of protons. In the process, they described the ion cyclotron resonance (ICR) mass spectrometer, which represented a closer step toward high-resolution MS. Furthermore, Wolfgang Paul and Helmut Steinwedel first described the quadrupole mass analyzer (QMS) in 1953. The QMS specified a type of dynamic mass analyzer that sorted out ions based on their stability within a quadrupolar field, achieved by using a combination of sinusoidal and static voltage potentials. In 1974, the Fourier transform ion cyclotron resonance (FT-ICR) mass spectrometer was developed by Melvin B. Comisarow and Alan G. Marshall. In FT-ICR MS, a Penning trap holds sample ions, which are excited by an oscillating electric field until they are separated. Subsequently, the Fourier transform (FT) converts the produced ion signal from its frequency domain to a mass spectrum [2].

2.1.2 Mechanism of Operation of Mass Spectrometry

All mass spectrometers consist of an ion source, a mass analyzer, and a detector operating under high vacuum (Fig. 2.2). The complete and most basic scheme of the MS technique converts the sample into gaseous ions that are then analyzed by their mass-to-charge ratio (m/z) and relative distribution. The process follows the steps of sample introduction, evaporation, and ionization or desorption/ionization while

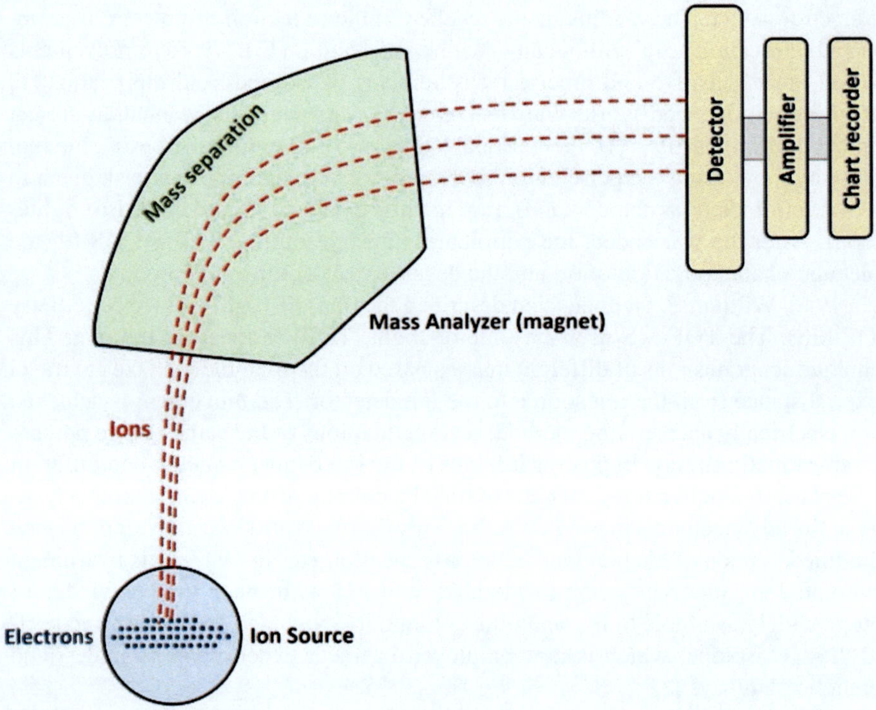

Fig. 2.2 General layout of a mass spectrometer and its main components: an ion source, a mass analyzer, and a detection system

each step requires careful optimization depending on the sample. Mass spectrometers have operated under total data system control since the 1990s, which is essential for data collection, customization of spectral plots, and in-depth data analysis [4].

2.1.2.1 Ion Sources

The production of gaseous ions of the sample is the primary step in the mass spectrometric characterization of the compounds. The ion source of the mass spectrometer produces an electron beam and accelerates it toward the mass analyzer. In this process, two planar electrodes are positioned in a 90° orientation to the axis of an electric field. The electrodes use a potential difference to provide a plasma discharge, which is a phenomenon of an ionized gas containing equal concentrations of positive ions and electrons. A hole in the cathode enables positive ions to appear collimated into a beam. It requires a setup that contains various electrode arrangements involving electron-emitting filaments that work with extensive ranges of voltages and pressures [5]. Several equations model the gas-phase ions' production process, such as the basic equation for electron ionization (Eq. 2.1).

$$M + e^- \rightarrow M^{\cdot+} + 2e^- \quad (2.1)$$

Radical cations with an odd value of electrons usually experience fragmentation to give either a molecule and a new radical cation or a radical and an ion with an even value of electrons. The fragmentation process is important due to the different chemical characteristics of both types of ions. Subsequently, the mass spectrometer separates ions according to their m/z and detects them proportional to their abundance, thus producing a mass spectrum of the molecule [6].

The ion source of the mass spectrometer ionizes the sample before analyzing it. MS allows a vast number of ionization techniques, although the user should always consider the internal energy transferred through the ionization process. For instance, some ionization methods result in extensive fragmentation due to a high energy transfer, while others may only generate ions of the molecular species due to a low energy transfer. The selection of the ionization technique should also depend on the physicochemical characteristics of the analyte, as some procedures such as the electron, chemical, and field ionization are only appropriate for gas-phase ionization and are restricted to volatile and thermally stable specimens [6].

2.1.2.2 Mass Analyzer

Gas-phase ions must be separated based upon their masses once their production process ends. However, their masses are unknown at this stage. Opportunely, the mass analyzer takes advantage of the mass-to-charge ratio (m/z) of ions and measures this physical property; therefore, the obtained m/z values are fractional parts of their actual masses. Since ion separation is a fundamental step of the process, several mass analyzers were introduced to measure the m/z of ions according to distinct principles. By way of illustration, an electrical mass analyzer uses kinetic energy, while a quadrupole mass analyzer uses m/z (trajectory stability), and an ion trap mass analyzer uses m/z (resonance frequency) to meet the separation function. Regardless of the followed principle, all mass analyzers use static or dynamic electric and magnetic fields. The difference in mass analyzers relies upon how such fields are used to achieve separation [6].

2.1.2.3 Detectors and Computers

After mass separation, the ions reach the ion detector, and it generates an electric current proportional to their abundance, thus transforming detectable ions into a functional signal. Similar to the other main components of the mass spectrometer, the choice of the detector largely depends on the performed analytical application. For instance, Faraday cup detectors measure the direct charge current generated by an ion that attacks a surface and is neutralized. Electron multipliers or electro-optical ion detectors operate when the ions hit a surface and produce a kinetic energy transference, producing secondary electrons that generate an electronic current if amplified. Signal amplification also represents an important step in some detectors such as the Fourier transform ion cyclotron resonance (FT-ICR) or orbitrap mass spectrometers. In these techniques, the ion detector consists of two metal plates inside the mass analyzer region, detecting ions through an image current obtained by the ion circuit that connects the plates. However, the number of ions leaving the mass analyzer usually is meager, and it cannot result in a clear signal unless a significant amplification is conducted. In contrast, the multiplication of the signal intensity can be carried out by a cascade effect in other detectors, except for the Faraday cup and image current detector [6].

2.1.3 Advantages and Disadvantages of Mass Spectrometry

Mass spectrometry is the gold standard characterization method. It allows a rapid and highly accurate analysis to obtain information related to molecular mass, isotope distribution patterns, identification of chemicals, and quantification of concentrations of several chemicals in mixed samples. In addition, this technique requires only a small sample quantity for examination. MS combines high selectivity and sensitivity to analyze samples with a diameter range from femtometers (10^{-15}) to attometers (10^{-18}) [7–9] thus facilitating the analysis of secondary metabolites. Since mass spectrometry is a very recurrent technique in the laboratory, there is a vast availability of MS platforms and technologies that provide various operational principles that can be applied. Hence, the possibility of expanding the number of identified molecules increases [8, 9]. Nonetheless, the application of this technique could be destructive for such samples [8].

Although a sample's structural information can be acquired indirectly by the characteristics of pattern and fragment ions [8, 9], it cannot be obtained directly from mass spectrometry. As stated before, the MS technique uses the sample's ionic form of molecules in a gaseous state, representing a limitation for samples with non-volatile compounds [7, 9]. Another disadvantage of using this technique is the elevated price of instrumentation and the need for specialized personnel to operate it [7]. The benefits and limitations of using MS as a material characterization technique are listed below (Table 2.1).

2.1.4 Applications of Mass Spectrometry

Mass spectrometry (MS) is a highly specific analytical method capable of acquiring quantitative and qualitative data of gaseous-, liquid-, or solid-state analytes. The scientific community employs MS for structure definition, identification, and quantification of molecules, which is helpful for diverse research fields such as pharmaceutics, biochemical genetics, environment, toxicology, polymers, food industry,

Table 2.1 Advantages and disadvantages of utilizing MS

Advantages	Disadvantages
Allows a rapid analysis [9]	Cannot record direct structural information [8, 9]
Allows a highly accurate analysis, given its dynamic range of four orders of magnitude [8, 9]	Requires the usage of the sample's ionic form of molecules in a gaseous state, hindering the analysis of non-volatile compounds [7, 9]
It is widely used for obtaining molecular mass, isotope distribution patterns, identification of chemicals, quantification of concentrations of several chemicals in mixed samples, and deduction of molecule structure [8, 9]	
Combines high sensitivity and selectivity to analyze samples with a diameter range from femtometers (10^{-15}) to attometers (10^{-18}) [7–9]	It is a destructive technique for samples such as metabolites [8]
Allows the analysis of secondary metabolites [8]	
Requires only a small amount of sample [8]	Has an elevated price of instrumentation [7]
There exist a wide array of MS platforms, contributing to augmenting the number of molecules that could be identified [8, 9]	Requires specialized personnel to operate the equipment [7]

MS mass spectrometry

metabolomics, and clinical testing. For instance, immunoassay-based tests are fundamental in molecular diagnostics, yet they present the hook effect and are of poor specificity at low concentrations, negatively affecting their performance. For this reason, it is recommended to use MS to identify small and large molecules (e.g., peptides and proteins) and determine their concentration in the specimen since this technique provides a high degree of sensitivity and rapid data acquisition. Moreover, it is a simple method that allows multi-analyte fingerprinting, non-destructive surface analysis, and spatially resolved analyte detection, among others. MS also enables early disease detection, monitoring treatment, and obtaining high-quality images for disease detection. Moreover, scientific advances have also provided solutions to those problems presented by the basic MS technique. Revolutionary ambient ionization methods such as the direct analysis in real-time mass spectrometry (DART-MS), the extractive electrospray ionization mass spectrometry (EESI-MS), and the desorption electrospray ionization mass spectrometry (DESI-MS) allow in situ analysis, accurate and fast surgical diagnosis, and little to no prior sample preparation needed [10–14].

In 2018, Zhu et al. used MS to reveal a novel glutamate (GLU) biosynthetic pathway and shed light on sunlight-induced neurobehavioral changes. Exposure to ultraviolet (UV) rays is frequently portrayed as the cause of skin diseases. However, moderate sunlight exposure is favorable for human health as UV exposure positively affects an individual's cognition, memory, and mood due to the concentration change

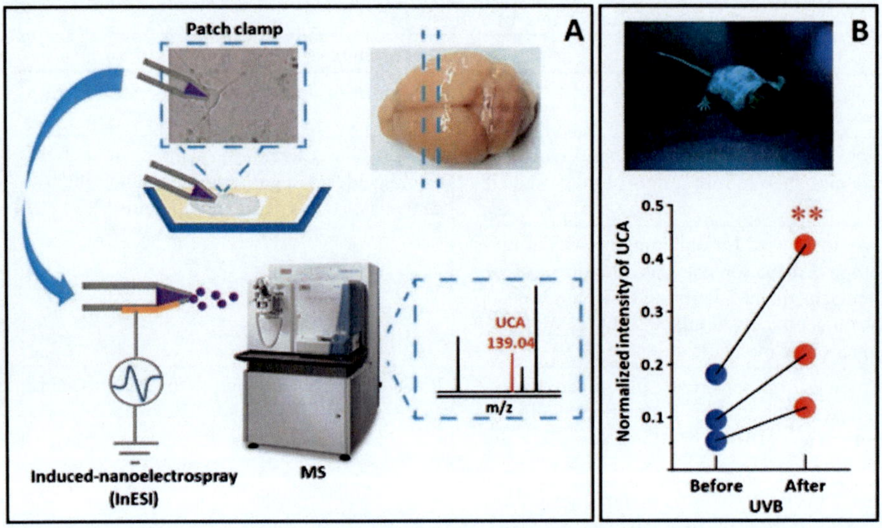

Fig. 2.3 Biological effect of UCA release due to UV exposure. **a** The scheme of the single-neuron mass spectrometer. **b** The UCA levels in mice before (blue dots, N = 3 mice) and after (red dots, N = 3 mice) 2 h of UVB exposure. Reproduced (or Adapted) with permission [15], ©2018, Elsevier

in molecules present in the blood or skin. In this regard, the scientific community has found that Urocanic Acid (UCA) concentration levels increase in the peripheral skin, blood, and urine after UV exposure. Nevertheless, the biological effects of higher concentrations of UCA in the human body remained unknown. Zhu et al. evaluated the neurobehavioral changes induced by UV exposure and developed a single-cell mass spectrometry method consisting of the union of a patch-clamp and MS technologies (Fig. 2.3a). Using a mouse model, the author discovered a notable increase in UCA levels in single neurons after moderate UV exposure (Fig. 2.3b). Subsequently, UCA passed the blood–brain barrier (BBB) to enter neurons and promoted the biosynthesis of GLU. In short, the UCA-GLU metabolic pathway in neurons was found to alter the mice's neuronal excitability, improving recognition memory and rotarod skill learning ability [15].

When it comes to the characterization of molecules at low concentrations within a sample, the study of metabolites represents a great analytical challenge since thousands of chemical species inhabit a cell at low concentrations. Therefore, high sensitivity is an essential feature of any technique employed for metabolite characterization. In this regard, the MS technique proves to be the most suitable choice since it owns high sensitivity, rapid analysis, and chemical specificity. Nevertheless, MS requires a laborious pre-treatment for tests under vacuum, manual assistance for instruments such as pipettes or capillaries, a relatively long time to obtain data results, and presents low sampling performance. In 2019, Cahill et al. proposed a single-cell printer liquid vortex capture mass spectrometry (SCP-LVC-MS) system to address such deficiencies. The proposed SCP-LVC-MS consisted of a single-cell separation

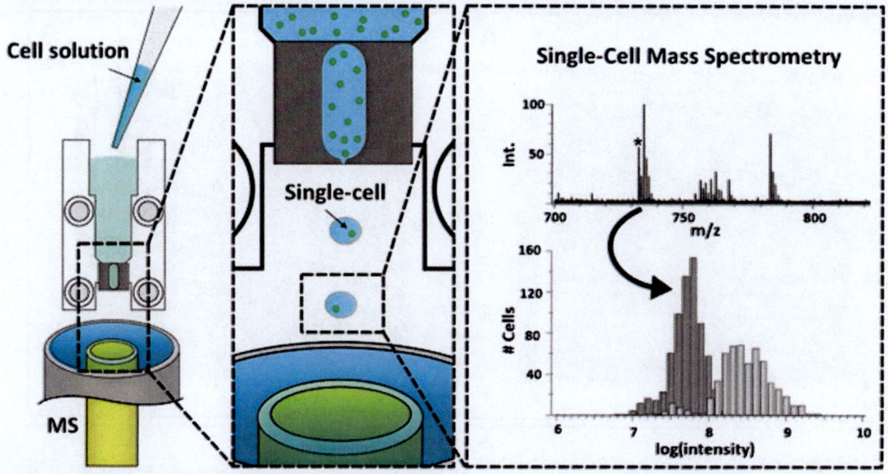

Fig. 2.4 Operation principle of the SCP-LVC-MS system. Reproduced (or Adapted) with permission [16], ©2019, American Chemical Society

principle via droplet expulsion (inkjet printer-like mechanism), and the characterization of single-cell was performed in a droplet (containing >1 or no cells) by liquid vortex capture mass spectrometry (LVC-MS) (Fig. 2.4). The system certainly improved the MS technique as the SCP-LVC-MS did not require a previous treatment in the analyte. In addition, the liquid vortex surface directed the analyte toward the ionization source in approximately 6 s, which resulted in a shorter sample-to-answer time [16].

Furthermore, identifying, localizing, and knowing the chemistry of nanomaterials and proteins in complex biological systems represent a challenging task. The standard procedure to obtain an image of the distribution of proteins is to label the analytes with fluorescent dyes (fluorescence microscopy). However, the diffraction limits the spatial resolution of this technique and compromises the image quality. A reliable and label-free alternative to examine proteins inside cells is the time-of-flight secondary ion mass spectrometry (TOF-SIMS) based on the m/z of the specimen. The TOF-SIMS allows observing amino acids and prosthetic groups while mapping the metal ions with high chemical specificity, surface sensitivity, and spatial resolution. In 2019, Leo et al. used this technique to detect the position of enzymes within cells exposed to silver nanowires (AgNWs). As a result, the authors successfully identified sensitive ion signals of various H2S-synthesizing enzymes within microglia including amino acid fragments and cystathionine-γ-lyase. This study was the first to simultaneously identify organic and inorganic compounds in a biological system (Fig. 2.5). It demonstrated that enzymes within the cells can be localized by imaging and that AgNWs are present both inside and outside the cell surface [17].

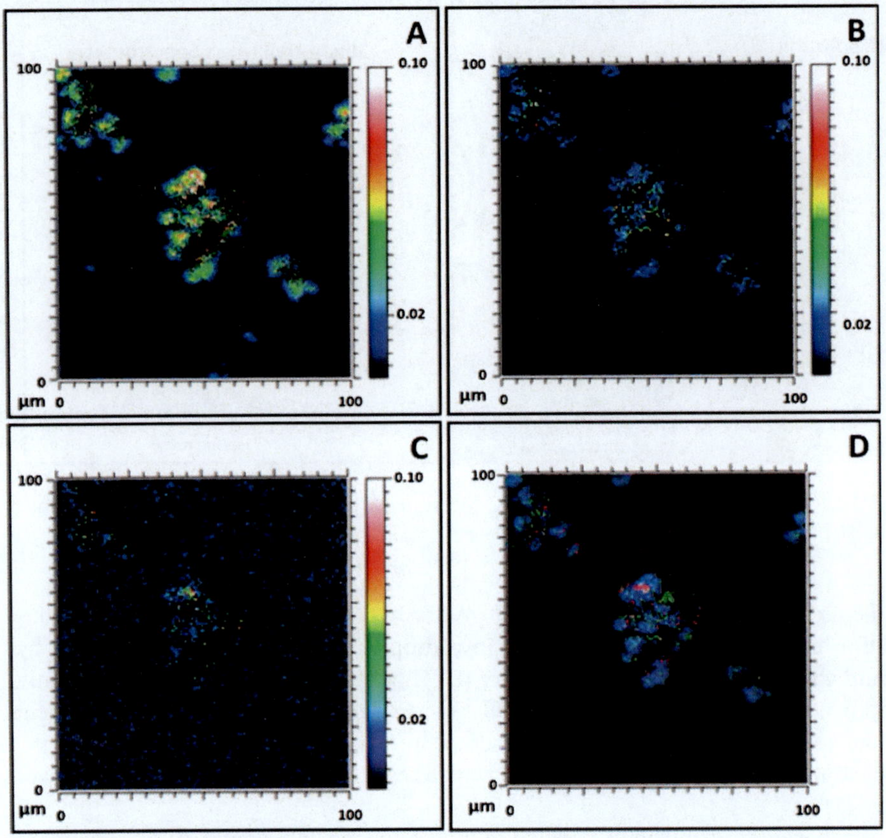

Fig. 2.5 Detection of the position of enzymes within cells exposed to AgNWs using the TOF–SIMS. **a** Protein fragments from cells. **b** Cystathionine fragments. **c** Silver ions. **d** Overlay of silver (red), cystathionine (green), and protein (blue). Reproduced (or Adapted) with permission [17], ©2019, American Chemical Society

2.2 Matrix-Assisted Laser Desorption Ionization Time-of-Flight Mass Spectrometry for Material Characterization

2.2.1 History of Matrix-Assisted Laser Desorption Ionization Time-of-Flight Mass Spectrometry

Mass spectrometry technology rapidly became popular in various research areas such as chemical and biological sciences. However, as the analysis took place through "hard ionization" exclusively, analyzing intact molecules became a relevant limitation of this technology. Responding to such needs, several scientists contributed to

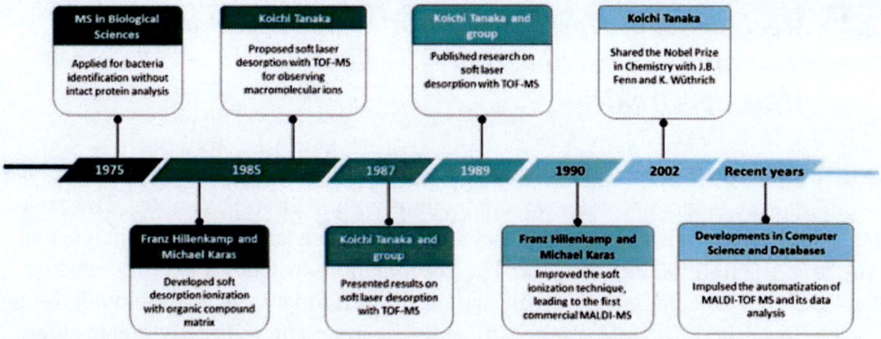

Fig. 2.6 History of MALDI-TOF spectrometry from 1975 to recent times

the further development of the MS technique in the 1980s [18, 19]. The following timeline (Fig. 2.6) illustrates the chronology of events that led to the invention of current matrix-assisted laser desorption/ionization time-of-flight (MALDI-TOF) spectrometry technology.

The history of MALDI-TOF spectrometry traces back to 1985 when Franz Hillenkamp and Michael Karas developed a soft desorption/ionization technique employed to preserve molecule fragmentations of an organic compound matrix of peptides. From this event, the scientific community adopted the term matrix-assisted laser desorption/ionization (MALDI) [18–20]. Koichi Tanaka, a Japanese electrical engineer, was interested in the development of ionization and data acquisition of the technique and gathered a TOF-MS research team at the Central Research Laboratory of Shimadzu Corporation. In the late 1985, Tanaka proposed a technique consisting of soft laser desorption and TOF-MS hardware to observe macromolecular ions. In 1987, the results of Tanaka and the TOF-MS research group were presented internationally and were subsequently published in 1989. In 2002, Tanaka, J. B. Fenn, and K. Wüthrich were awarded the Nobel Prize in Chemistry for their work on the soft desorption/ionization technique for mass spectrometric analyses of biological molecules [18, 19, 21, 22]. In the early 1990s, Hillenkamp and Karas sped up the commercialization of the MALDI mass spectrometer due to the improved soft ionization technique [18]. Today's technological advances in computer science and mass spectra databases allowed the automation of matrix-assisted laser desorption/ionization time-of-flight mass spectrometry (MALDI-TOF-MS) and its associated data analysis [19].

2.2.2 *Mechanism of Operation of Matrix-Assisted Laser Desorption Ionization Time-of-Flight Mass Spectrometry*

MALDI-TOF-MS is an attractive analytical technique to satisfy the need of the scientific community to identify unknown compounds within different samples. This technique applies the soft ionization of the analyzed specimen and directs the analyte ions to the time-of-flight (TOF) detector. The combination provides a simple, sensitive, reliable, and rapid examination of the sample. A pulsed laser beam, commonly of a wavelength in the ultraviolet range, strikes the sample and indirectly creates high-mass ions representing a characteristic fragment of the analyte. The matrix plays a crucial role in this technique since it allows the isolation of analyte molecules from each other by reducing the intramolecular forces through a process named matrix isolation, thus avoiding cluster generation. Also, the matrix absorbs most of the photon energy from the laser beam with high absorption coefficient and promotes the soft ionization of the specimen. Subsequently, these ionized analytes collide in an ion detector through a tube under vacuum to calculate the m/z depending on the TOF (Fig. 2.7) [18, 23–26].

The MALDI-TOF-MS system is composed of an ionization zone, an acceleration zone, and a separation zone. Within the ionization zone, the laser beam interacts with the sample to generate analyte ions. Subsequently, the ions are influenced by the acceleration zone's electrostatic field and increase their velocity toward the separation zone, as the latter is free from any electric field. In the last zone, an ion detector records the flight time, the number, and the intensity in which ions of the same mass impacts the detector. The m/z ratio of the data obtained by the MALDI-TOF-MS system is then calculated with Eq. 2.2.

$$\frac{m}{z} = 2eE\left(\frac{t}{d}\right)^2 \tag{2.2}$$

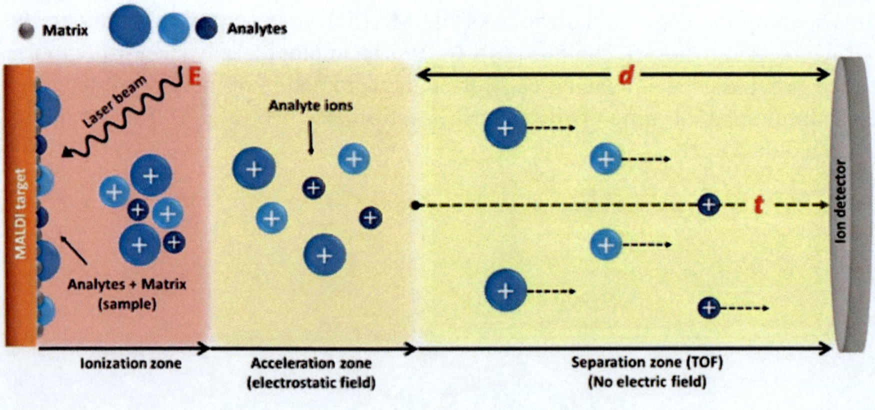

Fig. 2.7 Schematic representation of the MALDI-TOF-MS technique

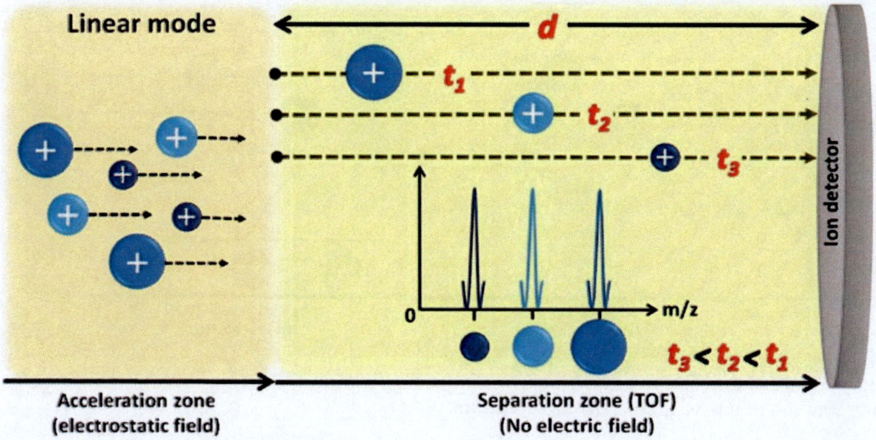

Fig. 2.8 MALDI-TOF-MS in the linear model. The time it takes for an ion to reach the ion detector relies on its mass

where *e* indicates the elementary charge, *E* represents the accelerating voltage from the laser beam, *t* corresponds to the TOF, and *d* expresses the distance of the separation zone. The m/z ratio value is directly proportional to the time-of-flight, meaning that smaller analytes (shorter m/z ratio) move faster (shorter TOF). In contrast, the larger ionized fragments of the analytes (longer m/z ratio) move slower (longer TOF). The time it takes for an ion to move across the separation zone and to reach the ion detector relies on its mass. As stated before, the TOF increases proportionally to ions' mass; hence, heavier ions generate a larger m/z value than lighter ones (Fig. 2.8). Regarding the m/z spectrum, individual mass peaks can be attributed to ions of different masses and represent relevant data regarding the fingerprinting of specimens [18, 24, 27].

The MALDI-TOF-MS operation principle states that the laser beam strikes all generated ions with the same kinetic energy. Therefore, ions of the same mass should take the same TOF to travel across the separation zone. Nonetheless, experimentally, the laser beam's kinetic energy strikes each discrete ion at a different intensity and alters the TOF. Specific mass ions produce a distribution of the m/z value instead of an individual value, thus reducing the resolution of the MALDI-TOF-MS spectrum. In consequence, the mass peak becomes wider and hinders the data interpretation (Fig. 2.9).

The reflection model improved the low resolution of the m/z ratio spectrum presented in the linear model. In the reflection model, the ion detector is installed in front of an ion reflector, which redirects the ions of the same mass and allows them to take an equal TOF after hitting the reflector. In another word, since the laser beam's energy is not equally distributed experimentally ($tx \neq ty$), the ionized analytes leave the acceleration zone taking a different TOF. Nevertheless, the ion reflector adjusts the velocity of the ions of the same mass after reflecting them ($t1 = t2$). Consequently,

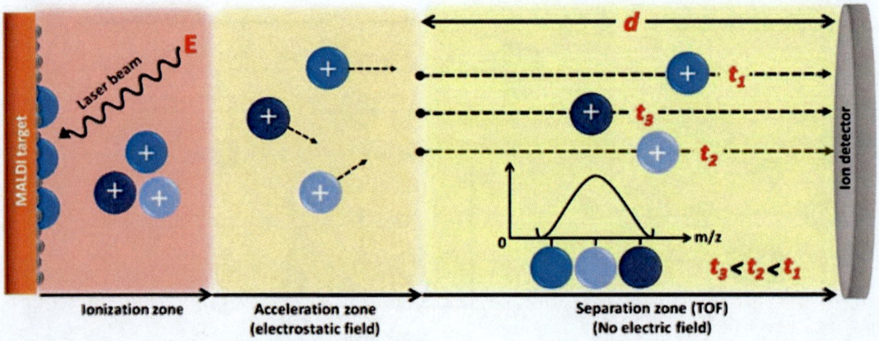

Fig. 2.9 Ions of the same m/z reach the ion detector at a different TOF in the linear model, reducing the resolution of the MALDI-TOF-MS spectrum

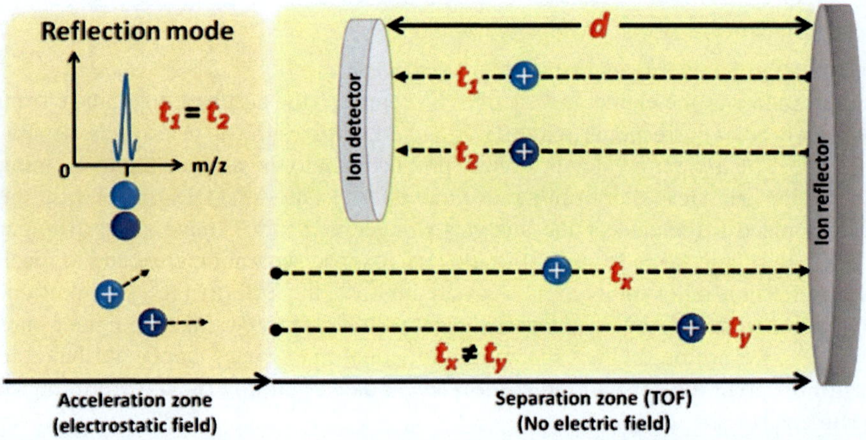

Fig. 2.10 MALDI-TOF-MS is in the reflection mode. The ion reflector improves the drawbacks presented in the linear mode

the ions reach the ion detector at a paired TOF, improving the mass peak's definition (Fig. 2.10) [18].

2.2.3 Advantages and Disadvantages of Matrix-Assisted Laser Desorption Ionization Time-of-Flight Mass Spectrometry

The MALDI-TOF technique has become one of the most common characterization techniques. Its principle of ion production from high molecular mass has expanded

MALDI-TOF-MS to several application fields, including epidemiologic studies, taxonomical classification, biotechnology, cell biology, and pharmaceutical research. In particular, the application of MALDI-TOF allows us to obtain biomarkers' profiles directly from unfractionated microorganisms such as viruses, spores, bacteria, or fungi. This characterization technique can perform bacterial and fungal species identification with a 95% accuracy and a turnaround time of 10 min [28]. Some applications of MALDI-TOF-MS are still in the development stage. For instance, the possibility of detecting bacterial resistance has aroused great interest in recent years. The overall process and usage of MADI-TOF are economical, although they can be costly for more novel applications where thorough optimization is needed. However, the demand for affordable instrumentation will increase with the number of applications. As a result, many manufacturers will begin to offer MALDI-TOF systems dedicated to more specific uses [29].

Nevertheless, the commercial MALDI platforms greatly rely on the current databases of molecules and organisms. Although the databases and software are continually updated, individual researchers hold limited access to such resources due to their high costs [26]. The MALDI-TOF technique also needs further standardization and optimization as it presents low analytical sensitivity when no prior sample culture preparation is performed. It is also not suitable for detecting low bacterial concentrations present in samples such as cerebrospinal fluids. In addition, MALDI-TOF cannot characterize distinct bacteria strains at a subspecies level, as strains within a single species are often highly comparable and members of a single species share similar MALDI-TOF spectra. Finally, it is not possible to completely implement MALDI-TOF in automated processes since several steps still need to be done manually [28] (Table 2.2).

Table 2.2 Advantages and disadvantages of MALDI-TOF

Advantages	Disadvantages
Possesses a turnaround time of 10 min, which is fast in comparison with other techniques [28]	Presents low analytical sensitivity when no prior sample culture preparation is performed [28]
Provides a 95% accuracy for bacteria species identification [28]	Is unable to characterize distinct bacteria strains on a subspecies level [28]
Allows obtaining biomarker profiles directly from unfractionated microorganisms [29]	Is not suitable for detecting a low number of bacteria present in samples [28]
Relies on existing databases, which are continually updated [26]	Is restricted to the limited access of software and databases due to their high costs [26]
Is an effective method for fungal identification, with high accuracy and low operating costs [29]	Needs several steps to be done manually; hence, it is not possible to implement the technique in an automated manner [28]

2.2.4 Applications of Matrix-Assisted Laser Desorption Ionization Time-of-Flight Mass Spectrometry

The MALDI-TOF-MS applies a soft ionization to safely provide molecular mass data of specimens susceptible to fragmentation, hence analyzing the sample without suffering great damage. This feature is helpful in polymer science since delicate materials need a comprehensive characterization of polymers prior to being used to develop biomaterials, biosensor systems, and imaging diagnostics. The MALDI-TOF-MS provides structural information, mass distribution data, and end-group analysis of synthesized polymers [30]. In addition, the rapidness, simplicity of use, reliability, eco-friendly, and low-cost features of the MALDI-TOF-MS make it an attractive option for the diagnosis of infections. The MALDI-TOF-MS technique is also broadly used in biochemical research to identify and quantify isolated biomolecules found at low concentrations in tissues, cells, and biological fluids (Fig. 2.11) [31]. In patients with the suspected microbial disease, the time spent on identifying the pathogen and detecting antimicrobial resistance is crucial to lead to more accurate, targeted, and timely treatment. The MALDI-TOF-MS enables the identification of pathogens in approximately 2 min, performs the analysis at a lower cost than genome sequencing or PCR, and accepts a considerable number of samples, such as cerebrospinal fluid, blood, or urine [32]. Through the mass spectrum obtained by the

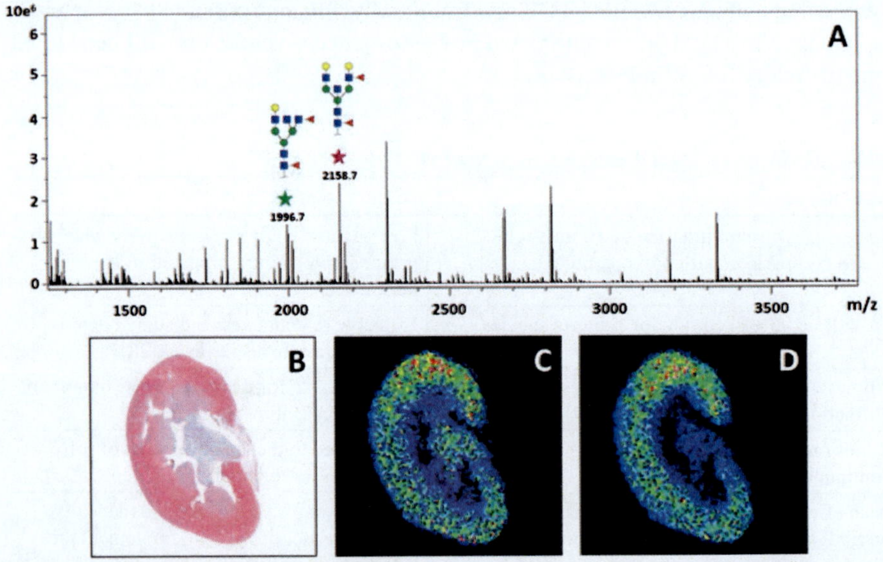

Fig. 2.11 Application of the MALDI-TOF technique on mouse kidneys to identify toxic and disease-causing molecules. **a** A MALDI spectra from the kidney tissue. **b** H&E staining; the labeled peaks correspond to N-glycans reported for the mouse kidney. **c** The ion Hex4dHex2HexNAc5 at m/z = 1996.7 is in the cortex and medulla. **d** The ion Hex5dHex2HexNAc5 at m/z = 2158.7 is more abundant in the cortex. Reproduced (or Adapted) with permission [31], ©2014, PLOS ONE

TOF data, a technique called peptide mass fingerprinting (PMF) allows the identification of different microorganisms when compared with a database, enabling the recognition of the family, genus, and species of unknown organisms.

The MALDI-TOF-MS is widely used in the biotechnology and pharmaceutical areas as well. It can detect biological and synthetic molecules with high accuracy. Moreover, it can analyze more than 100,000 samples in one-day profiling and differentiate various analytes [33]. The MALDI-TOF-MS also represents a novel approach to investigating unidentified micropollutants that are exposed to nature and endanger flora. This technique facilitates the quantification and identification of pharmaceutical waste, pesticides, and personal care products. Areas such as agriculture and food safety benefit from the powerful features of MALDI-TOF in the identification of these micropollutants [34].

The MALDI MS is also an effective mechanism for evaluating enzymatic activity. Feng et al. identified specific analytes via a mass spectrometric biosensing chip (MSC) based on an enzyme-specific recognition interface through fluorous–affinity interactions. The authors employed caspase-3 and protein kinase A (PKA) as targets to monitor the changes in the m/z ratio patterns and record the fingerprint of enzymes. The evaluation of MSC revealed four well-defined peaks associated with a splitting product (m/z = 1990), a phosphate hydrolysis product (m/z = 2070), a fluorous-tagged peptide (m/z = 2133), and a phosphorylation product (m/z = 2213) (Fig. 2.12). The MSC without target analytes resulted in no changes in the obtained mass spectrum, thus verifying the technique's specificity. The MSC allowed a more effective and simple quantification of enzyme activity compared with other methods.

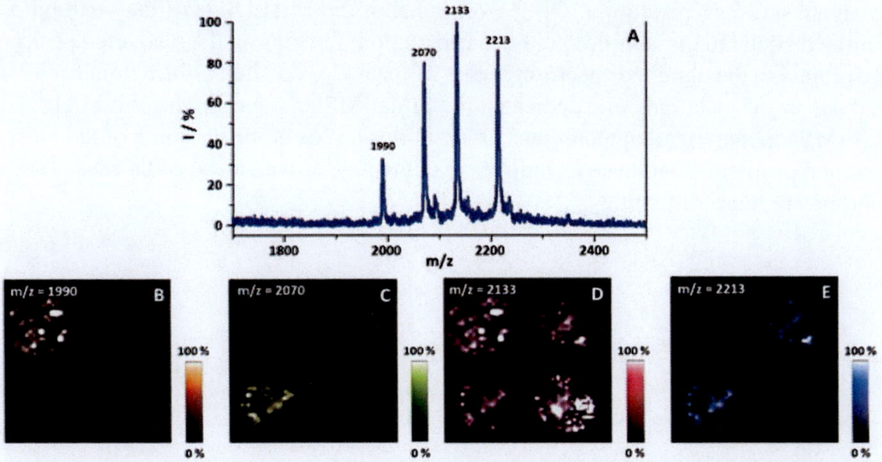

Fig. 2.12 Quantitation of enzyme activity via MSC. **a** Mass spectrum of MSC after enzymatic reactions of PKA and caspase-3. MS recording of MSCs upon reaction with **b** caspase-3, **c** PKA, **d** dual enzymes in assay buffer, and **e** assay buffer. Reproduced (or Adapted) with permission [35], ©2020, Elsevier

In addition, further using the MALDI MS enabled identifying the presence of enzyme inhibitors at good reproducibility and sensibility rates [35].

Polymeric materials hold properties such as flexibility and high resistance that are desirable for developing biomaterials and biosensor systems. Therefore, knowing the block copolymer, structure, end groups, and composition of new polymer materials is vital to leverage their potentials. Hosseini et al. synthesized a novel polymeric material composed of polymethyl methacrylate-co-methacrylic acid (PMMA-co-MAA) using different monomer ratios via free-radical polymerization. In this study, the MALDI-TOF-MS was used to present valuable information on the developed molecular mass distribution, end groups, and structures of the oligomer chains. One of the most significant advantages of this characterization technique is the capability of preserving the specimen's structure, thus observing the entire chain of oligomers in the mass spectrum. In this regard, the MALDI-TOF-MS provided clear evidence of the distribution of the monomer units in the synthesized material. The mass spectrum exhibited the ordered insertion of the methacrylic acid (MAA) segments as end groups in the methyl methacrylate (MMA) polymeric chain (Fig. 2.13) [30].

2.3 Troubleshooting of Mass Spectrometry Analysis Techniques

Some possible problems might be faced while performing MS or MALDI-TOF-MS techniques. On the one hand, the MS may present a loss of sensitivity, having no signal and low resolution. On the other hand, the MALDI-TOF may exhibit a saturated signal in the sample and the matrix region, flat tops on the recovered peaks, and dimers in the obtained spectrum. Table 2.3 provides detailed information for MS analysis in two sections, one dedicated to regular MS and another for the MALDI-TOF-MS characterization technique. Overall, this table comprises some of the most common problems that these techniques may display, as well as possible causes and solutions to these difficulties.

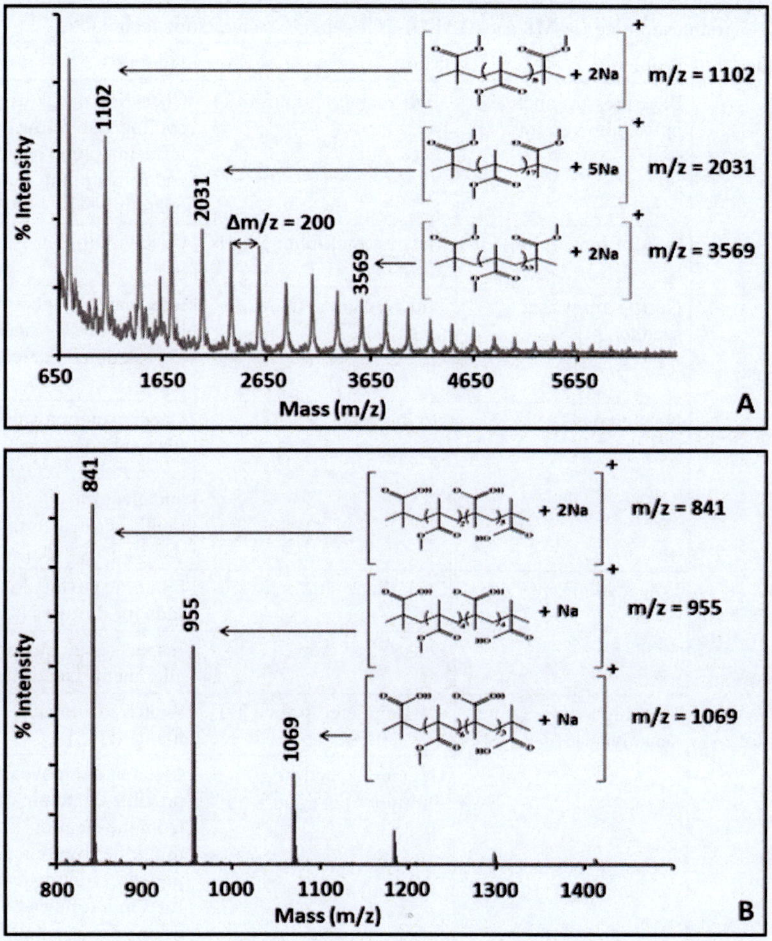

Fig. 2.13 MALDI-TOF-MS usage on polymeric material to obtain the distribution of the monomer units. The MALDI-TOF-MS spectrum of **a** PMMA and **b** PMMA-co-MAA. Reproduced (or Adapted) with permission [30], ©2014, Elsevier

Table 2.3 Troubleshooting for MS and MALDI-TOF-MS characterization techniques

Technique	Problem	Cause	Solution
MS	Base pressure increases in vacuum systems [4]	High vacuum pumps off [4]	Check ion gauge trip, cooling water flow, diffusion pump heater, and rough pump operation [4]
	Loss of sensitivity	Electron multiplier gain is down [4]	Check multiplier gain [4]
	Narrow mass range scanned [4]	Widen scanned mass range [4]	Ions may be higher or lower in mass. Check expected mass defects [4]
	No signal [4]	No ion beam [4]	Check isolation valve, filament continuity, accelerating voltage, and electron multiplier/detector, and the ion lens voltages [4]
	Low resolution [4]	Peak shape shows lift-off [4]	Examine the quadrupole rods for dirtiness [4]
		Peak shape skewed [4]	Inspect sector slits for alignment issues [4]
	Poor high-end sensitivity [36]	Contaminated source [36]	Switch to another filament [36]
		Degraded electron multiplier [36]	Clean and remove the possible contaminations from the electron multiplier by sonication preferably in methanol for 3 min followed by air drying for 30 min [36]
MALDI-TOF	Saturated signal in the sample and the matrix region [37]	High laser intensity [37]	Adjust the laser by using the slider controls [37]
	Poor crystallization on the sample plate [37]	Contaminated sample [37]	Clean the sample by washing, drop dialysis (floating membrane), or cation exchange [37]
	Flat tops on the recovered peaks [37]	Signal saturation due to high laser power[37]	Decrease laser setting until peak tops are sharp by using the slider controls [37]
	Round tops on peaks [37]	Saturated (flat top) and unsaturated (sharp top) scans [37]	

(continued)

Table 2.3 (continued)

Technique	Problem	Cause	Solution
		Laser power has changed during the acquisition of data [37]	Reacquire data using a laser setting that gives sharp peaks [37]. Note that data acquisition using different laser settings may have poor mass accuracy [37]
	Dimers, trimers, and tetramers in the spectrum [37]	High sample concentration [37]	Prepare the sample or matrix with an appropriate and diluted concentration [37]
	Dimer in the spectrum [37]	Signal saturation due to high laser intensity [37]	Adjust the laser by using the slider controls [37]
	Peaks are not symmetrical [37]	High laser intensity [37]	Decrease the laser intensity by using the slider controls [37]
		The sample contains more than one component [37]	Purify the sample before analyzing [37]
		The signal is saturating the detector [37]	Decrease laser intensity by using the slider controls [37]

MS mass spectrometry; *MALDI-TOF* matrix-assisted laser desorption/ionization time-of-flight

References

1. J.R. Yates, A century of mass spectrometry: from atoms to proteomes. Nat. Methods **8**(8), 633–637 (2011). https://doi.org/10.1038/nmeth.1659
2. S. Maher, F.P.M. Jjunju, S. Taylor, Colloquium: 100 years of mass spectrometry: perspectives and future trends. Rev. Mod. Phys. **87**(1), 113–135 (2015). https://doi.org/10.1103/RevModPhys.87.113
3. J. De Laeter, M.D. Kurz, Alfred Nier and the sector field mass spectrometer. J. Mass Spectrom. **41**(7), 847–854 (2006). https://doi.org/10.1002/jms.1057
4. J.H. Gross, *Mass Spectrometry*, 2nd edn. (Springer Berlin Heidelberg, Berlin, Heidelberg, 2011)
5. J.H. Beynon, Mass Spectrometry. https://www.britannica.com/science/mass-spectrometry
6. E. De Hoffmann, V. Stroobant, *Mass Spectrometry: Principles and Applications*, 3rd edn. (Wiley, 2007)
7. A. Scherl, Clinical protein mass spectrometry. Methods **81**, 3–14 (2015). https://doi.org/10.1016/j.ymeth.2015.02.015
8. A.-H.M. Emwas, The Strengths and Weaknesses of NMR Spectroscopy and Mass Spectrometry with Particular Focus on Metabolomics Research, in *Metabolomics. Methods in Molecular Biology*, vol. 1277, ed. by J.T. Bjerrum. (Humana Press, New York, NY, 2015), pp. 161–193
9. P.K. Ghosh, Introduction, in *Introduction to Protein Mass Spectrometry* (Academic Press, 2016), pp. 1–8
10. F.G. Strathmann, A.N. Hoofnagle, Current and future applications of mass spectrometry to the clinical laboratory. Am. J. Clin. Pathol. **136**(4), 609–616 (2011). https://doi.org/10.1309/AJCPW0TA8OBBNGCK

11. G.A. Harris, A.S. Galhena, F.M. Fernández, Ambient sampling/ionization mass spectrometry: applications and current trends. Anal. Chem. **83**(12), 4508–4538 (2011). https://doi.org/10.1021/ac200918u
12. B. Xi, H. Gu, H. Baniasadi, D. Raftery, Statistical analysis and modeling of mass spectrometry-based metabolomics data, in *Mass Spectrometry in Metabolomics. Methods in Molecular Biology (Methods and Protocols)*, vol. 1198, ed. by D. Raftery (Humana Press, New York, NY, 2014), pp. 333–353
13. J.R. Yates, C.I. Ruse, A. Nakorchevsky, Proteomics by mass spectrometry: approaches, advances, and applications. Annu. Rev. Biomed. Eng. **11**, 49–79 (2009). https://doi.org/10.1146/annurev-bioeng-061008-124934
14. E.J. Finehout, K.H. Lee, An introduction to mass spectrometry applications in biological research. Biochem. Mol. Biol. Educ. **32**(2), 93–100 (2004). https://doi.org/10.1002/bmb.2004.494032020331
15. H. Zhu et al., Moderate UV exposure enhances learning and memory by promoting a novel glutamate biosynthetic pathway in the brain. Cell **173**(7), 1716-1727.e17 (2018). https://doi.org/10.1016/j.cell.2018.04.014
16. J.F. Cahill, J. Riba, V. Kertesz, Rapid, untargeted chemical profiling of single cells in their native environment. Anal. Chem. **91**(9), 6118–6126 (2019). https://doi.org/10.1021/acs.analchem.9b00680
17. B.F. Leo et al., Label-Free time-of-flight secondary ion mass spectrometry imaging of sulfur-producing enzymes inside microglia cells following exposure to silver nanowires. Anal. Chem. **91**(17), 11098–11107 (2019). https://doi.org/10.1021/acs.analchem.9b01704
18. S. Hosseini, S.O. Martinez-Chapa, Principles and Mechanism of MALDI-ToF-MS Analysis, in *Principles and Mechanism of MALDI-ToF-MS Analysis. SpringerBriefs in Applied Sciences and Technology* (Springer, Singapore, 2017), pp. 1–19
19. R. Patel, MALDI-TOF MS for the diagnosis of infectious diseases. Clin. Chem. **61**(1), 100–111 (2015). https://doi.org/10.1373/clinchem.2014.221770
20. M. Karas, D. Bachmann, F. Hillenkamp, Influence of the wavelength in high-irradiance ultraviolet laser desorption mass spectrometry of organic molecules. Anal. Chem. **57**(14), 2935–2939 (1985). https://doi.org/10.1021/ac00291a042
21. Nobel Prize, The Nobel Prize in Chemistry 2002. https://www.nobelprize.org/prizes/chemistry/2002/summary/
22. K. Tanaka, The origin of macromolecule ionization by laser irradiation (Nobel lecture). Angew. Chemie Int. Ed. **42**(33), 3860–3870 (2003). https://doi.org/10.1002/anie.200300585
23. T.Y. Hou, C. Chiang-Ni, S.H. Teng, Current status of MALDI-TOF mass spectrometry in clinical microbiology. J. Food Drug Anal. **27**(2), 404–414 (2019). https://doi.org/10.1016/j.jfda.2019.01.001
24. R. Patel, A moldy application of MALDI: MALDI-ToF mass spectrometry for fungal identification. J. Fungi. **5**(1) (2019). https://doi.org/10.3390/jof5010004
25. M.W. Duncan, D. Nedelkov, R. Walsh, S.J. Hattan, Applications of MALDI mass spectrometry in clinical chemistry. Clin. Chem. **62**(1), 134–143 (2016). https://doi.org/10.1373/clinchem.2015.239491
26. N. Singhal, M. Kumar, P.K. Kanaujia, J.S. Virdi, MALDI-TOF mass spectrometry: An emerging technology for microbial identification and diagnosis. Front. Microbiol. **6**, 1–16 (2015). https://doi.org/10.3389/fmicb.2015.00791
27. A. Croxatto, G. Prod'hom, G. Greub, "Applications of MALDI-TOF mass spectrometry in clinical diagnostic microbiology. FEMS Microbiol. Rev. **36**(2), 380–407 (2012). https://doi.org/10.1111/j.1574-6976.2011.00298.x
28. S. Schubert, M. Kostrzewa, MALDI-TOF mass spectrometry in the clinical microbiology laboratory; beyond identification, in *Methods in Microbiology*, vol. 42, ed. by A. Sails, Y.-W. Tang. (Elsevier, 2015), pp. 501–524
29. L.F. Marvin, M.A. Roberts, L.B. Fay, Matrix-assisted laser desorption/ionization time-of-flight mass spectrometry in clinical chemistry. Clin. Chim. Acta **337**(1–2), 11–21 (2003). https://doi.org/10.1016/j.cccn.2003.08.008

30. S. Hosseini, F. Ibrahim, I. Djordjevic, M.M. Aeinehvand, L.H. Koole, Structural and end-group analysis of synthetic acrylate co-polymers by matrix-assisted laser desorption time-of-flight mass spectrometry: Distribution of pendant carboxyl groups. Polym. Test. **40**, 273–279 (2014). https://doi.org/10.1016/j.polymertesting.2014.09.017
31. T.W. Powers et al., MALDI imaging mass spectrometry profiling of N-glycans in formalin-fixed paraffin embedded clinical tissue blocks and tissue microarrays. PLoS ONE **9**(9), 1–11 (2014). https://doi.org/10.1371/journal.pone.0106255
32. G. Vrioni, C. Tsiamis, G. Oikonomidis, K. Theodoridou, V. Kapsimali, A. Tsakris, MALDI-TOF mass spectrometry technology for detecting biomarkers of antimicrobial resistance: current achievements and future perspectives. Ann. Transl. Med. **6**(12), 240–240 (2018). https://doi.org/10.21037/atm.2018.06.28
33. V. Greco et al., Applications of MALDI-TOF mass spectrometry in clinical proteomics. Expert Rev. Proteomics **15**(8), 683–696 (2018). https://doi.org/10.1080/14789450.2018.1505510
34. K.A. Maloof, A.N. Reinders, K.R. Tucker, Applications of mass spectrometry imaging in the environmental sciences. Curr. Opin. Environ. Sci. Heal. **18**, 54–62 (2020). https://doi.org/10.1016/j.coesh.2020.07.005
35. N. Feng, J. Hu, Q. Ma, H. Ju, Mass spectrometric biosensing: Quantitation of multiplex enzymes using single mass probe and fluorous affinity chip. Biosens. Bioelectron. **157**, 112159 (2020).https://doi.org/10.1016/j.bios.2020.112159
36. M.A. Ferry, Diagnosing and resolving mass spec problems (2017). [Online]. Available: https://www.spex.com/getmedia/5a104697-a29b-4ec1-b6cc-9b4150df64e0/Optimize_Issue_6_April_2017.pdf?ext=.pdf.
37. Applied Biosystems, Voyager^{TM} Biospectrometry^{TM} Workstation, (2001). [Online]. Available: http://www.chem.cmu.edu/cma/DE-STRManual.pdf

Chapter 3
Characterization Techniques for Topography Analysis

Hamed Hosseinian, Euth Ortiz Ortega, María José Rosales López, Andrea Rodríguez Vera, and Samira Hosseini

Abbreviations

3D	Three dimensional
AFM	Atomic force microscopy
Al	Aluminum
Au	Gold
CARS	Coherent anti-stokes Raman scattering
FAST	Fiber array spectral translation
LCIS	Liquid crystal imaging spectrometers
MEMS	Micro-electromechanical systems
NA	Numerical aperture
NS	Not specific
OBD	Optical beam deflection
PMMA	Poly(methyl methacrylate)
PA	Polyacrylic
PS	Polystyrene
RAMAN	Raman spectroscopy
S	Laser spot diameter
SEM	Scanning electron microscope
SERS	Surface-enhanced Raman spectrometry
Si	Silicon
Si_3N_4	Silicon nitride
SPM	Scanning probe microscopy
SSRS	Shifted subtracted Raman spectrometry
STM	Scanning tunneling microscopy
TEM	Transmission electron microscopy
TERS	Tip-enhanced Raman spectrometry

Hamed Hosseinian and Euth Ortiz Ortega—share the first authorship.

E. O. Ortega et al., *Material Characterization Techniques and Applications*,
Progress in Optical Science and Photonics 19,
https://doi.org/10.1007/978-981-16-9569-8_3

TiN	Titanium nitride
TiO_2	Titanium dioxide
λ	Wavelength

3.1 Atomic Force Microscopy for Material Characterization

3.1.1 History of Atomic Force Microscopy

Scanning probe microscopy (SPM) encompasses a comprehensive set of techniques dedicated to imaging and measuring properties of the surface of chemical and biological materials [1]. The first technique developed from SPM was the scanning tunneling microscopy (STM), introduced by Gerd Binnig and Heinrich Rohrer in 1982. Subsequently, Gerd Binnig, Calvin Forrest Quate, and Christoph Gerber modified STM four years later to produce a device capable of imaging atoms on solid surfaces, named atomic force microscopy (AFM). The AFM scans a small area over the surface of the sample and adjusts the height of the analyzed point. Consequently, the produced image of the height imitates the topography of the surface. Two factors promoted the use of AFM: the mass production of microfabricated cantilevers and the establishment of the optical beam deflection (OBD) method to detect cantilever deflection [2]. These two factors drove the commercialization of the AFM and enabled a broader scientific community to explore the technique and image animal cells, proteins, DNA, RNA, as well as lipids (Fig. 3.1) [3].

The ability to exclusively study solid-state samples was the most significant limitation of the AFM until 1989 when the technique was redefined to analyze liquid substances. The analysis of liquid samples revealed the possibility of creating three-dimensional (3D) images of the surfaces of non-conductive materials in vacuous and

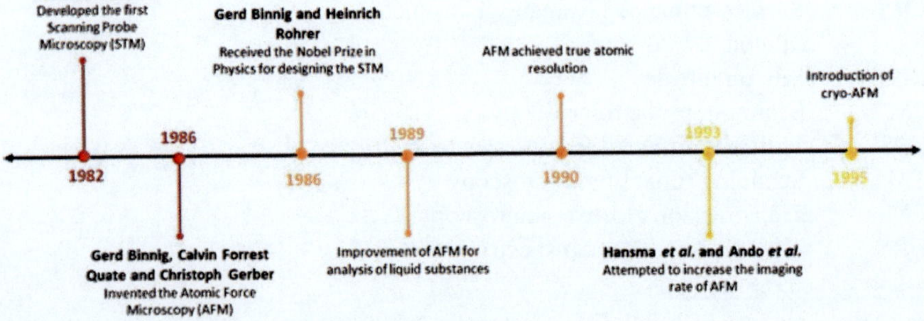

Fig. 3.1 History of AFM

non-vacuous (i.e., air or liquid) environments. In 1990, the AFM reached true atomic resolution and soon became the standard tool for nanometer-scale imaging of all types of surfaces [2]. However, it involved a rather slow process to capture dynamic biomolecular images. Hence, Hansma et al. and Ando et al. in 1993 attempted at increasing the imaging rate of AFM and published their results in 2002 [3].

Despite the great advancements in the field, AFM presented limitations for studying biological specimens including the softness and the thermal motion of flexible multidomain and subunit molecules at room temperatures. The invention of the cryo-AFM in 1995 represented a new mode of imaging biological macromolecules and addressed the previous challenges. The cryo-AFM significantly reduced sample contamination, provided thermal and mechanical stabilization to samples, and enabled an adjustable imaging temperature by operating in liquid nitrogen vapor. Today, most scientific fields use AFM to study animal cells, bacteria, and tumor cells, to mention a few. AFM instrumentation and specimen preparation protocols continue to evolve for the study of biological specimens, thus sustained efforts toward the development of the technique remain necessary [4].

3.1.2 Mechanism of Operation of Atomic Force Microscopy

AFM is a high-resolution technique derived from SPM. The AFM technique produces a topographical recording of the sample surface through the interaction between a tip and a specimen. The AFM equipment possesses a sharp probe tip that acts as the interacting probe and is mounted on a microcantilever. The most common form of a probe is a square-based pyramid or a cylindrical cone. Although cantilevers are primarily made of silicon nitride (Si_3N_4), the surface opposite to the probe tip is covered in a thin layer of either gold (Au) or aluminum (Al) to achieve a reflective surface. Moreover, the microcantilever or the sample surface is positioned above a piezoelectric crystal, thus conceding the probe the ability to move and scan the surface pixel by pixel. The probe is placed into and out of contact with the sample surface in a z-axis direction.

The process starts when the laser generates a light beam, and the reverse side of the microcantilever reflects it to a position-sensitive photodetector. Any deflection of the microcantilever causes the location of the laser spot on the photodetector to shift, hence, monitoring variations in the deflection. The photodetector is commonly a photodiode divided into four sections (Fig. 3.2). It calculates the deflection signal by subtracting the difference in the signal detected by the first and the second quadrants versus the one detected by the third and the fourth quadrants. Detecting lateral or torsional bending of the level is accomplished through the comparison of the signal strength detected by the second and third quadrants versus the first and fourth quadrants. The AFM equipment obtains the relative height information of the topographic features of the specimen through the scanning of the probe when it is in contact with the sample surface. Interactions such as electrostatic, magnetic, capillary, and van der Waals forces source the tip and sample interactions. Moreover, the AFM obtains

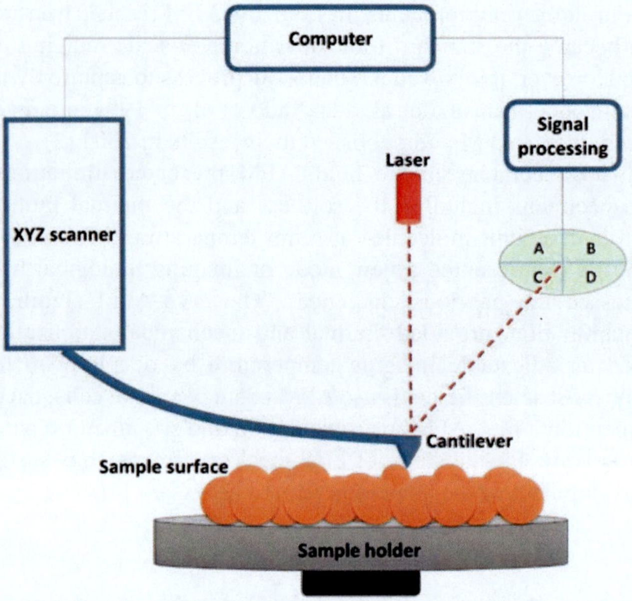

Fig. 3.2 Representation of the principles of AFM. The interactions between the probe tip and the sample's surface mediate the topographical images of the specimen

the topographical images by operating under contact mode, non-contact mode, or tapping mode.

3.1.2.1 Contact Mode Imaging

Contact mode imaging is the most accessible mode of AFM operation. It acquires its name due to the constant contact between the probe and the sample. Moreover, the simplicity of contact mode operation represents an advantage for imaging films and relatively flat surfaces. This technique is bifurcated into scanning at a constant force and at a variable force. In the constant force mode, the deflection and force of the cantilever are constant due to a feedback mechanism. The z-height is modified to cause a return to the original deflection when the cantilever is deflected. The alteration in the z-position is monitored, and the obtained information is plotted as a function of the x and y axes to depict a topographical recording of the sample surface. Conversely, the feedback mechanisms are interchanges in variable force imaging to maintain a constant z-height. This approach is only suitable for rather smooth samples with low-lying surface features. Although variable force imaging is not appropriate for all surfaces, it can provide images with a sharper resolution than constant force mode [1].

3.1.2.2 Tapping Mode Imaging

In tapping mode imaging, the cantilever can oscillate at a value near to its resonant frequency. The amplitude of oscillation is restricted due to the repeated engagement and disengagement actions of the probe with the surface. The oscillatory amplitude of the cantilever varies when facing a differing topography during the sample scanning process. Finally, a topography image is captured through a feedback mechanism to adjust the z-height of the piezoelectric crystal and preserve constant amplitude. In this regard, as the probe scans across the surface, the lateral forces are significantly decreased compared with the contact mode [1].

3.1.3 Advantages and Disadvantages of Atomic Force Microscopy

The AFM is mainly distinguished due to its high atomic image resolution (1–10 nm) of 3D surface topography. Additionally, the ability of this technique to operate in air, liquid, and vacuum has enabled its wide application in medicine, engineering, and natural sciences [5–7], thus allowing imaging biological samples in their natural environment [7]. Moreover, the AFM equipment is small, requires simple sample preparation, and presents a low cost of operation compared to other microscopy techniques such as the scanning electron microscope (SEM) or transmission electron microscopy (TEM) [6, 7].

However, the limited operation range in the spatial directions of the cantilever piezoelectric crystal device could cause tip breakage [5, 7]. In addition, the imaging process of surface roughness beyond a threshold becomes challenging due to the 100 μm limitation in scan range in the x and y axes and the 10 μm of scan range in the z-axis [7]. Further advantages and disadvantages could be attributed to the particular behaviors of the different operation modes of AFM. For instance, the non-contact mode does not damage the sample, while contact and tapping modes cause mild to severe damage to samples. Noteworthy, the non-contact mode can achieve high scan speed, while the tapping mode can perform at a slower scan speed. Lastly, the usage of non-contact mode decreases image resolution compared to the rest of the AFM modes (Table 3.1) [6, 7].

3.1.4 Applications of Atomic Force Microscopy

AFM is a near-field scanning tool used for monitoring the interactions between a sharp tip of a flexible cantilever and a sample. This analytical technique moves accurately on the three axes (x, y, z), thus reaching a spatial resolution above the

Table 3.1 Advantages and disadvantages of employing AFM

Advantages	Disadvantages
High atomic image resolution (1–10 nm) of 3D surface topography [5–7]	Risk of tip breakage [5–7]
Operates in air, liquid, and vacuum [5–7]	Scan range of 100 μm in the *x* and *y* axes [7]
Small size of equipment [6]	Scan range of 10 μm in the *z*-axis [7]
Low-cost operation [6, 7]	
Simple sample preparation [6, 7]	Poor image resolution using non-contact mode [6, 7]
No damage to sample using non-contact mode [6, 7]	Damage to samples using contact and tapping modes [6, 7]
High scan speed using contact mode [6, 7]	Slow scan speed using non-contact mode [6, 7]

subnanometer. In addition, AFM operates under liquid, vacuum, and air conditions and involves rapid and minimal specimen preparation (Fig. 3.3) [8, 9]. It is widely applied in research disciplines such as piezoelectric materials [10], microelectromechanical systems (MEMS) [11, 12], molecular biology [13], thin-film coating [14], energy materials [15, 16] semiconductors [17], cell biology [18], tribology [19], surface chemistry identification [20], and polymer science [21, 22]. AFM can

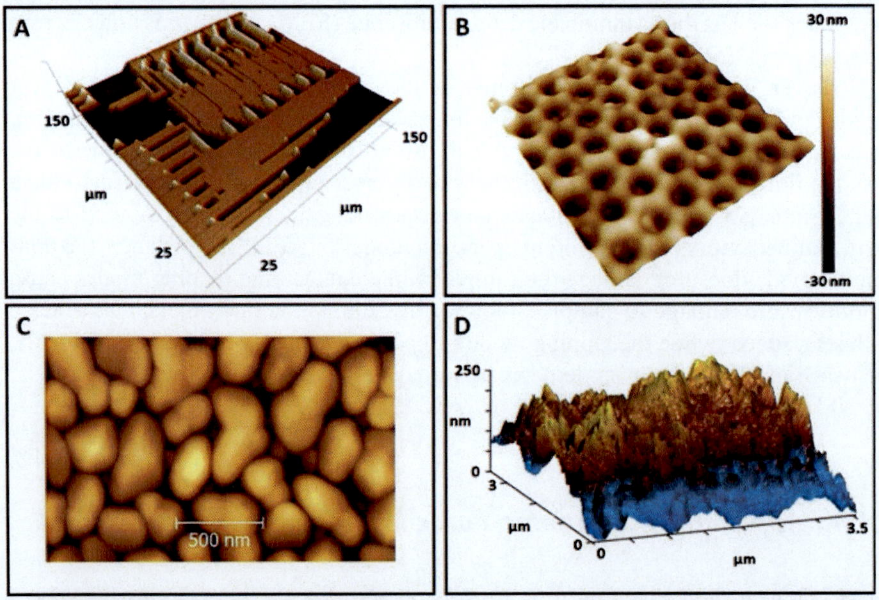

Fig. 3.3 Images obtained through AFM. **a** Component of a Pentium processor core, © 2009, Januszartur. **b** Nanostructured transistor for a transparent glucose sensor, © 2016, Oregon State University. **c** Magnetron sputtered TiN layer, © 2013, Piret Pikma. **d** TiO_2 film, © 2012, V. Dallacasa

analyze the topography of inorganic and biological materials, including DNA, RNA, viruses, bacteria, proteins, tissues, and organs. Moreover, this technique can estimate the bond strength between biomolecules, determine the mechanical properties, and monitor the protein dynamics [23]. Furthermore, functionalizing the AFM tip to record the interactions between material interfaces and biomolecules at a single-molecule level enables the extension of this technology's scope toward biomolecular imaging, biosensing, protein–surface interactions, cell mechanics, biomolecule unfolding mechanism, and molecule–molecule interaction [9, 24].

The development of effective and rapid cancer diagnostic methods is crucial for the early detection of the disease and the subsequent treatment to counteract it. Lekka et al. presented an attractive detection technique for breast and prostate cancer cells in different tissues. The diagnosis method relied upon AFM and studied the elastic properties of living cells by obtaining Young's modulus value of every cell type. The system consisted of a sensitive probe tip that indented the cell, producing a deviation of the cantilever. A laser-photodiode system measured such deflection. The healthy cells (rigid) generated large deflections and short indentations, while the cancer cells (soft) caused small deflections with high indentations (Fig. 3.4a). The force versus indentation curve calculated the cell rigidity, usually acquired by measuring the deflections caused by the sample's rigid or flexible surface. The study used glass for reference since it is a material that does not deform easily and produces

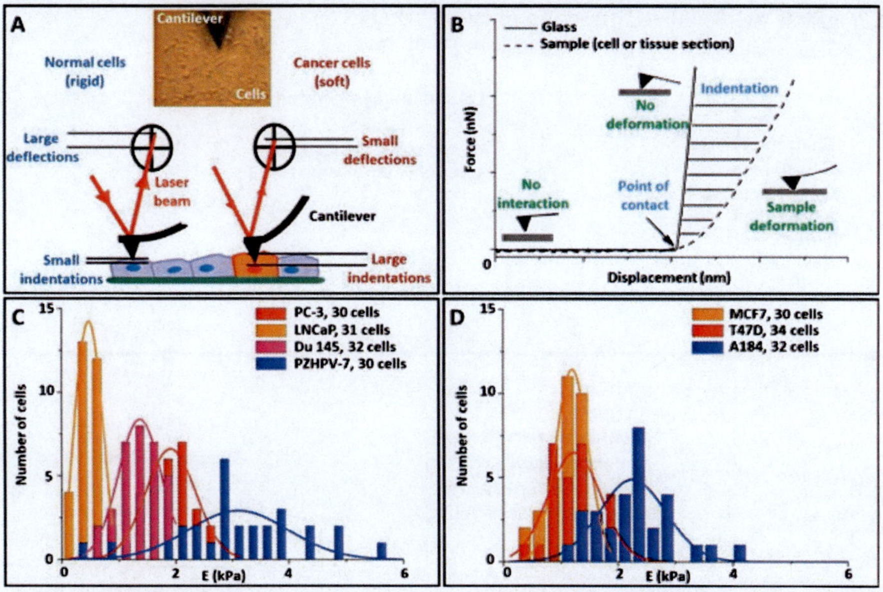

Fig. 3.4 **a** Determination of Young's modulus by AFM for cell indentation. **b** Young's modulus distributions with fitted Gaussian functions obtained for **c** breast cancer cell (MCF7 and T47D), and **d** prostate cancer cell compared to the healthy breast cell line. Reproduced (or adapted) with permission [25], ©2012, Elsevier

a straight inclined line. Similarly, the displacement of the sample's line relies on the amount of indentation (Fig. 3.4b). The results showed that healthy cells (PZHPV-7 and A184) have a higher Young's modulus value than breast (T47D and MCF7) and prostate (Du 145, LNCaP, and PC-3) cancer cells. The main benefit of this diagnostic technique is its simplicity and the requirement of only a few tens of cells to calculate Young's modulus with high accuracy and reliability (Fig. 3.4c and d) [25].

3.2 Raman Spectroscopy for Material Characterization

3.2.1 History of Raman Spectroscopy

Chandrasekhara Raman is known as the pioneer of the Raman scattering of light and the creator of Raman spectroscopy (RAMAN). A decisive moment in his scientific career came in a voyage to Europe while observing the deep blue color of the Mediterranean Sea. A few years earlier, Lord Rayleigh had laid the groundwork for the blue color of the sky, explaining that such color relied on the scattering of sunlight by the gaseous atmosphere. Consequently, Chandrasekhara Raman was interested in experimentally and theoretically studying light scattering in transparent media. In 1922, he explained energy transfer between light quanta and molecules under his monograph "Molecular Diffraction of Light" (Fig. 3.5).

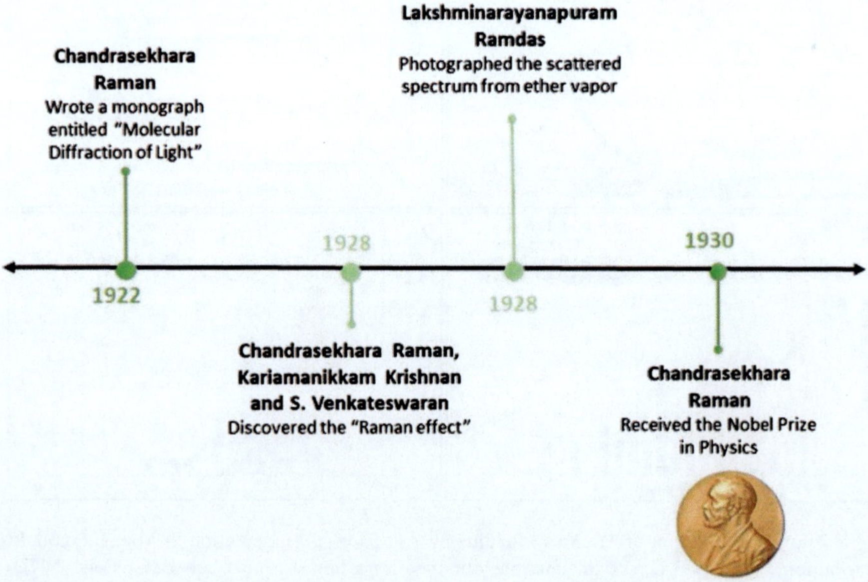

Fig. 3.5 History of RAMAN

In 1928, Chandrasekhara Raman, Kariamanikkam Krishnan, and S. Venkateswaran examined the scattered light track through a direct vision spectroscope and determined that the modified scattering was separated from the scattering corresponding to the incident unmodified light [26]. Such discovery was described as a new radiation, named later as the Raman effect. The observation of the Raman effect represented a milestone in many scientific fields since it occurs in a vast quantity of liquids, gases, vapors, and crystals. Later that year, Lakshminarayanapuram Ramdas first photographed the scattered spectrum from ether vapor [26].

Chandrasekhara Raman received numerous awards for his remarkable developments on light scattering and the discovery of the Raman effect, such as the Nobel Prize in Physics in 1930. He founded the Raman Research Institute to continue contributing to crystal physics, crystal optics, and mineralogy and remained the Director of the institution until his death on November 21st, 1970 [27].

Following the course of research in recent years, various Raman microscopes have been developed since the early 2000s, in addition to the progress on ordinary Coherent anti-Stokes Raman scattering (CARS) microscopes. Furthermore, the spatial resolution tip has been improved with the use of metallic and near-field optics. Reports on microscopes using other nonlinear Raman processes were initiated in 2008 [28].

3.2.2 *Mechanism of Operation of Raman Spectroscopy*

RAMAN is a helpful imaging tool in providing multidimensional analytical knowledge on the sample's surface. The principal characteristics of this technique are low sensitivity to water, minimal requirements for specimen preparation, and high selectivity. The light scattering process occurs when a photon interacts with the chemical bonds of a molecule, creating vibrational motions that cause the photon to be emitted in every direction. Rayleigh scattering is the phenomenon where the emitted photons have identical frequencies as the incident photons. In contrast, Raman scattering takes place when the photons are emitted at a different frequency. The Stokes Raman scattering results when the frequency of the emitted photon is lower than the incident photon, while the anti-Stokes Raman scattering takes place if the frequency of the photon increases when interacting with the molecule (Fig. 3.6a). RAMAN enables the analysis of samples' molecular structure and chemical composition since this technique combines digital imaging technology and spectroscopy. RAMAN's scope on imaging consists of several Raman spectra obtained from a data cube also called a data hypercube, formed by individual squares representing one pixel. The spatial dimensions are expressed by "X" and "Y," while the spectral dimension is the wavelength (λ) (Fig. 3.6b) [29–32].

Raman images are produced by laser scanning the specimen surface and simultaneously collecting the spectra. The scanning methods are classified into point-by-point scan and line scan. The data cube is created in the point-by-point method by collecting the Raman spectra at each square in which the area of interest is divided, thus obtaining high-resolution maps. This process reaches spatial resolutions from

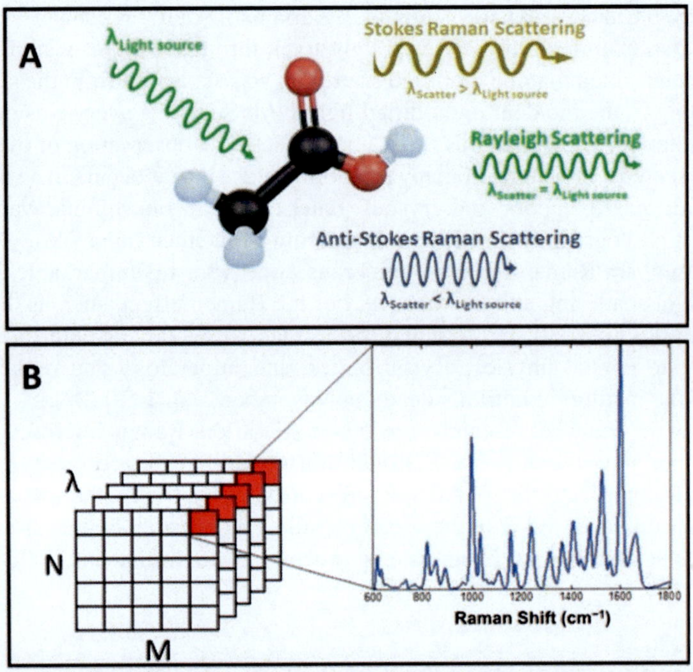

Fig. 3.6 **a** Light scattering process. **b** Data cube was obtained by Raman imaging. Reproduced (or adapted) with permission [30] © 2019, Journal of the Brazilian Chemical Society

2 μm to 25 μm and spectral resolutions in the range of 1–4 cm^{-1}. Moreover, the equipment works in a confocal configuration to reduce the fluorescence background and secondary scattering effects (Fig. 3.7a). Contrarily, the line scan method involves scanning the area of interest using the incident laser by creating vertical or horizontal lines through cylindrical optics or moving mirrors. Each scan line produces a spectrum; therefore, the spatial resolution is determined by the number of rows, width, and height of the area of interest (Fig. 3.7b) [29, 30, 32].

Wide-field techniques illuminate the entirety of the sample with a light source and analyze the discrete frequencies of light over time using a wide-aperture optical filter. In most cases, liquid crystal imaging spectrometers (LCIS) provide diffraction-limited spatial resolutions and spectral resolutions that compete with single-stage dispersive monochromators. LCIS can operate in a spectrum range from 400 to 2500 nm and offers large out-of-passband rejection performances and wide-open spectral ranges (Fig. 3.7c). The fiber array imaging Raman, dimension-reduction array, fiber image compression, and fiber array spectral translation (FAST) have become the most used imaging techniques during the last decade. The fibers array enables data collection in a single frame delimited by two spatial dimensions and one spectral dimension. This mechanism is possible since the Raman light captured from the sample is focused on a two-dimensional array of optical fibers. Finally,

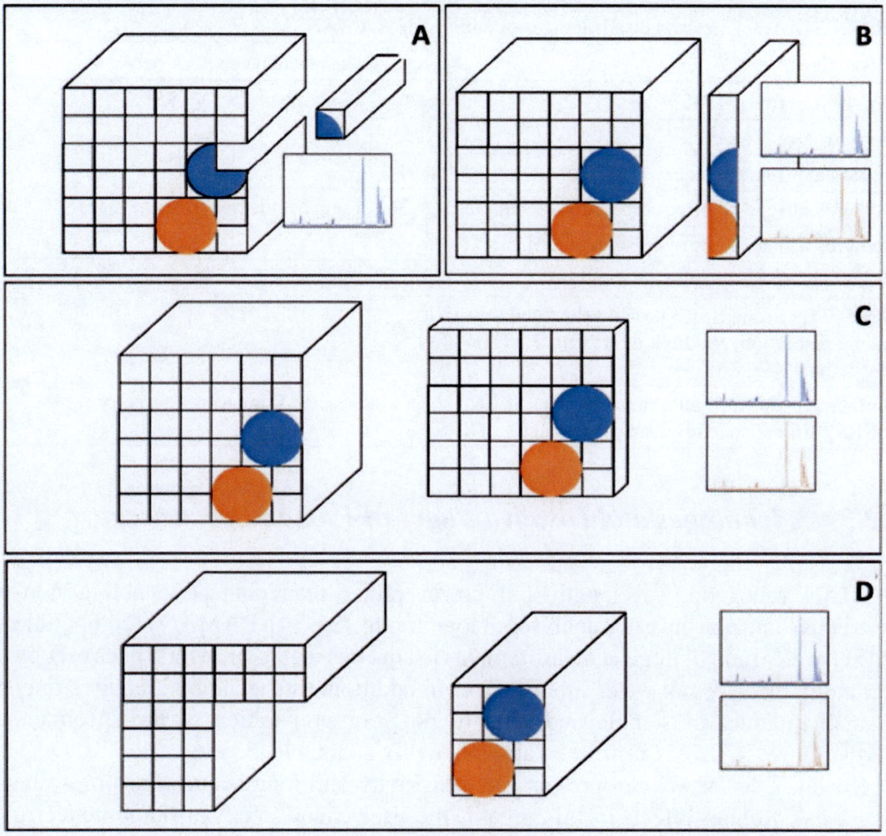

Fig. 3.7 Scanning and wide-field Raman imaging methods. **a** Point-by-point scanning process. **b** Line scanning process. **c** Operation mechanism of LCIS. **d** FAST function principle

specialized software reconstructs the spatial and spectral information in a data cube (Fig. 3.7d) [29].

The area of interest is not completely scanned if the spot dimension is larger than the resolution; therefore, the resolution should be equal to the spot dimension. Nevertheless, the airy disk equation (Eq. 3.1) states that the resolution cannot be larger than half the spot dimension:

$$S = 1.22\frac{\lambda}{\mathrm{NA}} \tag{3.1}$$

where S is the laser spot diameter, λ represents the laser wavelength, and NA depicts the numerical aperture of the laser. For instance, a green laser with a 100X objective (NA of 0.9) produces a spot dimension of approximately 720 nm; thus, the maximum resolution is approximately 360 nm [29, 32].

Table 3.2 Advantages and disadvantages of employing RAMAN

Advantages	Disadvantages
Non-destructive [34, 35]	Low sensitivity [34, 35]
Non-invasive [34, 35]	
Weak water interference [33–36]	
Requires little to no prior preparation of samples [33–36]	Long acquisition times [34, 35]
Requires a small quantity of sample [36]	
Highly adaptable instrumentation [34]	Fluorescence interference [34, 35]
SERS, TERS, and SSRS provide extended sensitivity, spatial resolution, and high fluorescence adaptability, respectively [33–36]	

Surface-enhanced Raman spectrometry (SERS); shifted subtracted Raman spectrometry (SSRS); tip-enhanced Raman spectrometry (TERS)

3.2.3 Advantages and Disadvantages of Raman Spectroscopy

RAMAN relies on wavelength and power rather than contact, enabling non-destructive and non-invasive analyses of topography [33, 34]. RAMAN is compatible with chemical and biological measurements in aqueous media since this spectroscopy technique offers weak water interference. In addition, this technique requires only a small quantity of sample and little to no prior preparation of the specimen. Furthermore, RAMAN instrumentation is highly adaptable [33–36].

However, RAMAN can present low sensitivity and long acquisition times due to a weak Raman effect. In this regard, the user cannot analyze fluorescence in samples since fluorochromes interfere with the Raman effect [33, 34]. Despite the mentioned shortcomings, advanced RAMAN techniques, including the surface-enhanced Raman spectrometry (SERS), the tip-enhanced Raman spectrometry (TERS), and the shifted subtracted Raman spectrometry (SSRS), overcome the limitations of the conventional RAMAN through extended sensitivity, spatial resolution, and high fluorescence compatibility, respectively [33, 35] (Table 3.2).

3.2.4 Applications of Raman Spectroscopy

RAMAN is frequently used for topography studies and mapping chemical compounds in irregular surfaces. The applications of this technique have significantly increased over time since it also enables the physicochemical characterization of materials. In this regard, RAMAN holds an additional advantage over other microscopy techniques that only study the morphology of the samples. RAMAN is a non-destructive imaging tool that can analyze samples in the presence of water and identify inorganic and organic compounds by distinguishing Raman signals.

Therefore, RAMAN can be employed in the food research domain for the recognition of pathogenic microorganisms, in geology for the identification of terrestrial and extraterrestrial materials, in environment research for studying the effects of contamination in the atmosphere, in pharmaceutics for the development of drugs, art industry for the mapping of pigments, forensic analysis for fingerprint detection, and agriculture for studying pesticides and agricultural products. Nevertheless, the majority of investigations are focused on materials science (e.g., analysis of new catalysts, carbon compounds, and polymers) and biomedical sciences (e.g., studies of diseases and cells) [30, 37–41].

Plastic pollution within our environment is a severe concern of current society as the industry and households are the principal sources of plastic disposal. Microplastics are plastic particles from 1 μm to 5 mm, while nanoplastics have a diameter size of < 1000 nm. Microplastics and nanoplastics can enter living organisms via inhalation or ingestion, resulting in adverse effects such as the accumulation of plastic in organs generating metabolism problems caused by the hydrophobic nature of these polymers. Therefore, there is a pressing need to understand the risk of microplastics and nanoplastics in human health and the environment. However, effective methods for the identification of nanoplastics are yet to be developed. Sobhani et al. proposed RAMAN to recognize and visualize nanoplastics of 600 nm and below from a vehicle's clear coating made from polyacrylic (PA). Polystyrene (PS) individual nanoparticles were clearly identified using glass support, and aggregates of nanoplastic were observed as blue circles (Fig. 3.8a and b). The difference between individual and aggregated nanoplastics was defined by the intensity of the color (Fig. 3.8c). The Raman spectra confirm the presence of plastic since the Raman signal of PS corresponds to 1000 cm^{-1} (Fig. 3.8d) [42].

RAMAN is also a novel method for identifying human breast cancer by detecting abnormal mass tissue from a biopsy. Abramczyk et al. examined fresh tissue samples containing tumor mass and healthy tissue, staining the specimens with biomarkers and studying them anew. Consequently, the non-stained and stained sample images were compared to analyze the similarities. The optical biopsy images acquired through RAMAN showed an arbitrary distribution of biochemical compounds, such as carotenoids and fatty acids, through the breast tissue (Fig. 3.9). Moreover, the non-cancerous tissue presented a significantly higher quantity of adipose cells, stained in blue, than the cancerous tissue. This study suggested that RAMAN imaging can serve as an effective and sensitive diagnostic method for cancer [43].

RAMAN imaging can analyze bone growth and monitor bone tissue composition. The bone sample could be analyzed in a hydrated state (native form) or dehydrated poly(methyl methacrylate) (PMMA)-embedded tissues. In this technique, the polar groups have a weaker contribution to Raman spectra than non-polar groups due to a difference in polarizability. Figure 3.10a depicts the optical image of bone tissue, while the following figures represent well-identified Raman spectra of bone. Figure 3.10B portrays the mineral:matrix ratio, from the ν_1 PO_4 or ν_2 PO_4 peak area and the amide I, amide III, CH_2, proline and hydroxyproline, and phenylalanine peak areas. Figure 3.10c describes the mineral maturity/crystallinity through ν_1 PO_4 peak broadening, and Fig. 3.10d pictures the carbonate:phosphate ratio from the area of the

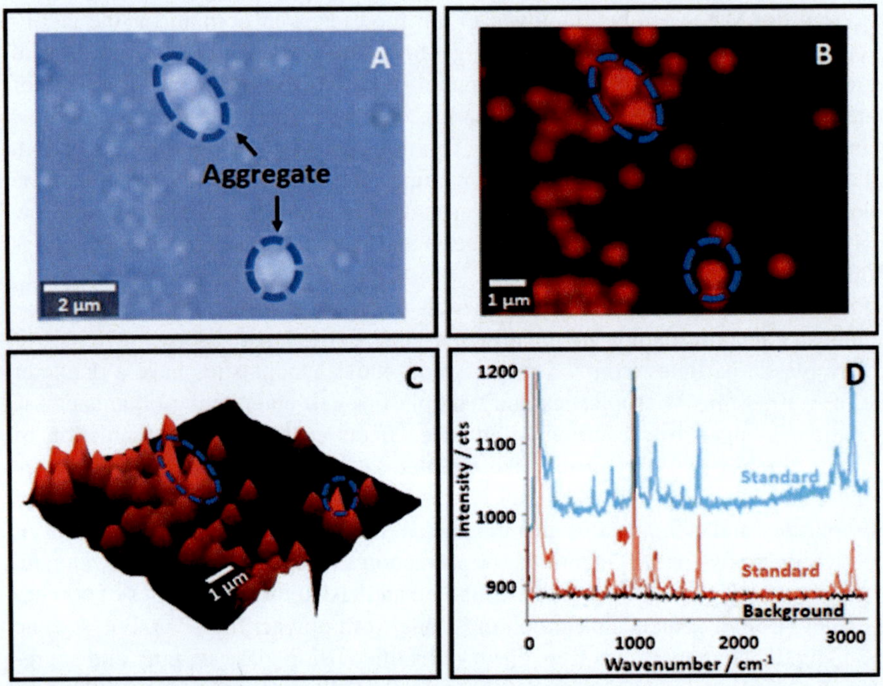

Fig. 3.8 Identification and visualization of nanoplastics (600 nm). **a** Microscopic image (100X). **b** and **c** Raman mapping images. **d** Raman spectra. Reproduced (or adapted) with permission [42], ©2020, Elsevier

composite ν_1 CO_3–ν_3 PO_4:ν_1 PO_4 peaks [44]. This information provides a valuable set of data with regard to the age, drug treatment, and disease stage of the patient.

3.3 Troubleshooting of Topography Analysis Techniques

AFM requires a precise selection of the cantilever based upon the anticipated elastic response of the material; otherwise, the user may face difficulty in mapping the local elasticity of the surfaces with various elastic characteristics. In addition, cantilevers may follow a different method of mapping depending on their use in a liquid or gaseous environment. In the air, the user should use acoustic excitations to excite the cantilever, while in water, the user should employ a measurement buffer or a photothermal excitation to prevent mechanical resonance [45]. Moreover, the image resolution can be increased by connecting a carbon nanotube to the end of the tip. The accuracy can be easily examined prior to each experiment by scanning a calibration grid with grooves.

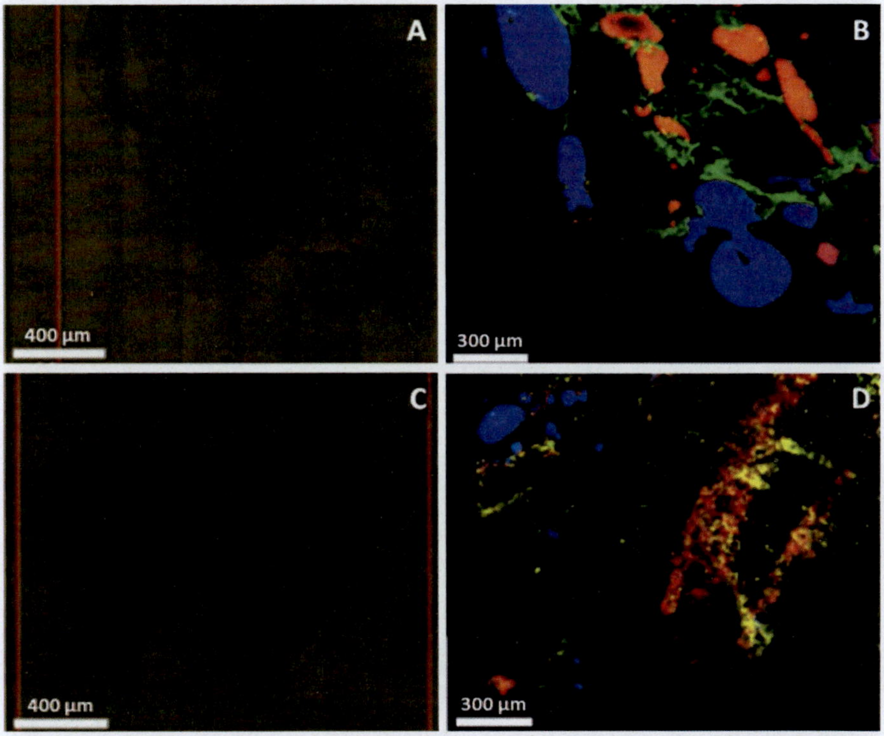

Fig. 3.9 Non-stained stitching image of **a** thin layer of non-cancerous tissue as opposed to (**c**) thin layer of cancerous tissue. The stained Raman image of **b** previous non-cancerous tissue and **d** previous cancerous tissue. Reproduced (or adapted) with permission [43], ©2012, Elsevier

If the photodetector has too much light or is saturated, changing the photodetector gain and switching to low gain should solve the problem. Samples can also signify a source of error; for instance, AFM cannot analyze the specimen if it is not securely attached to the stage. Moreover, scanning contaminated samples can lead to receiving false feedback and obtaining an opaque image [47]. Similarly, an essential component for AFM, such as the probe, must be handled with care ensuring that it possesses a sharp tip with a high aspect ratio in order to obtain high-resolution imaging. Since the cantilever measures the interaction force, the probe always exerts some force on the sample. Such behavior can cause distortion or damage to the sample since multiple biological samples are soft and delicate. Furthermore, if the probe is not seated in the holder correctly, the user should readjust the probe. Noticing if the probe is broken or damaged is challenging since there is no direct visual manner to examine the tip condition. However, a recommendation is to analyze the scans to assess the usability of the probe and decide if it needs replacement. Additionally, the selection of objective must not affect AFM quality, and the use of oil or water immersion lenses should prevent possible vibration into the setup when performing simultaneous light microscopy and AFM imaging [48].

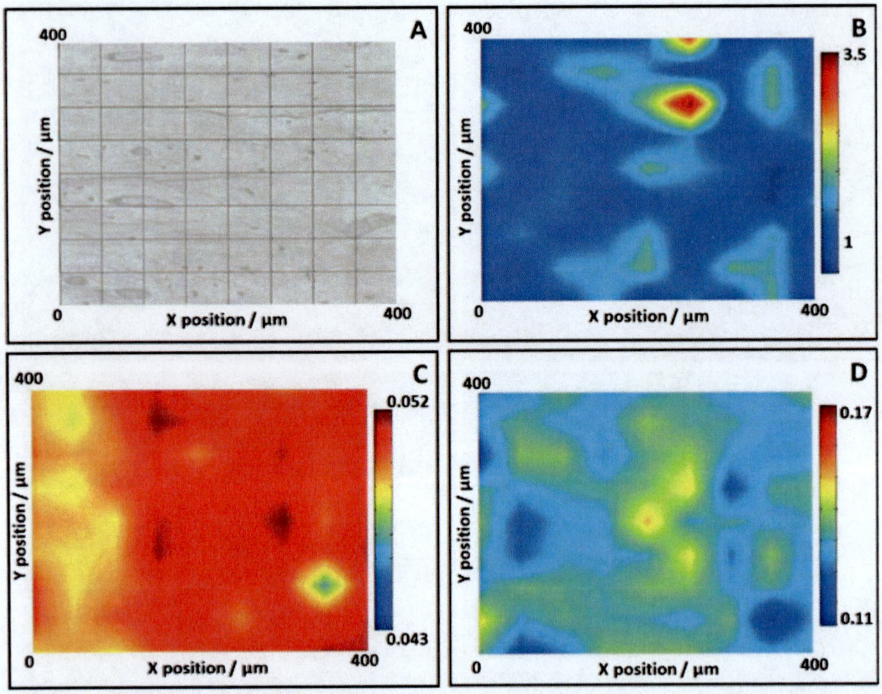

Fig. 3.10 Multiple examples of optical images taken from human proximal femoral cortical bone tissue. Red regions correspond to different tissue components. Reproduced (or adapted) with permission [44], ©2020, Elsevier

The user may also experience problems using the AFM software, such as a laser that does not turn on and a non-responsive photodetector. For the first case, the user should exit the software, power cycle the equipment, and restart the program. In the second case, it is recommended to turn off the software and restart the software computer and the program. The different modes of the AFM technique also present individual errors that are important to consider. For instance, not all specimens can be firmly attached to a solid support in the AFM contact mode; therefore, the user should employ the tapping mode for a more dynamic method.

Users of RAMAN commonly face problems related to the inadequate regulation of laser power intensity. The laser intensity may be lower than expected due to low laser power; hence, it is necessary to revise the laser's power and check if any filter is on the path. Another frequent scenario is sample burning as a result of high laser power, which is solved by reducing the power [50]. In addition, the majority of problems of the RAMAN technique involve a relatively low signal-to-noise ratio, possibly caused by a low value of integration time and an inadequate focus. Augmenting the integration threshold time value for signal collection and carefully refocusing may be the solution for these problems, respectively [49, 50].

Moreover, it is common that the software used for analysis and data acquisition does not detect the Raman unit. The cause may be that the Raman unit is not on or that the USB cable of the Raman unit is incorrectly connected. This problem can be fixed by turning on the Raman unit before opening the software and verifying connections [49]. Furthermore, having an excessive fluorescent sample is inconvenient since RAMAN presents interference with fluorescence. Fluorescent specimens may occur due to a dry sample or its need for bleaching. In this regard, the user should wet the sample, or to pre-run the operation with a short integration time if necessary [50]. Table 3.3 shows the detailed information of such errors, causes, and solutions for common imaging problems presented in AFM and RAMAN techniques for topography analysis.

Table 3.3 Troubleshooting for AFM and RAMAN in topography analysis

Technique	Problem	Cause	Solution
AFM	Topography is not stable [45]	Tip and sample contact are not optimized [45]	Increase the scan window to get a stable image, then go back to the initial scan size [45]
	High-resolution images cannot be recorded [45]	The tip is contaminated [45]	Expose the sample tip to ultraviolet light for 30 min [45]
	There is a poor contrast in the first or in the second mode channels [48]	The quality of both modes' observables is affected by the imaging parameters [48]	Optimize the signal-to-noise ratio to obtain high-resolution contrast. Reach a balance between contrasts [48]
	Poor image resolution [46]	NS	Attach a carbon nanotube to the end of the tip [46]
	The specimen cannot be firmly attached to a solid support in the contact mode [46]	NS	Switch to dynamic or tapping mode [46]
	The laser does not turn on, or the photodetector is non-responsive [47]	NS	Exit the software and restart the computer and the program [47]
	AFM probe is broken or damaged [47]	NS	Analyze scans to assess the usability of the probe. Replace if necessary [47]
	The photodetector is saturated with light [47]	NS	Change the photodetector gain switch to low gain [47]

(continued)

Table 3.3 (continued)

Technique	Problem	Cause	Solution
	False feedback and opaque image [47]	NS	Scan a non-contaminated area of the sample [47]
	Distortion or damage occurred to the sample [48]	NS	Use a sharp tip that has a high aspect ratio to record the topography with higher accuracy [48]
RAMAN	The software does not detect the Raman unit [49]	The Raman unit is not on [49]	Turn on Raman unit before opening software [49]
		USB cable of Raman unit is not connected [49]	Verify connections [49]
	Low intensity [50]	Low laser power [50]	Check the laser power and possible filters in the path [50]
	Sample burning [50]	High laser power [50]	Lower the laser power [50]
	Low signal-to-noise ratio [49, 50]	Inadequate focus [49, 50]	Refocus carefully [49, 50]
		Integration time value is set at a low threshold [49, 50]	Increase integration threshold time value for signal collection [49, 50]
	Excessive fluorescence [50]	Dry sample [50]	Wet the sample [50]
		Bleaching [50]	Pre-run with short integration time [50]

AFM atomic force microscopy; *NS* not specific; *RAMAN* Raman spectroscopy

References

1. T. Glatzel et al., Advanced atomic force microscopy techniques. Biophys. Rev. **12**(4), 893–894 (2012). https://doi.org/10.1002/jmr.857
2. T. Glatzel, H. Hölscher, T. Schimmel, M.Z. Baykara, U.D. Schwarz, R. Garcia, Advanced atomic force microscopy techniques. Beilstein J. Nanotechnol. **3**(1), 893–894 (2012). https://doi.org/10.3762/bjnano.3.99
3. Y.F. Dufrêne et al., Imaging modes of atomic force microscopy for application in molecular and cell biology. Nat. Nanotechnol. **12**(4), 295–307 (2017). https://doi.org/10.1038/nnano.2017.45
4. P. Parot et al., Past, present and future of atomic force microscopy in life sciences and medicine. J. Mol. Recognit. **20**(6), 418–431 (2007). https://doi.org/10.1002/jmr.857
5. B. Prats-Mateu, N. Gierlinger, Tip in–light on: Advantages, challenges, and applications of combining AFM and Raman microscopy on biological samples. Microsc. Res. Tech. **80**(1), 30–40 (2017). https://doi.org/10.1002/jemt.22744
6. M.K. Khan, Q.Y. Wang, M.E. Fitzpatrick, Atomic force microscopy (AFM) for materials characterization. Mater. Charact. Using Nondestruct. Eval. Methods 1–16 (2016). https://doi.org/10.1016/B978-0-08-100040-3.00001-8

7. A.T. Luís, D. Hlúbiková, V. Vaché, P. Choquet, L. Hoffmann, L. Ector, Atomic force microscopy (AFM) application to diatom study: Review and perspectives. J. Appl. Phycol. **29**(6), 2989–3001 (2017). https://doi.org/10.1007/s10811-017-1177-4
8. S. Liu, Y. Wang, Application of AFM in microbiology: A review. Scanning **32**(2), 61–73 (2010). https://doi.org/10.1002/sca.20173
9. Q. Li, T. Zhang, Y. Pan, L.C. Ciacchi, B. Xu, G. Wei, AFM-based force spectroscopy for bioimaging and biosensing. RSC Adv. **6**(16), 12893–12912 (2016). https://doi.org/10.1039/c5ra22841g
10. X. Xu et al., An improved AFM cross-sectional method for piezoelectric nanostructures properties investigation: Application to GaN nanowires. Nanotechnology **22**(10) (2011). https://doi.org/10.1088/0957-4484/22/10/105704
11. C. Steffens, F.L. Leite, C.C. Bueno, A. Manzoli, P.S.D.P. Herrmann, Atomic Force Microscopy as a tool applied to nano/biosensors. Sensors (Switzerland) **12**(6), 8278–8300 (2012). https://doi.org/10.3390/s120608278
12. H. Liu, B. Bhushan, Nanotribological characterization of molecularly thick lubricant films for applications to MEMS/NEMS by AFM. Ultramicroscopy **97**(1–4), 321–340 (2003). https://doi.org/10.1016/S0304-3991(03)00058-5
13. A. Wu, Z. Li, L. Yu, H. Wang, E. Wang, A relocated technique of atomic force microscopy (AFM) samples and its application in molecular biology. Ultramicroscopy **92**(3–4), 201–207 (2002). https://doi.org/10.1016/S0304-3991(02)00133-X
14. J. Qi, G.E. Thompson, Comparative studies of thin film growth on aluminium by AFM, TEM and GDOES characterization. Appl. Surf. Sci. **377**, 109–120 (2016). https://doi.org/10.1016/j.apsusc.2016.03.115
15. A. Halama, B. Szubzda, G. Pasciak, Carbon aerogels as electrode material for electrical double layer supercapacitors—Synthesis and properties. Electrochim. Acta **55**(25), 7501–7505 (2010). https://doi.org/10.1016/j.electacta.2010.03.040
16. Y. Qiu, X. Zhang, S. Yang, High performance supercapacitors based on highly conductive nitrogen-doped graphene sheets. Phys. Chem. Chem. Phys. **13**(27), 12554–12558 (2011). https://doi.org/10.1039/c1cp21148j
17. R.A. Oliver, Advances in AFM for the electrical characterization of semiconductors. Reports Prog. Phys. **71** (7) (2008). https://doi.org/10.1088/0034-4885/71/7/076501
18. M. Favre et al., Parallel AFM imaging and force spectroscopy using two-dimensional probe arrays for applications in cell biology. J. Mol. Recognit. **24**(3), 446–452 (2011). https://doi.org/10.1002/jmr.1119
19. D. Martínez-Martínez, L. Kolodziejczyk, J.C. Sánchez-López, A. Fernández, Tribological carbon-based coatings: An AFM and LFM study. Surf. Sci. **603**(7), 973–979 (2009). https://doi.org/10.1016/j.susc.2009.01.043
20. E. Ionization, M. Spectrometry, D. Chimica, U. Parma, P. Area, Effects of Tailored Surface Chemistry on Desorption, pp. 1311–1319 (2015). https://doi.org/10.1021/jasms.8b05078
21. B.J. Haupt, T.J. Senden, E.M. Sevick, AFM evidence of Rayleigh instability in single polymer chains. Langmuir **18**(6), 2174–2182 (2002). https://doi.org/10.1021/la0112514
22. S. Zou, H. Schönherr, G.J. Vancso, Stretching and rupturing individual supramolecular polymer chains by AFM. Angew. Chemie **117**(6), 978–981 (2005). https://doi.org/10.1002/ange.200460963
23. S. Kasas, N.H. Thomson, B.L. Smith, P.K. Hansma, J. Miklossy, H.G. Hansma, Biological applications of the AFM: From single molecules to organs. Int. J. Imaging Syst. Technol. **8**(2), 151–161 (1997). https://doi.org/10.1002/(SICI)1098-1098(1997)8:2%3c151::AID-IMA2%3e3.0.CO;2-9
24. K.C. Neuman, A. Nagy, Single-molecule force spectroscopy: Optical tweezers, magnetic tweezers and atomic force microscopy. Nat. Methods **5**(6), 491–505 (2008). https://doi.org/10.1038/nmeth.1218
25. M. Lekka et al., Cancer cell detection in tissue sections using AFM. Arch. Biochem. Biophys. **518**(2), 151–156 (2012). https://doi.org/10.1016/j.abb.2011.12.013

26. R.S. Krishnan, R.K. Shankar, Raman effect: History of the discovery. J. Raman Spectrosc. **10**(1), 1–8 (1981). https://doi.org/10.1002/jrs.1250100103
27. R. Singh, C. V. Raman and the discovery of the Raman effect. Phys. Perspect. **4**(4), 399–420 (2002). https://doi.org/10.1007/s000160200002
28. H. Kano, H. Segawa, P. Leproux, V. Couderc, Linear and nonlinear Raman microspectroscopy: History, instrumentation, and applications. Opt. Rev. **21**(6), 752–761 (2014). https://doi.org/10.1007/s10043-014-0123-9
29. S. Stewart, R.J. Priore, M.P. Nelson, P.J. Treado, Raman imaging. Annu. Rev. Anal. Chem. **5**, 337–360 (2012). https://doi.org/10.1146/annurev-anchem-062011-143152
30. H. Mitsutake, R.J. Poppi, M.C. Breitkreitz, Raman imaging spectroscopy: History, fundamentals and current of the technique. J. Braz. Chem. Soc. **30**(11), 2243–2258 (2019)
31. Witec, Topographic confocal Raman imaging of archaeological samples, vol. 49, no. 0 (2014)
32. N. Bost, F. Westall, F. Foucher, G. Guimbretière, Petrographical and mineralogical applications of Raman mapping, in *Raman Spectroscopy and Applications*, ed. by K. Maaz. INTECH (2017)
33. J.R. Beattie, J.J. McGarvey, A.W. Stitt, Raman spectroscopy for the detection of AGEs/ALEs, in *Methods in Molecular Biology*, vol. 965, ed. by L. Galluzzi, I. Vitale, O. Kepp, G. Kroemer. Springer (2013), pp. 297–312
34. K. Eberhardt, C. Stiebing, C. Matthäus, M. Schmitt, J. Popp, Advantages and limitations of Raman spectroscopy for molecular diagnostics: an update. Expert Rev. Mol. Diagn. **15**(6), 773–787 (2015). https://doi.org/10.1586/14737159.2015.1036744
35. D. Chenery, H. Bowring, Infrared and Raman spectroscopic imaging in biosciences. Spectrosc. Eur. **15**(4), 8–14 (2003) [Online]. Available: https://www.spectroscopyasia.com/system/files/pdf/IR_15_4.pdf
36. E.E. Lawson, H.G.M. Edwards, A.C. Williams, B.W. Barry, Applications of Raman spectroscopy to skin research. Ski. Res. Technol. **3**(3), 147–153 (1997). https://doi.org/10.1111/j.1600-0846.1997.tb00179.x
37. H. Sato, T. Tanaka, T. Ikeda, S. Wada, H. Tashiro, Y. Ozaki, Biomedical applications of a new portable Raman imaging probe. J. Mol. Struct. **598**(1), 93–96 (2001). https://doi.org/10.1016/S0022-2860(01)00809-2
38. T. Yaseen, D.W. Sun, J.H. Cheng, Raman imaging for food quality and safety evaluation: Fundamentals and applications. Trends Food Sci. Technol. **62**, 177–189 (2017). https://doi.org/10.1016/j.tifs.2017.01.012
39. M.D. Schaeberle, C.J. Lau, P.J. Treado, C.G. Karakatsanis, Raman chemical imaging: Noninvasive visualization of polymer blend architecture. Anal. Chem. **67**(23), 4316–4321 (1995). https://doi.org/10.1021/ac00119a018
40. N. Gierlinger, M. Schwanninger, The potential of Raman microscopy and Raman imaging in plant research. Spectroscopy **21**(2), 69–89 (2007). https://doi.org/10.1155/2007/498206
41. C. Otto, C.J. De Grauw, J.J. Duindam, N.M. Sijtsema, J. Greve, Applications of micro-Raman imaging in biomedical research. J. Raman Spectrosc. **28**(2–3), 143–150 (1997). https://doi.org/10.1002/(sici)1097-4555(199702)28:2/3%3c143::aid-jrs70%3e3.0.co;2-9
42. Z. Sobhani, X. Zhang, C. Gibson, R. Naidu, M. Megharaj, C. Fang, "Identification and visualisation of microplastics/nanoplastics by Raman imaging (i): Down to 100 nm. Water Res. **174** (2020). https://doi.org/10.1016/j.watres.2020.115658
43. H. Abramczyk, B. Brozek-Pluska, J. Surmacki, J. Jablonska-Gajewicz, R. Kordek, Raman 'optical biopsy' of human breast cancer. Prog. Biophys. Mol. Biol. **108**(1–2), 74–81 (2012). https://doi.org/10.1016/j.pbiomolbio.2011.10.004
44. E.A. Taylor, E. Donnelly, Raman and Fourier transform infrared imaging for characterization of bone material properties. Bone **139**, 115490 (2020). https://doi.org/10.1016/j.bone.2020.115490
45. S. Benaglia, V.G. Gisbert, A.P. Perrino, C.A. Amo, R. Garcia, Fast and high-resolution mapping of elastic properties of biomolecules and polymers with bimodal AFM. Nat. Protoc. **13**(12), 2890–2907 (2018). https://doi.org/10.1038/s41596-018-0070-1
46. C.S. Goldsbury, S. Scheuring, L. Kreplak, Introduction to atomic force microscopy (AFM) in Biology. Curr. Protoc. Protein Sci. (SUPPL.58), 1–19 (2009). https://doi.org/10.1002/0471140864.ps1707s58

47. B. Contents, AFM—Atomic Force Microscope Description Lab Manual, pp. 1–55
48. JPK Instruments AG, "The NanoWizard® AFM Handbook (2005), p. 40
49. N.E. Leadbeater, J.R. Schmink, Use of Raman spectroscopy as a tool for in situ monitoring of microwave-promoted reactions. Nat. Protoc. **3**(1), 1–7 (2008). https://doi.org/10.1038/nprot.2007.453
50. N. Gierlinger, T. Keplinger, M. Harrington, Imaging of plant cell walls by confocal Raman microscopy. Nat. Protoc. **7**(9), 1694–1708 (2012). https://doi.org/10.1038/nprot.2012.092

Chapter 4
Characterization Techniques for Chemical and Structural Analyses

Euth Ortiz Ortega, Hamed Hosseinian, María José Rosales López, Andrea Rodríguez Vera, and Samira Hosseini

Abbreviations

2D	Two-dimensional
3D	Three-dimensional
7-APTES	(7-aminoproyl)triethoxysilane
Ag/rGO	Silver nanoparticles over reduced graphene oxide
Al	Aluminum
anti-TG	Antithyroglobulin
anti-TPO	Antithyroid peroxidase
APTES	Aminosylane
AuNPs	Gold nanoparticles
BE	Binding energy
$C_4H_4O_3$	Succinic anhydride
CA	Cellulose acetate
CT	Computed tomography
DPPH	1,1-diphenyl picrylhydrazyl
DRIFT	Diffuse reflectance infrared Fourier transform
DRIFTS	Diffuse reflectance infrared Fourier transform spectroscopy
EBFCs	Enzymatic biofuel cells
EDC	Carbodiimide
fcc	Face-centered cubic
FID	Free induction decay
FFT	Fast Fourier transform
FTIR	Fourier transform infrared spectroscopy
FTS	Fourier transform spectroscopy
GO*x*	Glucose oxidase
H	Hydrogen
H_2O_2	Hydrogen peroxide
He	Helium

E. O. Ortega et al., *Material Characterization Techniques and Applications*, Progress in Optical Science and Photonics 19,
https://doi.org/10.1007/978-981-16-9569-8_4

IR	Infrared
KBr	Potassium bromide
KCl	Potassium chloride
KE	Kinetic energy
KOH	Sodium hydroxide
Mg	Magnesium
MRI	Magnetic resonance imaging
MWCNTs	Multi-walled carbon nanotubes
Ne	Neon
NEP	Noise equivalent power
$-NH_2$	Amino groups
Ni	Nickel
NMR	Nuclear magnetic resonance
NS	Not specific
OPPs	Organophosphorus pesticides
PBS	Phosphate buffered saline
PCA	Principal component analysis
PS	Photosensitivity
Pt	Platinum
Pt-Ni	Platinum-nickel
RAMAN	Raman spectroscopy
RRS	Resonance Raman spectroscopy
S/N	Signal-to-noise
SERS	Surfacc-enhanced Raman scattering
SIMCA	Soft independent modeling of class analogy
SNR	Signal-to-noise ratio
SOC	State of charge
SW	Spectral width
SSRS	Shifted subtracted Raman spectroscopy
TERS	Tip-enhanced Raman spectroscopy
TMB	3,30,5,50-tetramethylbenzidine
Trp	Tryptophan
TRS	Transmission Raman spectroscopy
Tyr	Tyrosine
UV	Ultraviolet
UV-Vis	Ultraviolet-visible
V	Vanadium
XPS	X-ray photoelectron spectroscopy
XRD	X-ray diffraction
$Zn(NO_3)_2 \cdot 6H_2O$	Zinc nitrate hexahydrate
ZnO	Zinc oxide
λ	Wavelength
Φ_{spec}	Spectrometer work function

4.1 Raman Spectroscopy for Material Characterization

4.1.1 History of Raman Spectroscopy

Chandrasekhara Raman is, indeed, the protagonist of the history of Raman spectroscopy (RAMAN) (Fig. 4.1). He was a famous Indian physicist whose most notable work was the discovery of the Raman effect. In 1871, Lord Rayleigh published a paper describing the Rayleigh scattering phenomenon, which stated that the gases in the atmosphere cause the scattering of sunlight, giving the sky its blue color and the deep sea its dark blue by the reflection of the blue hue of the sky. Encouraged by this statement, Chandrasekhara Raman later began a series of experimental and theoretical studies on the scattering of light in transparent media. Through his monograph entitled "Molecular Diffraction of Light" in 1922, he explained that energy could be transferred between light quanta and molecules.

In 1928, Chandrasekhara Raman, Kariamanikkam Krishnan, and Sundaram Venkateswaran examined ether vapor at high temperatures and analyzed the scattered track through a direct vision spectroscope [1]. The experiment showed that

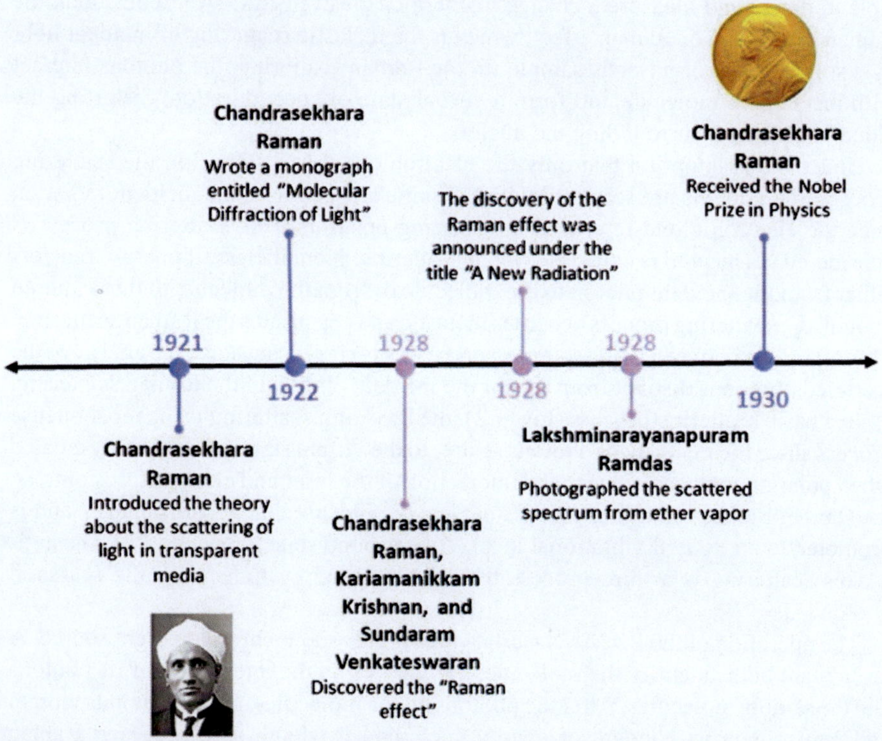

Fig. 4.1 History of RAMAN

the modified scattering was separated from the classical scattering in the spectrum, demonstrating a change of wavelength in scattering that they announced as "A New Radiation". The Raman effect, or new radiation, was observed in many liquids, gases, vapors, and crystals, and its use was quickly acknowledged in chemistry and physics. In fact, by the end of 1929, the number of published papers on the Raman effect had grown to over 200 [2]. In 1930, Chandrasekhara Raman's discovery of the effect named after him, and his study on the scattering of light granted him the Nobel Prize in Physics. Since Raman scattering is the principle of RAMAN, Raman spectroscopy's invention is solely associated with the discovery of the Raman effect.

4.1.2 Mechanism of Operation of Raman Spectroscopy

Three scenarios could occur upon light's interaction with matter: The sample may absorb, scatter, or not interact with the photons. A sample molecule may absorb a photon if the energy of an incident photon matches an energy gap between the molecule's ground and excited state, thus boosting the molecule to a higher excited energy state. A wide range of absorption spectroscopy techniques employs this principle to detect and measure a change in energy. However, RAMAN relies upon the Raman scattering, or Raman effect, which is the inelastic scattering of incident light by a solid, liquid, or gaseous sample. In the Raman scattering, the photons interact with the sample molecule and form a virtual state of short duration, distorting the cloud of electrons surrounding the nuclei.

Under the assumption that only the electron cloud is distorted in the scattering process, the photons are scattered with the same frequency as the incident radiation once the electron cloud returns to the starting position. This scattering process is considered elastic and is called Rayleigh scattering. Nonetheless, if energy transfers either from the incident photon to the molecule or from the molecule to the scattered photon, the scattering process becomes inelastic and constitutes the Raman scattering. The difference between both scattering processes is significant since the energy of the scattered photon is distinct from that of the incident photon in the Raman scattering.

In a basic scattering process (Fig. 4.2), the Rayleigh scattering is the most intense process since the majority of photons scatter to the virtual state, which is only created when polarization occurs due to the interaction of the laser and electrons. In contrast, the Raman Stokes scattering process depicts a molecule that absorbs energy and is promoted to an excited vibrational level of the ground state known as "n". The anti-Stokes scattering is a third process that involves energy transfer to the scattered photon [3].

The operation of the RAMAN starts with a highly monochromatic light source. A laser beam helps identify the molecules' properties by the interaction of its photons and the sample molecules. After the photons hit the molecules, the atom bonds vibrate and emit a new wavelength of energy, including Rayleigh scattering and Raman scattering. This step is essential since the change between the laser wavelength and the emitted new energy wavelength reveals information about the molecules within

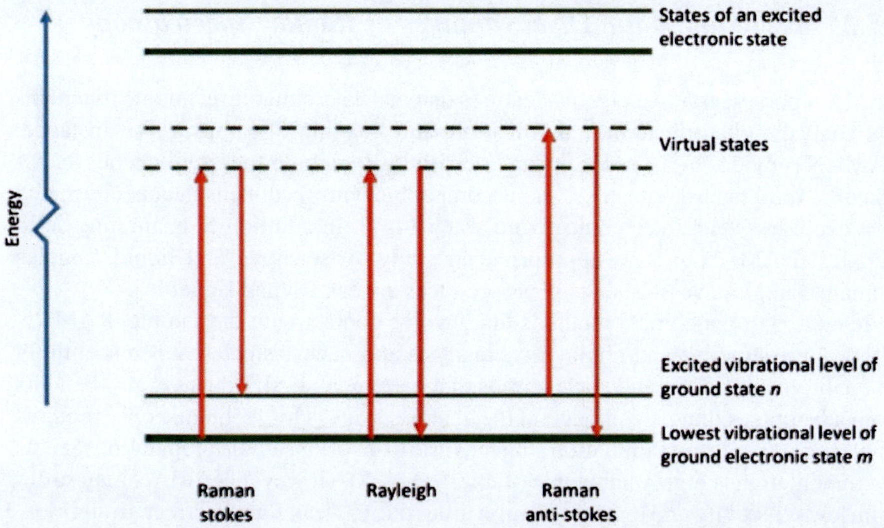

Fig. 4.2 Schematic diagram of the Rayleigh, Raman Stokes, and Raman anti-Stokes scattering processes. The excited vibrational level "*n*" is distinct by one vibrational unit from the lowest energy vibrational state "*m*". The upward arrow illustrates the excitation energy, while the downward arrows represent the scattered energy [3]

the sample. Subsequently, the Rayleigh filter allows only the Raman scatterings to pass through. The emitted energy contained in the Raman scattering segregates into a light spectrum by a diffraction grating. Finally, a detector analyzes the obtained spectrum and converts it into a readout on a computer (Fig. 4.3). Since a particular vibration characterizes each molecule, the readout obtained by RAMAN shows peaks in the Raman spectrum, representing the different vibrations of the intermolecular bonds and obtaining a comprehensive analysis of the sample.

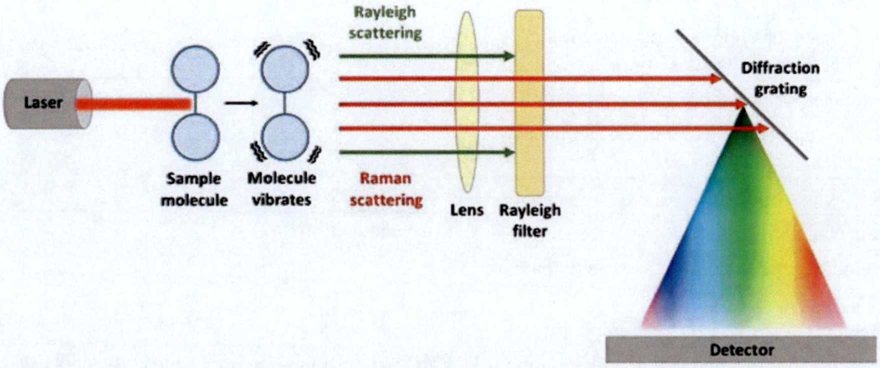

Fig. 4.3 Illustration of the basic principle of RAMAN. The photons of the laser beam vibrate the molecules, starting the scattering process and from which their diffracted beams reach the detector

4.1.3 Advantages and Disadvantages of Raman Spectroscopy

RAMAN possesses advantageous features that make it a suitable technique for chemical analysis, quantification, classification, and imaging of samples. For instance, RAMAN provides high specificity and sensitivity, facilitates the simultaneous detection of several macromolecules, and is compatible with specimens in aqueous media as it excludes weak interference from water [4–7]. In addition to examining solid samples, RAMAN is also appropriate to study air-sensitive and liquid samples contained in glass vessels since glass produces a weak Raman signal [6].

In contrast to analytical methods that involve contact with the sample, RAMAN offers nondestructive and noninvasive analysis approaches since it depends entirely on the produced power and wavelengths of a specimen [4, 5]. RAMAN also benefits from advantages that extend beyond the analysis stage. This technique only requires small amounts of the examination subject, little to no prior sample preparation, and its instrumentation is highly adaptable (Table 4.1) [4–7]. However, RAMAN may suffer from low sensitivity and long acquisition times due to weak Raman effect. In addition, fluorescence may interfere with the Raman effect, and therefore, samples containing fluorochromes cannot be subjected to analysis [4, 5].

In an attempt to overcome the disadvantages of the conventional RAMAN, other specialized techniques such as the surface-enhanced Raman scattering (SERS), the tip-enhanced Raman spectroscopy (TERS), and the shifted subtracted Raman spectroscopy (SSRS) offer extended sensitivity, superior spatial resolution, and allow the analysis of the fluorescent samples, respectively. Similarly, the resonance Raman

Table 4.1 Advantages and disadvantages of RAMAN

Advantages	Disadvantages
It is appropriate for chemical analyses, quantification, classification, and imaging of samples [4, 5]	Samples may present a weak Raman effect [4, 5]
Possesses high specificity [5]	
Facilitates the simultaneous detection of several macromolecules [5]	
Excludes weak interference from water [4–7]	Low sensitivity [4, 5]
Studies air-sensitive and liquid samples contained in glass vessels [6]	
It is nondestructive [4, 5]	
It is noninvasive [4, 5]	Long acquisition time [4, 5]
Requires little to no prior sample preparation [4–7]	
Requires small amounts of sample [6]	
Includes highly adaptable instrumentation [4]	Fluorescent samples are not subject to analysis [4, 5]

spectroscopy (RRS) and the transmission Raman spectroscopy (TRS) provide higher specificity and the ability to examine larger volumes (Table 4.1) [4, 7].

4.1.4 Applications of Raman Spectroscopy

The different applications of RAMAN are very promising due to the technique's noninvasive and high specificity characteristics. RAMAN's operation mechanism is based on the production of various Raman spectra representing the vibration of each chemical bond or functional group in a molecule. In this regard, medical diagnostics using RAMAN can discriminate between benign and malignant cancer tissues [8]. In addition, further analysis of distinctive spectral characteristics and intensity differences can differentiate cancer types into normal, benign, and malign. Oncology research has a particular interest in RAMAN as a method for categorizing dead tissue in breasts [9] and intraoperative margin assessment for breast cancer (Fig. 4.4). Using Raman spectra can also shorten the operative time by monitoring sentinel lymph nodes and surgical margins for breast-conserving surgery. In 2014, Kast et al. presented the clinical applications of RAMAN on prostate cancer, such as screening, biopsy, margin assessment, treatment efficacy, and other future areas of research with a focus on multiplexing Raman spectroscopy with other modalities [10]. The results align with similar reviews indicating successful clinical proof-of-concept and the opportunity to implement surgical RAMAN fiber optic probe studies for evaluating various types of cancer in vivo such as bladder, breast, lung, brain, skin, and cervical [11]. Likewise, the neurology domain works with RAMAN for brain surgery. In 2012, Kircher et al. located a whole-brain tumor before surgery and during surgery in mice by using a triple-modality of magnetic resonance imaging (MRI), photoacoustic imaging, and surface-enhanced Raman scattering (SERS) [12].

Raman-based techniques are helpful at every stage of a biotechnological manufacturing process [14–16]. Several process parameters, such as the reactor feed quality,

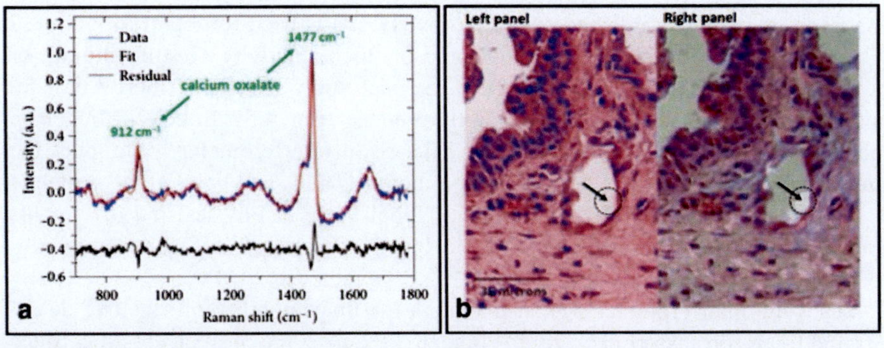

Fig. 4.4 **a** Raman spectrum of a breast lesion. **b** The histopathology of breast lesions. Reproduced (or adapted) with permission [13], ©2017, Pin Gao et al., Hindawi

inoculum age, and harvest point, affect the yield and the quality of recombinant protein production. However, the batch-to-batch variability of the raw materials and feed media is a difficult parameter to control and can majorly affect the outcome of the bioprocess. In 2010, Li et al. used Raman spectroscopy and chemometric data analysis to characterize a cell-media complex from a commercial recombinant protein manufacturing process [17]. RAMAN, the main component analysis (PCA), and a soft independent modeling of class analogy (SIMCA) were used to analyze the aqueous cell-media complex and assess its composition and quality. The results showed effective discrimination and identification of all the components of the complex and demonstrated that variability in each batch could be monitored and controlled.

The RAMAN method is also vastly applied in the forensic field, particularly the surface-enhanced Raman scattering (SERS). The high sensitivity of SERS is ideal for analyzing biological samples such as fingerprints, DNA, RNA, and urea from blood and sweat. Kamińska et al. applied SERS to examine antigen–antibody interactions in different blood groups samples [18]. The authors also used a PCA, which showed 96 to 98 percent accuracy for the correct choice of blood groups. Moreover, Zhang et al. used SERS and Raman mapping to investigate single latent fingerprints [19]. Raman active reporter molecules were embedded in silver nanoparticles to allow multiplexing. Subsequently, electrostatic interactions allowed the attachment of antibodies to lysozymes, cotinine, and immunoglobulin G. This specificity of antibody together with multiplexing facilitated a high-speed mapping and sensitive SERS analysis.

4.2 Fourier Transform Infrared Spectroscopy for Material Characterization

4.2.1 History of Fourier Transform Infrared Spectroscopy

The Fourier transform infrared spectroscopy (FTIR) is used to obtain the infrared (IR) spectrum of absorption, emission, and photoconductivity of solid, liquid, or gaseous samples. The history of FTIR dates back to 1880 when Albert Abraham Michelson invented the "Michelson interferometer" (Fig. 4.5). In 1887, Michelson and Edward Williams Morley used the Michelson interferometer to conduct the famous experiment that rejected the ether hypothesis and shed light on the principle of relativity [20]. Michelson received the Nobel Prize in Physics in 1907 for his notable work on the measurement of the speed of light and his contributions to optical precision instruments.

Albeit the interferometer's great potential, the unavailability of sensitive detectors and Fourier transform algorithms hindered the advance of the technique. Since the invention of the interferometer, scientists developed handmade high-resolution Fourier transform infrared spectrometers. However, the arrival of commercially

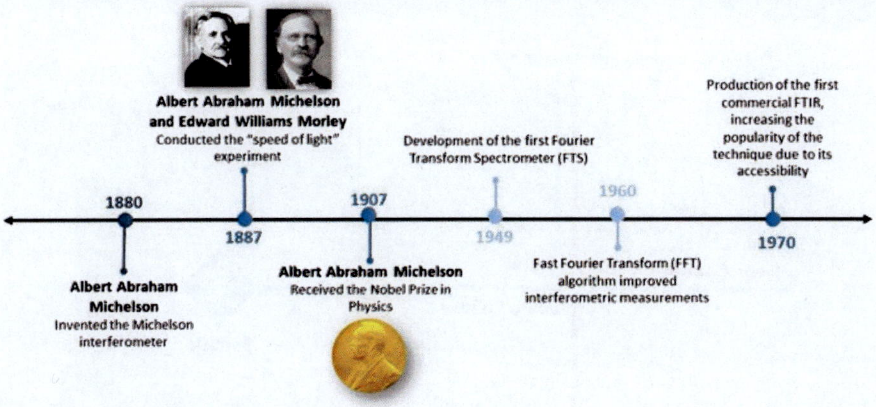

Fig. 4.5 History of FTIR

available high-resolution instruments allowed scientists to focus on research and theoretical techniques rather than instrument development [21].

In 1949, despite the time-consuming task of calculating Fourier transforms, the scientific community developed the first Fourier transform spectroscopy (FTS) equipment and generated the first Fourier transform spectrum. Shortly after, the 1960s witnessed many advances in the theory of interferometric measurements and their use in physical systems due to the introduction of the fast Fourier transform (FFT) algorithm. The FFT algorithm allowed Fourier transforms to be easily and effectively computed [20]. However, the computational task of calculating Fourier transforms was still an obstacle in the way of achieving high-resolution FTIR until the 1980s. The improvement of FTIR was tightly connected to the evolution of fast computers and algorithms [21].

The first commercialized FTIR went on sale by 1970 and became a very accessible and popular technique among scientists, yet the equipment's large size and high cost were limiting factors [20]. Additionally, prior to 1980, the resolution of most Fourier transform spectrometers was minimal and hardly superior to the traditional grating instruments [21]. Fast computers now aid FTIR in performing Fourier transforms in a matter of seconds. Such progress in technology lowered the cost, allowed availability, and improved the capabilities of the FTIR technique [20].

4.2.2 Mechanism of Operation of Fourier Transform Infrared Spectroscopy

FTIR is a robust procedure to determine functional groups and molecular bonds in a material. Spectrometry techniques use spectrophotometers to determine the absorption spectrum of a compound in a similar fashion as the FTIR equipment provides

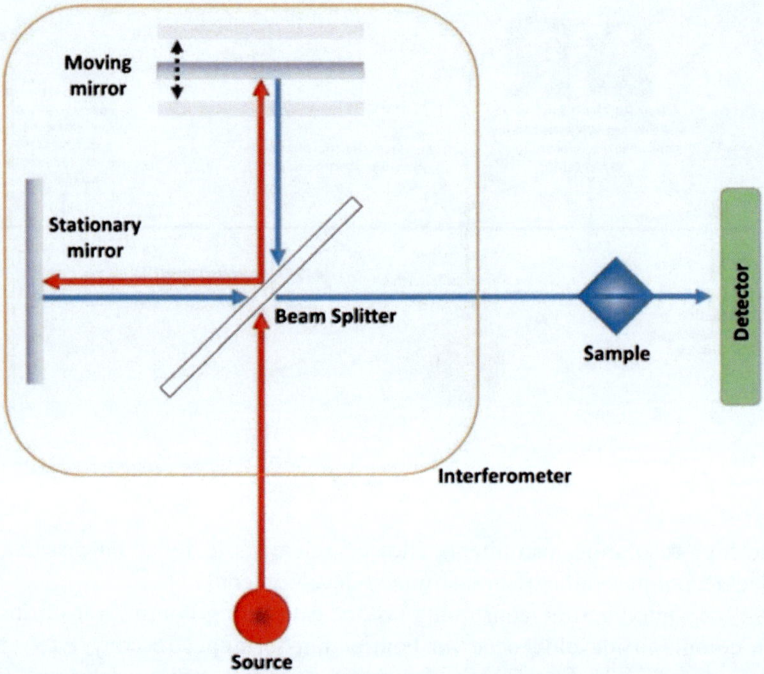

Fig. 4.6 Schematic illustration of a FTS setup based on Michelson interferometer. A beam of IR irradiation passes through the interferometer and is redirected to the sample. The detector receives the interferogram signal and translates it into the sample spectrum via Fourier transformation software

its infrared (IR) spectrum (Fig. 4.6). The FTIR apparatus has a glowing black-body source that emits a beam of IR irradiation, which enters the interferometer and initiates the spectral encoding. The interferometer creates an interferogram depicting the constructive and destructive interference of the recombination of beams with different lengths. Subsequently, the beam is redirected to the samples compartment to interact with the specimen, where the latter absorbs specific frequencies of energy obtained from the interferogram of the sample. Finally, the detector receives the interferogram signal for all frequencies and simultaneously acquires a background signal. After the interferogram subtracts the background spectrum from the sample spectrum, the desirable spectrum is obtained by a Fourier transformation software [22].

The Fourier transform infrared spectrometer requires an IR radiation source, an IR radiation detector, and an interferometer. The principal source of IR radiation is heat or thermal radiation. Therefore, the IR radiation source of FTIR is frequently electrically heated to reach an incandescent state. Among the three subregions that compose the IR spectrum, the FTIR operates in the mid-IR region as it is composed of fundamental absorption bands and provides molecular identification. The widely known mid-IR sources are the Globar, the Nernst glower, and the tungsten glower

[23]. The source may either have an adjustable power supply to vary the temperature or an adjustable aperture to change the intensity of light entering the interferometer. In contrast with dispersive IR spectrometers generating a slit-shaped light beam, the FTIR instrument produces a round-shaped beam image to benefit from the source energy [24].

The IR radiation detector measures the radiation intensity and translates it into an amplifiable electrical signal to produce a spectrum. The performance of an IR detector is evaluated according to its photosensitivity (PS), noise equivalent power (NEP), and detection ability. The PS is the output signal per watt of incident radiation, while the NEP is the amount of incident radiation equivalent to the detector's intrinsic noise, and the detection ability is the PS per unit of the active area of the detector [23]. The NEP indicates the detection limits which are usually proportional to the active area of the detector.

The modern FTIR spectrometer relies on the Michelson interferometer, consisting of a movable mirror, a fixed mirror, and a beam splitter. The incoming infrared beam is collimated and directed through a beam splitter, which acts as a partially reflecting mirror and transmits half the radiation to a fixed mirror and the remaining half to a movable mirror. The two beams are reflected in the beam splitter to recombine, focus, and ultimately direct them to the detector.

4.2.3 Advantages and Disadvantages of Fourier Transform Infrared Spectroscopy

The FTIR technique benefits from the infrared (IF) spectra since such spectrum can provide extensive sample information. In this regard, the peak positions of the sample spectra reveal the structure of the molecules, while the peak intensities expose the concentration of molecules, and the peak widths determine the hydrogen bonding and pH [25]. The technique can also examine nearly all samples since numerous molecules have strong absorbances in the mid-IR radiation range at which FTIR operates [25, 26].

The FTIR has other extensive qualities that differentiate it from other spectroscopy techniques, such as its operation simplicity [25–27], low cost [25], high reproducibility, and accurate measurements [27]. Moreover, the conventional FTIR technique can even detect micrograms (10^{-6}) of the analyte within a sample, making it highly sensitive. This feature is possible due to the sensitivity of detectors employed, the high optical throughput that outcomes in low noise levels, and the fast scans that allow the diminishment of random measurement noise to any expected level [25, 27]. Furthermore, the FTIR does not require calibration as it employs self-calibration equipment by implementing an internal wavelength calibration standard using a Helium–Neon (He–Ne) laser.

The mechanism of operation of FTIR represents an advantage in itself as this method provides a short response time due to the simultaneous measurement of all

Table 4.2 Advantages and disadvantages of FTIR

Advantages	Disadvantages
Can examine nearly all samples [25, 26]	Inability to detect chemical species that do not vibrate [25, 26]
Can provide extensive sample information [25]	
Short response time [25–27]	Difficulty of compound differentiation in mixtures [25, 26]
Simple operation [25–27]	
Relatively inexpensive [25, 26]	Incompatibility with water as it causes interference [25]
High sensitivity [25, 27]	
Self-calibrating instrumentation [27]	Single beamed instrument [27]
High reproducibility and accurate measurements [27]	

frequencies. However, the technique cannot detect noble gases, monoatomic ions, and homonuclear diatomic molecules since these chemical species do not vibrate, which is necessary for FTIR's operational process. Evaluating the sample spectra' peaks to differentiate a component in a mixture is also restricted, considering that associating peaks to specific molecules is more challenging due to the augmented difficulty that a complex sample spectrum entails [25, 26].

Additionally, the interference of water is a relevant drawback to consider. Water produces wide and intense peaks that can cover the spectra of solutes dissolved in this solvent and can also disintegrate compounds used for sample preparation, thus interfering in the obtained sample spectrum [25]. In comparison with different dispersive instruments with double beams, FTIR instruments have only a single beam representing a limitation for this technique [27]. Table 4.2 provides information on the advantages and disadvantages of FTIR as a chemical and structural analyses technique.

4.2.4 Applications of Fourier Transform Infrared Spectroscopy

As an accessible, accurate, and efficient technique, FTIR is utilized to identify polymeric, inorganic, and organic species in samples of known or unknown chemical compositions. The industry and the scientific community commonly employ this analytical tool due to its advantageous features. For instance, the FTIR technique needs minimum or no sample preparation, requires only a minimum amount (1–5 μl) of sample to conduct the analysis, and is a nondestructive procedure since the infrared light does not damage the sample. These attractive features place FTIR as an essential technique for multiple fields such as environment [28, 29], forensics [30, 31], pharmaceutical [32, 33], food quality control [34, 35], polymer science [36, 37], plastics development [38, 39], environment quality control [40, 41], and sensing and biosensing [42, 43].

Breast cancer is one of the leading causes of death in women, and it is often detected at late stages since this type of cancer tends to be asymptomatic in the early stages. The treatment for breast cancer can be highly effective, especially when the disease is timely diagnosed. However, X-ray mammography, magnetic resonance imaging (MRI), and other currently available clinical methods are regularly used once the condition is advanced. FTIR is a noninvasive and reliable medical diagnosis technique enabling the analysis of easy-collecting samples such as saliva, cells, tissues, or even hair. In 2019, Sitnikova et al. conducted a study on 66 breast cancer patients and 86 healthy donors to identify breast cancer from blood serum employing FTIR coupled with multivariate methods. The authors observed that the significant difference in the FTIR spectra of cancer patients and healthy donors corresponds to nucleic acids with modified functional groups, specifically, oscillatory deformations in C–H bonds and stretching vibrations of PO_2. Such mutations were detected in the 1450–1250 cm^{-1} and 3100–3000 cm^{-1} spectral zones (Fig. 4.7) [44].

A malignant brain tumor is another deadly cause for a patient. The available clinical methods for early brain cancer diagnosis are expensive, discouraging those from low-income backgrounds from receiving an adequate diagnosis and associating symptoms such as dizziness and headaches with less dangerous diseases. Gray et al. successfully detected brain tumors by analyzing biological species present in blood serum through FTIR (Fig. 4.8a). Despite the complicated biological composition of blood serum, the combination of the FTIR spectra and machine learning algorithms allowed the authors to discriminate between non-cancer and brain cancer specimens (Fig. 4.8b). The technique reached a high specificity (91.5%) and sensitivity (92.8%).

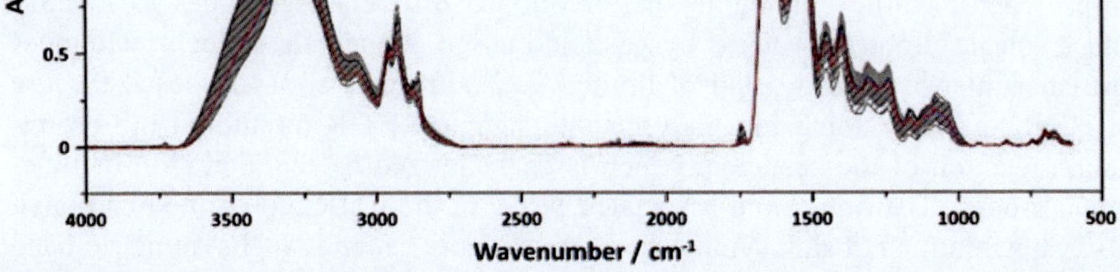

Fig. 4.7 FTIR spectra of blood serum of a group of patients with breast cancer and a group of healthy donors. Reproduced (or adapted) with the permission [44], ©2020, Elsevier

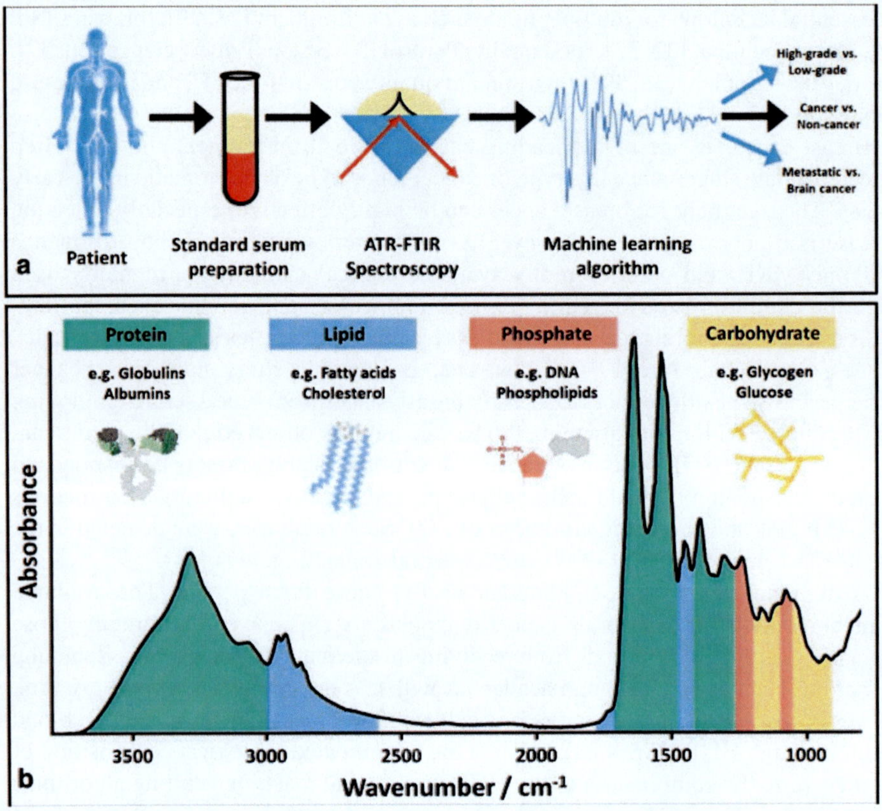

Fig. 4.8 **a** The general procedure to identify brain tumors through ATR-FTIR and machine learning algorithms. **b** The ATR-FTIR spectra of human blood serum and associated compounds. Reproduced (or adapted) with permission from [45], ©2018, BMJ Publishing Group Ltd

Proteins, lipids, phosphates, and carbohydrates adjust their absorption spectra in the presence of cancer cells due to their chemical nature (bond vibrations) [45].

Pesticides are a significant concern to the ecosystem due to the negative impact they have on the ecosystem as well as human life. Flonicamid is a chemical compound used to exterminate insects from vegetables, apples, and cottonseed crops. The scientific community has developed different techniques to identify and eradicate problems caused by pesticide usage. Nonetheless, such techniques are labor-intensive, costly, and of limited access. Kumar et al. proposed the use of FTIR to detect flonicamid in vegetable samples. FTIR monitored the bromination reaction of the pesticide, which is characterized by the alteration of functional groups. The flonicamid-associated peaks in the FTIR represented carbonyl, C–C stretching, NH band, symmetrical stretching C–N, and N–H symmetric band at 785.61 cm^{-1}, 1035.81 cm^{-1}, 1321.34 cm^{-1}, 1417.46 cm^{-1}, and 1654.85 cm^{-1}, respectively (Fig. 4.9a). These functional groups formed a complex during the chem-

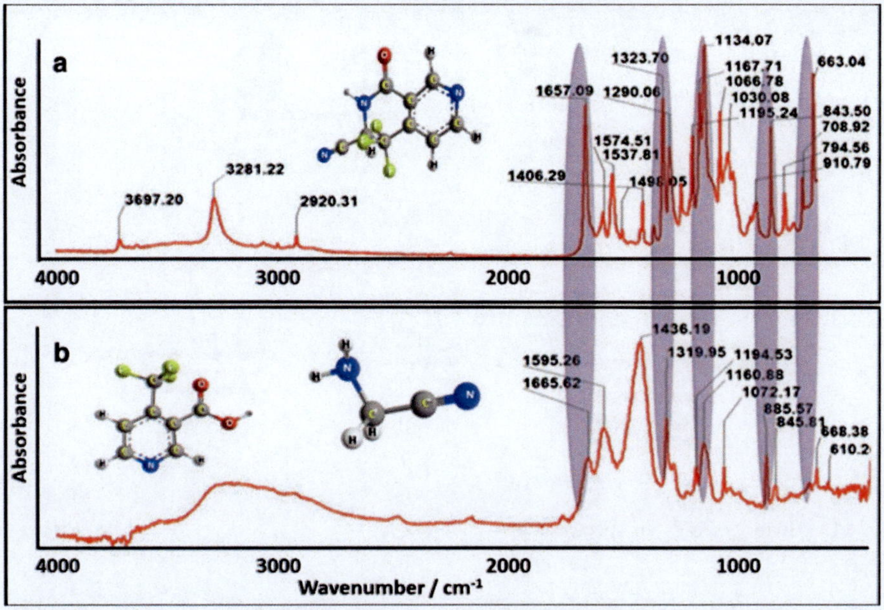

Fig. 4.9 FTIR spectra of **a** flonicamid and **b** flonicamid with sodium hydroxide. Reproduced (or adapted) with the permission [46], © 2020, Elsevier

ical process, thus shifting the bands (Fig. 4.9b). The change in the absorbance indicated the presence of the pesticide in the analyzed sample [46].

4.3 Ultraviolet–Visible Spectroscopy for Material Characterization

4.3.1 History of Ultraviolet–Visible Spectroscopy

In 1801, the German physicist Johann Wilhelm Ritter discovered ultraviolet (UV) radiation when he split sunlight through a prism and irradiated silver chloride with each produced color. Subsequently, Ritter extended the electromagnetic spectrum beyond the visible region and named "ultraviolet" to the region beyond the violet (Fig. 4.10) [47]. The scientific community was eager to unravel knowledge on the UV technique until the last decades of the nineteenth century. W. N. Hartley was the pioneer in the study of the UV spectra of organic compounds and attempted to correlate the photographed UV absorption spectra of organic compounds with certain functional groups.

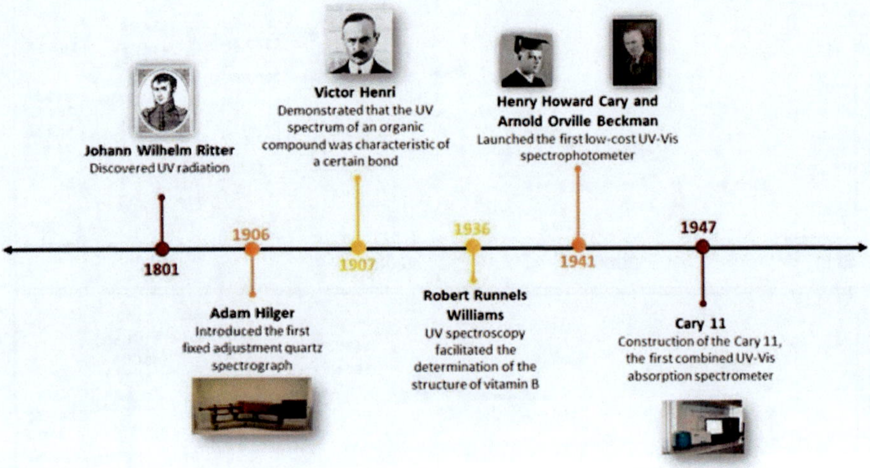

Fig. 4.10 History of UV–Vis spectroscopy

At the time, UV spectroscopy relied upon quartz spectrographs that photographically recorded spectra. However, the development of the equipment and its applications continued to evolve. In 1906, Adam Hilger introduced the first fixed adjustment quartz spectrograph and made progress in other complementary elements to this tool including the sector photometer and the photoelectric "Spekker" photometer. A year later, the French chemist Victor Henri followed Hartley's previous work and took it to the next stage. Henri confirmed that the UV spectrum of an organic compound was characteristic of a specific bond rather than a function of the whole molecule, which was an essential feature for the operation of UV spectroscopy.

The quantitative determination of vitamin A in the 1930s was a critical factor that led to the development of ultraviolet–visible (UV–Vis) spectrophotometers [47]. Shortly after in 1936, Robert Runnels Williams determined the structure of thiamine via UV spectroscopy. This technique became widely popular among organic chemists, and such tendency only increased when Henry Howard Cary and Arnold Orville Beckman introduced the first low-cost UV–Vis spectrophotometer in 1941. The technological development of UV spectroscopy never ceased. The first combined UV–Vis absorption spectrometer, the Cary 11, was constructed in 1947 [48]. By the 1950s, many other companies were producing UV–Vis spectrophotometers, and the instruments have become larger, more complex, and rather pricier.

4.3.2 *Mechanism of Operation of Ultraviolet–Visible Spectroscopy*

Quantitative absorption spectroscopy aims to calculate the amount of analyte in a sample solution. The determination relies on the capacity of an analyte to absorb specific amounts of light from a reference beam that passes through the sample solution. Analytes naturally absorb radiation in the UV–Vis range; however, certain analytes may not behave the same. In such cases, the analyte must undergo chemical modification to convert into a species that absorbs radiation of a suitable wavelength. Chemically modified or not, the analyte in the sample solution affects the amount of radiation absorbed by the analyte or transmitted through the solution. Therefore, the relative transmittance or absorbance of the solution can be used as an indicator of analyte concentration [49].

In contrast with the vibrational transitions that characterized the previously reviewed spectroscopy techniques, the molecules subjected to UV–Vis spectroscopy undergo electronic transitions in the UV or visible region. In UV–Vis spectroscopy, the molecules containing non-bonding electrons absorb UV or visible light energy and are excited to higher anti-bonding molecular orbitals. This technique is the basic standard method to quantitatively determine analyte concentrations in a solution based on the Beer–Lambert law (Eq. 4.1) [50]:

$$A = \left(\frac{I_0}{I}\right) = \varepsilon cl \tag{4.1}$$

where A is the measured absorbance, I_0 is the intensity of the incident light at a given wavelength, and I is the transmitted intensity. However, the second form of the equation enables the linear relationship between concentration and absorbance of a solution. In the second form, the L is the path length through the sample, C is the concentration of the absorbing species, and ε is the extinction coefficient of the studied species and wavelength. In quantitative analysis, ε is employed at the maximum absorption wavelength to minimize errors in the absorbance curve caused by instrumental wavelength uncertainty [50].

The operation of a UV–Vis spectrophotometer starts when the integrated lamp emits UV or visible light. Both wavelengths are different and thus produced with different lamps. For instance, a deuterium lamp emits ultraviolet light, and a tungsten lamp generates visible light. The monochromator is the next component in the operation mechanism of a UV–Vis spectrophotometer. A monochromator uses two slits separated by a prism or a diffraction grating to produce a light beam with an extremely narrow bandwidth. In another word, it can be considered as the light of a single color. The initial light beam passes through the first slit of the monochromator and is refracted into a rainbow of colors when striking the prism. Each wavelength moves to a different place in space; however, only one wavelength goes through the second slit. The next element of the UV–Vis spectrometer is the beam splitter which divides a beam of light into two parallel light beams of equal intensity. The light

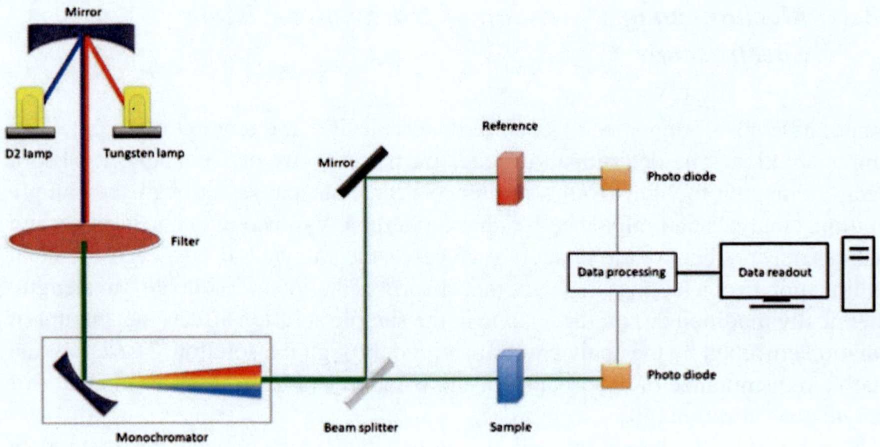

Fig. 4.11 Schematic diagram of a double beam UV–Vis spectrophotometer

beams strike two cuvettes containing the sample and the sample reference, and the beams strike the detector as they exit the sample containers. Finally, the detector converts the impact of photons into electrical current via computational programs (Fig. 4.11).

4.3.3 Advantages and Disadvantages of Ultraviolet–Visible Spectroscopy

UV–Vis spectroscopy is a low-cost, nondestructive, and self-calibrating technique [51, 52]. The procedure does not require specialized operation training, holds high sensitivity, and is a relatively rapid technique for structural and chemical quantitative and qualitative analysis. In addition, this technique effectively quantifies a species concentration in a liquid sample, assesses reaction kinetics, and is highly selective toward nanoparticles. Such features contribute to the versatility of this method, hence bringing the UV–Vis technique to a number of applications in research and industrial settings.

On the other hand, the narrow wavelength range, compound interference, and difficulty in analyzing complex samples are a few limitations of UV–Vis spectroscopy. Although the UV–Vis light spectrum is 100–800 nm and the UV–Vis spectrometers can operate wavelengths from 200 to 1100 nm, a higher quality of results is obtained inside a narrower interval from 220 to 780 nm. In addition, the subproducts in the UV–Vis spectra that are originated from the solute–solvent interaction could cause interference while utilizing UV–Vis spectroscopy. The technique is also limited to the sample's complexity since quantifying an analyte in a complex mixture is rather challenging (Table 4.3) [51].

Table 4.3 Advantages and disadvantages of UV spectroscopy

Advantages	Disadvantages
Simple operation [51, 52]	Distorted results outside the wavelength range [51]
High sensitivity [51]	
Rapid measurement [52]	
Quantitative and qualitative analysis [51]	
Quantifying the analyte concentration in a liquid sample [51]	Interference from the solute–solvent reactions [51]
Assesses reaction kinetics [51]	
High selectivity to nanoparticles [52]	
Versatility [51, 52]	Increased difficulty in the quantification of mixtures [51]
Inexpensiveness [51, 52]	
Nondestructive [51]	
Self-calibrating instrumentation [51, 52]	

4.3.4 *Applications of Ultraviolet–Visible Spectroscopy*

Spectroscopy techniques utilize the interaction of light with matter to characterize an unknown specimen. In UV–Vis spectroscopy, the amount of light absorbed by the sample allows the determination of the physical and chemical properties of the specimen. UV–Vis spectroscopy requires a homogeneous liquid analyte to accurately quantify cells, proteins, ions, and bioactive molecules. The liquid state of the sample is essential since the existence of solid particles involves the predominance of a light scattering process rather than an absorption process. The scientific community widely uses the UV–Vis spectroscopic technique to analyze samples due to its low cost, accessibility, short response time, sensitivity, portability, simple operation, and real-time and in situ detection. The UV–Vis technique is widely applied in various fields, including bioprocesses [53], detection of environmental pollutants [54], pharmaceutical industry [55], batteries [56], food processing [53], portable sensing devices [57], nanoparticles characterization [58], and in-line process analytical technology [59]. UV–Vis spectroscopy can monitor antioxidants, phenolic compounds, and sugar to improve the quality of food products [53]. In addition, this technique can effectively analyze the characteristics of plasmonic nanoparticles since their shape and size determine the wavelength at which they absorb light. For that reason, UV–Vis spectroscopy can obtain data on the size, size distribution, and shape of metal nanoparticles, in addition to their stability when embedded within a polymeric material [60].

The electrolyte concentration is significant in an electrochemical energy system to avoid performance losses. In this regard, a method to monitor the state-of-charge

(SOC) is important for battery performance since the SOC indicates the ion levels in the electrolyte. Since UV–Vis spectroscopy is independent of the temperature or electrochemistry of the analyte, it is an accurate technique to predict the SOC. However, this technique is limited by the amount of electrolyte concentration it can precisely detect. A vanadium redox battery employs vanadium ions to produce energy, yet a high dose of vanadium ions in the electrolyte significantly reduces the signal-to-noise ratio (SNR) since the Lambert–Beer law does not apply to high concentrations [61]. Furthermore, UV–Vis spectroscopy can study changes in the protein structure as the peptide bonds absorb UV light from 180 to 230 nm. The aromatic amino acids present an absorption peak from 230 to 300 nm, and their absorption spectrum is strongly affected by pH. The phenol from Tyrosine (Tyr) is sensitive to environmental pH changes. Therefore, the absorption spectrum presents shifts depending on the reactions between Tyr and the surrounding environment [62]. Moreover, the UV–Vis spectroscopic technique can also identify pesticides in a simple, fast, and sensitive manner. Identifying organophosphorus pesticides (OPPs) is essential to prevent their negative effects on humans and the environment, and the UV–Vis technique can achieve this by implementing metal nanoparticles. For instance, gold nanoparticles (AuNPs) modified with ytterbium agglomerate the metallic nanoparticles when interacting with the OPPs. Consequently, the absorption peak at 520 nm decreases corresponding to the existence of the pesticides [63].

A healthy person has a 4 mM glucose concentration, while a diabetes patient can surpass the 11 mM breach of glucose level. Hence, constant monitoring of glucose concentration helps generate a timely diagnosis and avoid health complications. Tran et al. developed a colorimetric glucose biosensor based on the colorimetric reaction between silver nanoparticles over reduced graphene oxide (Ag/rGO) and 3,30,5,50-Tetramethylbenzidine (TMB), using hydrogen peroxide (H_2O_2) from the glucose oxidation catalyzed by glucose oxidase (GO_x) (Fig. 4.12a). The analyte emits a blue hue to indicate the presence of glucose, and the coloration becomes more intense when glucose levels rise. The absorption peak close to 550 nm shifted with the increase of glucose, and the calculated limit of detection was 40 μM (Fig. 4.12b). The biosensor demonstrated the capability to measure the glucose concentration in real samples from human serum diluted in phosphate buffered saline (PBS) (Fig. 4.12c) [64].

Phytochemicals have medical attributes that are beneficial to humans. There are two general categories of phytochemicals: the primary metabolites in charge of plant growth and the secondary metabolites responsible for plant care. Flavonoids and phenolic compounds are secondary metabolites with advantageous pharmacological properties, such as antioxidant and anti-carcinogenic. Kumar et al. studied the antioxidant effect of phenolic compounds isolated from the plant *Dillenia pentagyna* using UV–Vis spectroscopy and acid ascorbic as reference. The UV–Vis spectra showed that the free radical activity was inhibited by 76% (Fig. 4.13a) and 94% (Fig. 4.13b) in the presence of 10 μg ml^{-1} of leaf extract and ascorbic acid, respectively. The polyphenols from phenolic and flavonoid compounds manifested antioxidant activity deactivating free radicals and breaking down peroxides. These plant extracts showed

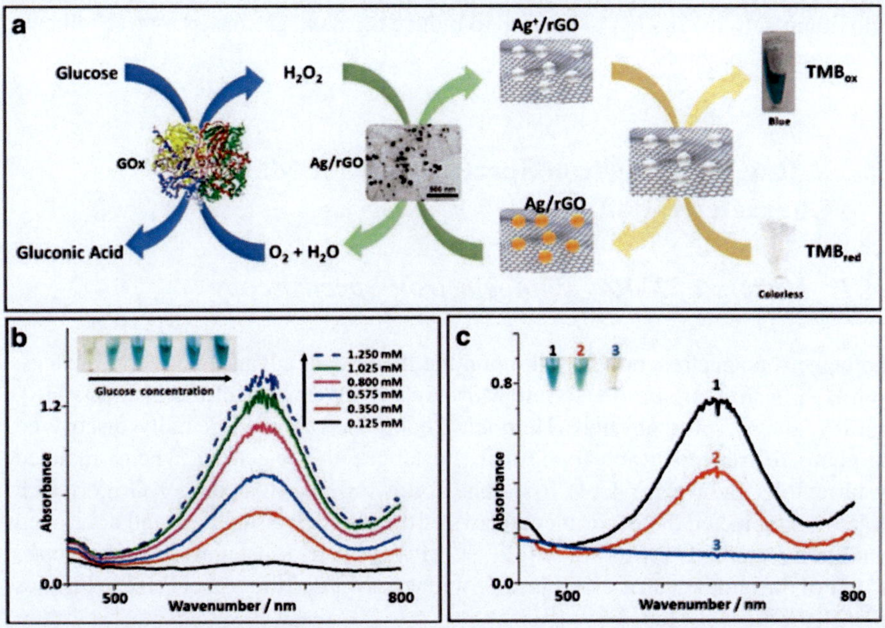

Fig. 4.12 **a** The principle of operation of the glucose biosensor based on Ag/rGO. UV–Vis spectra of AgNPs@rGO/GO$_x$/TMB **b** at different glucose concentrations and **c** at different samples. Reproduced (or adapted) with permission [64], © 2020, Elsevier

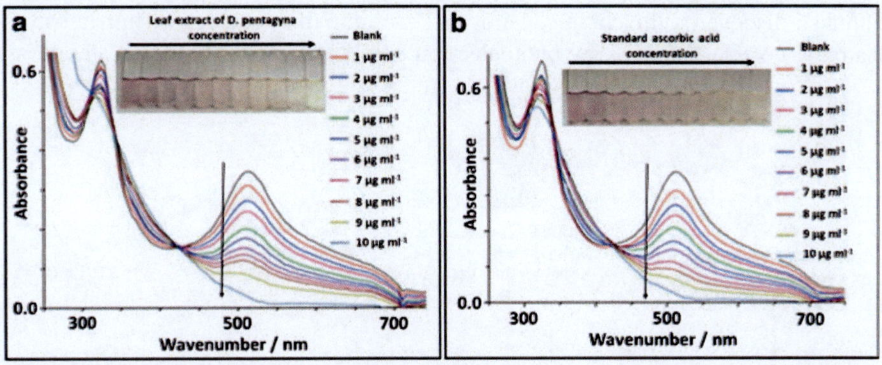

Fig. 4.13 The UV–Vis spectra at different concentrations (from 1 to 10 μg ml^{-1}) of **a** leaf extract of *D. pentagyna* and **b** standard ascorbic acid for DPPH free radical scavenging activity. Reproduced (or adapted) with permission [65], © 2020, Elsevier

antioxidant activity and could be used to reduce the damage caused by free radicals in the human body [65].

4.4 X-Ray Photoelectron Spectroscopy for Material Characterization

4.4.1 History of X-Ray Photoelectron Spectroscopy

Photoelectron spectroscopy is built upon the theoretical advances of early physics, from Bohr's structure of the atom to Moseley's X-ray spectra of the elements [66]. In 1887, the German physicist Heinrich Rudolf Hertz experimentally discovered the photoelectric effect and found that substances emit electrons when irradiated by ultraviolet light (Fig. 4.14) [67]. The following year, Alexander Grigorievich Stoletov established the experimental laws of the photoelectric effect and developed quantitative methods for its study [67]. However, Albert Einstein further developed the law of the photoelectric effect in 1905 and was awarded the Nobel Prize in Physics in 1921 for his work [66].

In 1954, E. Sokolowski, C. Nordling, and Kai Siegbahn introduced the earliest iron-free double-focusing spectrometer, which was the first high-resolution electron spectrometer with a ten-fold resolving power of contemporary iron instruments for determining slow forming electrons. E. Sokolowski, C. Nordling, and Kai Siegbahn also discovered the chemical shift in 1956, initially detecting chemical shifts of atomic core levels in X-ray photoelectron spectra of copper oxides. Later, they

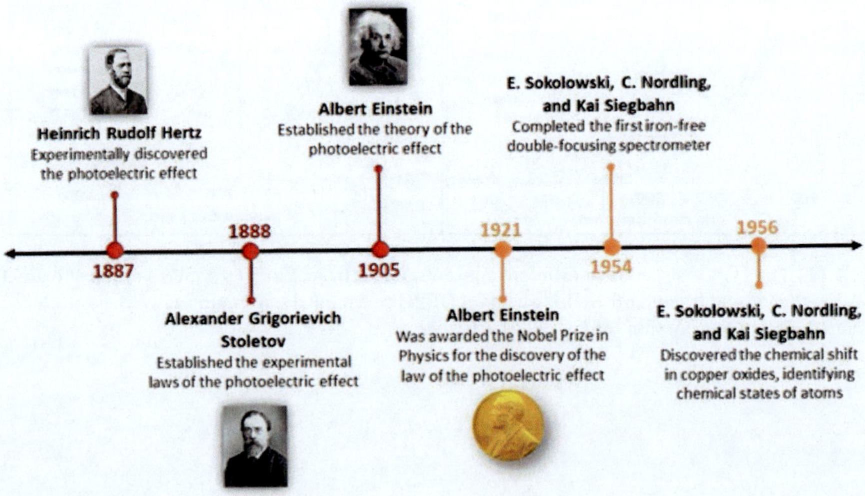

Fig. 4.14 History of the XPS technique

distinguished the chemical states of non-metal atoms and the oxidation states of metal atoms by analyzing the fine structure of their X-ray photoelectron spectroscopy (XPS) spectra [67]. In 1964, Kai Siegbahn, C. Nordling, and Hagstrom observed the spectra of specific substances and uncovered that multiple photoelectron lines were associated with the location where solely one line was anticipated to appear [66].

Finally, after the invention of commercialized spectrometers, XPS was used for the analysis of solid surfaces. Kai Siegbahn wrote the first monograph in electron spectroscopy. The Journal of Electron Spectroscopy and Related Phenomena, a specialized international journal, has been published since 1972. Today, several manuals are available for teaching XPS to science enthusiasts in the areas of physics, chemistry, and material science [67].

4.4.2 Mechanism of Operation of X-Ray Photoelectron Spectroscopy

The XPS process is divided into three stages. The X-ray photoemission process initiates when an X-ray photon is absorbed and transfers its energy to a core electron. The target atom replies to this excitation by emitting a photoelectron, thus creating an empty core state. Subsequently, the photoelectron moves to the material's surface and escapes into the vacuum to meet the electron detector that measures its energy [68].

During this process, X-rays radiate to the sample with energies lower than 6 keV, for which the output energy of the emitted photoelectrons is later evaluated. The emitted electron originates from a complete transmission of the X-ray energy to a core level electron, as shown mathematically in Eq. 4.2. The equation represents an equivalence between the X-ray (hυ) energy and the binding energy (BE) of the electron, taking into consideration the electron kinetic energy (KE) and the spectrometer work function (Φ_{spec}) [69].

$$hv = \mathrm{BE} + \mathrm{KE} + \Phi_{\mathrm{spec}} \tag{4.2}$$

Equation (4.2) can be reordered to obtain Eq. (4.3) in order to define the binding energy of an electron, whereby the terms on the right side of the equation are measured in the XPS experiment:

$$\mathrm{BE} = hv - \mathrm{KE} - \Phi_{\mathrm{spec}} \tag{4.3}$$

Throughout the analysis, photons of predetermined energy levels radiate the sample resulting in the photoelectric effect. A segment of the electrons that are at the sample surface leaves the sample for the vacuum chamber for the analyzer slit of the spectrometer. The electron current is measured according to the quantity of

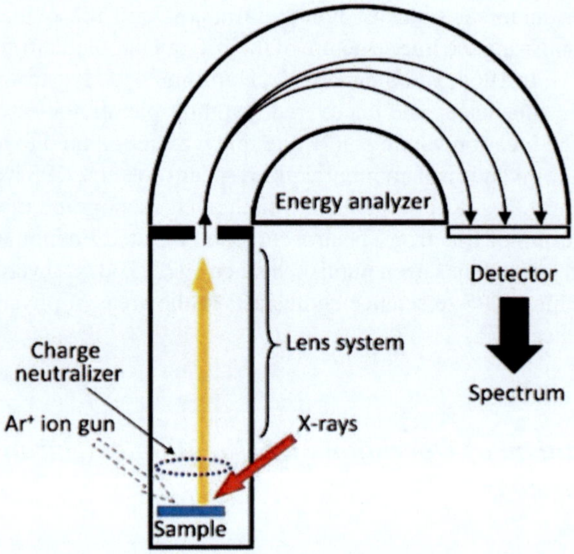

Fig. 4.15 Illustration of a photoelectron spectrometer equipped with a hemispherical electron energy analyzer. Reproduced (or adapted) with permission [69], ©2020, Elsevier

electrons per unit of time as a function of their energy. The XPS spectra depict the intensity vs. energy plots [69].

A typical schematic view of XPS is shown in Fig. 4.15. A light source directs a photon beam to the sample surface. The energy of the photons excites the electrons at the surface of the sample, and they consequently escape from it. Subsequently, the lens system gathers a proportion of these emitted electrons and transfers them into the detectors through the hemispherical analyzer to finally draw the spectrum. The instrument's performance is dependent primarily on the lens functions, and the efficiency can be severely impaired without properly tuning the lenses.

4.4.3 Advantages and Disadvantages of X-Ray Photoelectron Spectroscopy

XPS provides an efficient quantitative analysis through the estimation of atom percentage per element [70–74]. XPS is employed to provide chemical bonding information of surfaces of a wide range of materials since the technique can analyze most elements [70–72]. Such materials include pure metals, metal alloys, polymers, ceramics, organic and biological samples, thin films, and biomaterials [72]. In addition, the XPS technique is nondestructive to samples and possesses an ultrasensitivity which marks XPS as a highly sophisticated surface analytical tool [70, 72].

Table 4.4 Advantages and disadvantages of XPS

Advantages	Disadvantages
Efficient quantitative analysis [70–74]	Cannot analyze H and He [70–72]
	Samples require homogeneous thickness (~10 nm) [71, 72]
Analyzes most elements [70–72]	Large analysis area (≈1 cm^2) [70, 72]
Nondestructive [70, 72]	Expensive instrumentation [70, 72, 74]
Ultra-sensitivity [70, 72]	Low resolution (≈0.1–1.0 eV) [70, 72]
Provides chemical bonding information of surfaces [70, 72]	Long time of analysis (0.5–8 h) [70, 72]
	High vacuum is needed [70, 72, 73]

He helium; *H* hydrogen

However, a significant disadvantage to highlight is the inability of the technique to study hydrogen and helium surfaces [70–72]. Additionally, the sample must meet a requirement of at least 10 nm of homogenous thickness to obtain consistent peak intensities and an accurate XPS spectrum [71–73]. Samples must also completely cover the size beam, leading to an extensive analysis area of approximately 1 cm^2 [70, 72]. Likewise, the high price of instrumentation and low resolution represent other downsides of XPS [70, 72, 74]. In addition, XPS analysis is rather time-consuming. Based on the content of the sample, the analysis may last from half an hour to eight hours [70, 72]. Furthermore, a high vacuum environment is desirable to acquire the highest electron count throughout the spectra acquisition (Table 4.4) [70, 72, 73].

4.4.4 Applications of X-Ray Photoelectron Spectroscopy

It is known that XPS is an excellent spectroscopic technique of ultrasensitive nature for characterizing the surface chemistry of a sample. This method can analyze both organic and inorganic materials since it is a nondestructive technique, thus enabling the study of polymers, metals, glasses, ceramics, and semiconductors. This spectroscopic technique can acquire quantitative and qualitative information from a sample and identify the properties of new materials including biocompatibility and adhesion. Hence, this is an attractive tool for the fields of catalysis [75], solid-state nanostructures [76], biosensors [77], polymers [78], bioengineering [79], space [80], biomaterials [81], energy production [82], and biology [83]. For instance, when a polymer suffers a surface modification due to a physical or chemical treatment, XPS provides information on the functional groups created after the treatment and their nature [84].

In addition, XPS enables obtaining information regarding the chemical states and the chemical distribution to determine the stability or chemical reactivity in a sample under analysis [82].

In 2020, Pushkarev et al. developed a highly sensitive glass biochip for antibodies biosensing to detect Antithyroglobulin (anti-TG) and Antithyroid Peroxidase (anti-TPO) in human serum. Figure 4.16a represents the stages for the construction of the biochip. At the first stage, the glass was cleaned with piranha solution and modified

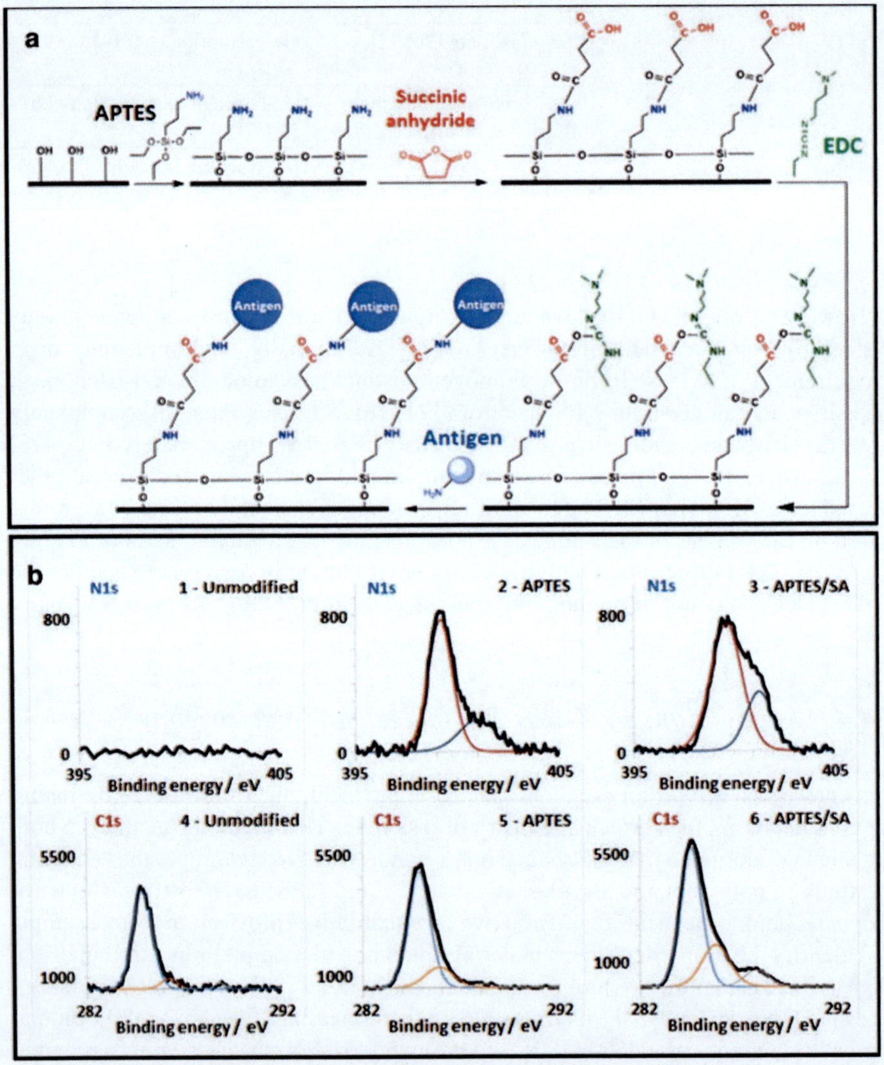

Fig. 4.16 **a** The steps of attaching the antigen to the glass surface. **b** The XPS spectra of the biochip surface. Reproduced (or adapted) with permission [85], © 2020, Elsevier

using Aminosylane (APTES) to introduce amino groups (-NH_2) on the glass surface. Subsequently, a succinic anhydride ($C_4H_4O_3$) treatment formed carbonyl groups coupled with the –NH_2 groups. The carbodiimide (EDC) provided o-acylisourea active ester before the antigen attachment. Finally, peptide bonds between the amino group of the antigen and the functionalized glass surface formed the functional glass biochip. The XPS analysis (Fig. 4.16b) followed every stage in the construction of the biosensing device. The obtained data validated the formation of amino groups (N1s peaks at 400 eV) after the APTES treatment in the first stage. Furthermore, the XPS spectra identified the presence of carboxyl groups on the glass surface due to the emergence of peaks at 288 and 286 eV, attributed to C=O and C–O bonds, respectively. The glass biochip demonstrated that the detection performance remains practically constant after one month of storage [85].

4.5 Diffuse Reflectance Infrared Fourier Transform Spectroscopy for Material Characterization

4.5.1 History of Diffuse Reflectance Infrared Fourier Transform Spectroscopy

The diffuse reflectance infrared Fourier transform spectroscopy (DRIFTS) is applied to study powder samples and polymer fibers and films. In 1976, Ronald R. Willey presented the optical design of an instrument that aimed to extend the diffuse reflectance technique into the infrared region (Fig. 4.17), since his FTIR spectrometer was particularly designed to perform infrared diffuse reflectance measurements. In the 1980s, the DRIFTS technique expanded, allowing its development and subsequent application in a wide range of samples and fields. Hans Maulhard and Dietmar Kunath used DRIFTS with their customized optical device and a commercialized version of FTIR to investigate a silane coupling agent on glass fibers. In 1984, Martin T. McKenzie et al. examined the coupling agent (7-Aminoproyl)triethoxysilane (7-APTES) on E-glass fibers, hence introducing the usage of potassium bromide (KBr) over-layers to obtain diffuse reflectance infrared Fourier transform (DRIFT) spectra of polymer fibers [86].

4.5.2 Mechanism of Operation of Diffuse Reflectance Infrared Fourier Transform Spectroscopy

The DRIFTS technique relies upon scattering radiation within a sample to study powders and rough solid surfaces. Two scenarios could occur when directing incident light to a sample: The specimen may produce a singular reflection from the surface

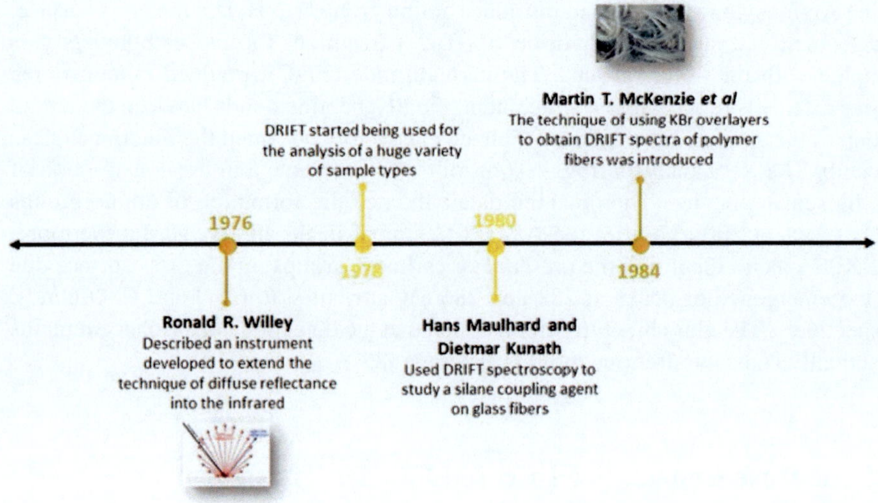

Fig. 4.17 History of DRIFTS

or produce a multiplied reflection. The second scenario originates the diffusely scattered light over a wide area that DRIFTS uses for the analyzing process. The DRIFTS equipment is equipped with collection optics that reject the specularly reflected radiation and receive the highest achievable level of diffused reflected light. Theories describing diffuse reflectance usually consider plane-parallel surfaces within the sample and ubiquitous light scattering [87]. Therefore, Eq. (4.4) defines the observed reflectance spectrum using the Kubelka–Munk equation representing a ratio of the single beam spectra of the sample diluted by KBr powder against pure powder KBr.

$$f(R_\infty) = \frac{(1 - R_\infty)^2}{2R_\infty} = \frac{K}{s} = \frac{C}{k} \tag{4.4}$$

where K is the absorption coefficient, and s is the scattering coefficient of the sample material. The reflectance, depicted as R_∞, denotes no radiation reaching the back surface of a sample due to its thickness. Such factors predict a linear relationship between K and the maximum value of $f(R_\infty)$, considering that s is constant. Although the scattering coefficient depends on the particle size and packing, it should stay constant for reproducible outcomes [87]. Finally, C is the sample concentration, and k relates to the sample's particle size and molar absorptivity (Eq. 4.5).

$$k = s/2.303e \tag{4.5}$$

The particle size, sample packing, and dilution are a few factors that should be considered in order to obtain a high-quality DRIFT spectrum. For instance, the particle size of the sample should be small (50 μm or less) and uniform, resulting in narrower bandwidths and more accurate relative intensities. In addition, the dilution of the sample in a non-absorbing matrix powder, such as KBr or potassium chloride (KCl), ensures deeper penetration of the incident light into the sample, thus increasing the diffusely scattered light. The matrix powder should contain a ~5% relative quantity of finely powdered sample. Homogeneity of the samples is also a fundamental factor in obtaining a high-quality spectrum. Furthermore, the sample packing should be loose to augment the infrared (IR) beam penetration. In spite of following the previous preventive measures, the diffuse reflectance spectrum may vary from that acquired through a transmission IR analysis. A Kubelka–Munk conversion, however, would compensate for a few of these drawbacks [87].

The DRIFTS technique possesses an IR source that directs IR beams to a sample and penetrates in its particles. Nevertheless, the IR beam can either reflect off the next particle or be transmitted through the next particle (Fig. 4.18). Although the IR energy reflecting off the surface is typically lost, the randomly oriented scattering light phenomenon is isotropic and generates a diffuse reflectance [88]. Subsequently, a spherical mirror, focusing onto the detector, collects the scattered IR energy. The detected IR light is partially absorbed by the sample particles, enabling the detector to bring the sample information.

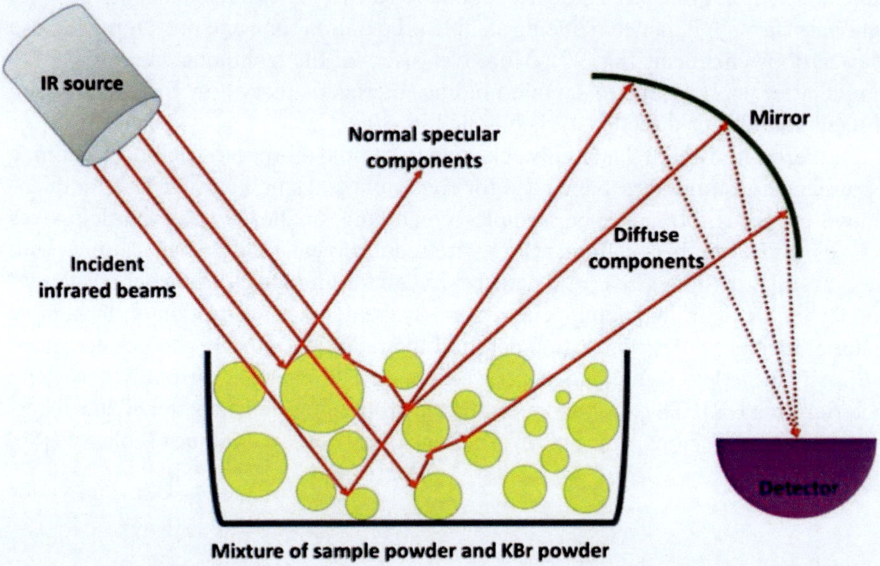

Fig. 4.18 Schematic illustration of DRIFTS. The incident IR beams reflect from the sample, enabling the detector to receive diffused beams and provide sample information

Table 4.5 Advantages and disadvantages of DRIFTS

Advantages	Disadvantages
Obtains chemical and structural information from all types of solid surfaces [89]	Consistent and small particle size is required [86]
Highly sensitive [86]	
Rapid [89, 90]	Limited depth penetration [86]
No laborious sample preparation is required [86, 90]	Distortions due to strongly absorbing indexes of materials [86, 89, 92]
Nondestructive [86, 89–91]	

4.5.3 Advantages and Disadvantages of Diffuse Reflectance Infrared Fourier Transform Spectroscopy

DRIFTS obtains chemical and structural information from all classes of solid substrates, both in compressed and powdered forms [89]. In addition, this technique offers favorable attributes, including high sensitivity and rapid analysis of samples [86, 89, 90]. Unlike Fourier transform infrared spectroscopy (FTIR), the subjected samples to DRIFTS do not require to form a pellet translucid enough to let light pass through it, hence reducing sample preparation time and preventing the use of hazardous chemicals [86, 90]. Moreover, users of the technique can retrieve the sample after analysis due to sample-holding microcells that allow the examination of fixed volumes or direct analysis on-site [86, 89–91].

Furthermore, DRIFTS is highly sensitive to sample morphology, although it may represent a constraint due to the need for consistent and small particle size for quantitative purposes. For instance, samples containing smaller average particle sizes (<10 μm) yield higher-quality spectra with reduced peak widths than samples with large average particle sizes (>90 μm) [86]. Another drawback of this characterization technique is limited depth penetration. As the average absorption of the sample material increases, the penetration depth of the radiation into the powder decreases [86]. Albeit the low signal and sloping baselines features of the strongly absorbing materials that DRIFTS can analyze, strong absorbtion indexes in matter create distortions in spectral regions, thus limiting the sensitivity of the technique (Table 4.5) [86, 89, 92].

4.5.4 Applications of Diffuse Reflectance Infrared Fourier Transform Spectroscopy

The industrial and scientific community encountered in DRIFTS an opportunity to overcome the drawbacks of FTIR. In addition, this technique offers unique advantages such as improving the signal-to-noise (S/N) ratios, allowing the usage of the visible and UV spectrum, analyzing gas-surface interactions, identifying and quantifying functional groups on the sample surface, providing high sensitivity and resolution, and representing a nondestructive and low time-consuming method. The mentioned factors secure the DRIFTS as a desirable and valuable technique for different applications in characterization of inorganic and organic materials, pharmaceutical quality control, food research, clinical investigations, the study of soil organic matter, and the study of polymers [86, 93–98]

DRIFTS can study the composition and structure of a dried bacterial sample. For instance, DRIFTS identified the effect of the nutritional stress on cultivated biomass of *A. brasilense Sp7* cells through a significant change in the spectral region attributed to vibrations of peptide groups [99]. This technique also provides a simple and reliable instrument for monitoring the degree of curing in films and epoxy composites by following the signals corresponding to C–O–C deformation at 4528 cm^{-1} [100]. Additionally, quantitatively detecting pharmaceutical products in a solvent-free, quick, sensitive, reliable, and simple way is possible using DRIFTS [101]. Moreover, DRIFTS can identify different binders and colorants in modern reverse glass, mural, canvas, and wall paintings without causing any damage [102].

The diamond surface modified with DNA provides a functionalized surface to identify biological contaminants, direct monitoring of DNA structures, and gene expression studies. Ushizawa et al. immobilized DNA over a diamond surface, covalently bonding a phosphate group from DNA (5′-end) with thymidine (3′-end) previously anchored to the diamond surface. Simultaneously, thymine residues and adenine from DNA formed a hydrogen bond (Fig. 4.19a). From DRIFTS data, the authors concluded that the chemisorbed species, thymidine, and DNA formed ester linkages with the diamond surface. In the thymidine case, bands attributed to thymine residue appeared at 1460, 1650, and 1703 cm^{-1}. Meanwhile, PO_2^- stretching arose at 1100 cm^{-1}, corresponding to DNA moiety (Fig. 4.19b) [103].

The functional groups on the carbon surface play a crucial role in the final properties of carbon materials and their performance in several technical applications such as energy storage, wastewater purification, catalysis, and energy generation. The chemical composition of the surface of carbon materials continuously changes due to the environmental conditions surrounding them, for instance, humidity or pH. Nonetheless, the above represents an advantage since the carbon materials can be subjected to physical or chemical processes, as condensation, to favor certain functional groups. Herold et al. functionalized the surface of oxidized polymer-derived carbon using different physicochemical methods and used DRIFTS to identify the selectively created functional groups. Thermal desorption at 510 °C favored the formation of lactones with an absorption band appearing at 1771 cm^{-1}. Carboxylic acids appeared via a hydrolysis treatment in sodium hydroxide (KOH) and were detected through their absorption band at 1758 cm^{-1}. Finally, anhydrous groups formed by applying a condensation method at 315 °C after the hydrolysis treatment, and two absorption bands arose at 1852 and 1792 cm^{-1} (Fig. 4.20) [104]. Therefore, DRIFTS is an excellent characterization tool to identify different chemical species in a reliable, easy, and low time-consuming manner.

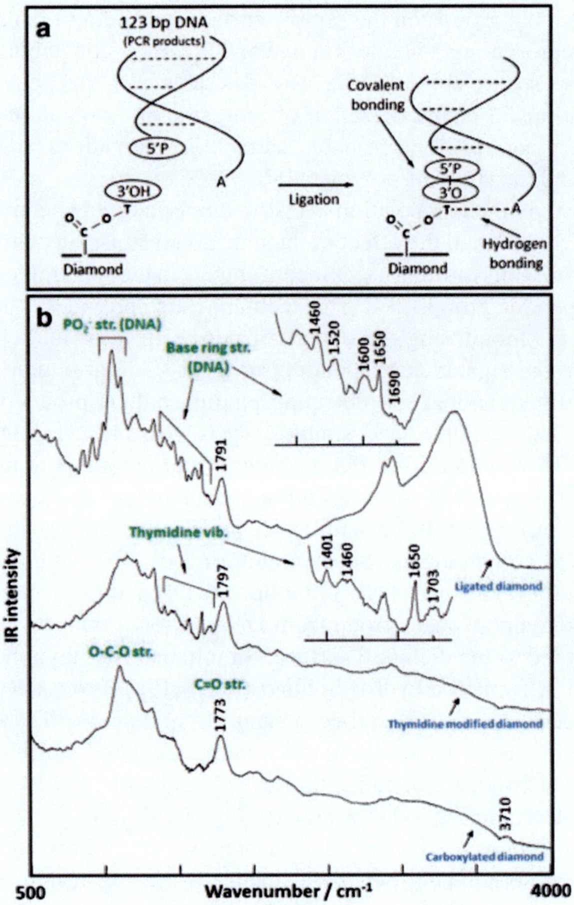

Fig. 4.19 **a** The immobilization of DNA on a diamond surface via a ligation process. **b** The DRIFTS spectra of several treated carbon surfaces. Reproduced (or adapted) with the permission [103], © 2002, Elsevier

4.6 X-Ray Diffraction for Material Characterization

4.6.1 History of X-Ray Diffraction

The history of X-ray diffraction (XRD) began in 1912 when Max von Laue and Paul Ewalds introduced the idea through an experiment that evidenced the existence of space lattices in crystals and the "wave nature" of X-rays (Fig. 4.21). On June 8, 1912, Max von Laue introduced his findings of X-ray diffraction in crystals as in a three-dimensional diffraction grating [105].

Shortly after, William Henry Bragg and Lawrence Bragg relied on the discovery to confirm Barlow's hypothetic model of rock salt, thus initiating the X-ray diffraction

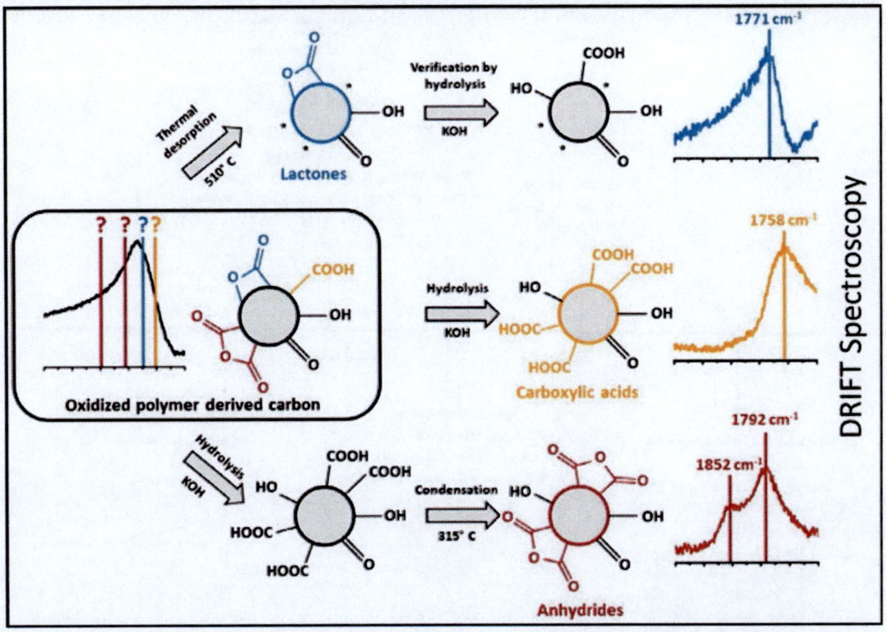

Fig. 4.20 The general diagram to achieve selectively created functional groups on the surface of carbon materials through heat treatment and hydrolysis methods. Reproduced (or adapted) with permission [104], © 2021, Elsevier

analysis of a single crystal [106]. Furthermore, William Henry Bragg contributed to the history of XRD by constructing a precise X-ray spectrometer and introducing the ionization chamber for more accessible quantitative measurements of X-ray intensities [107].

A year after his great discovery, Max von Laue gave birth to two new sciences: X-ray crystallography and X-ray spectroscopy. Max von Laue received the Nobel Prize in Physics in 1914 for discovering X-ray diffraction by crystals. William Henry Bragg and Lawrence Bragg received this prestigious prize in 1915 for their works in analyzing crystal structure utilizing X-rays [105]. In 1955, Alexander Kitaigorodsky published a monograph on the theory of XRD to identify new structures which are still used today. A considerable number of new structures are identified every year, which gathers a wide repository of materials' datasets for crystallography, crystal chemistry, quantum chemistry, and other techniques aiming to understand and model the organization of the matter at an atomic level [106].

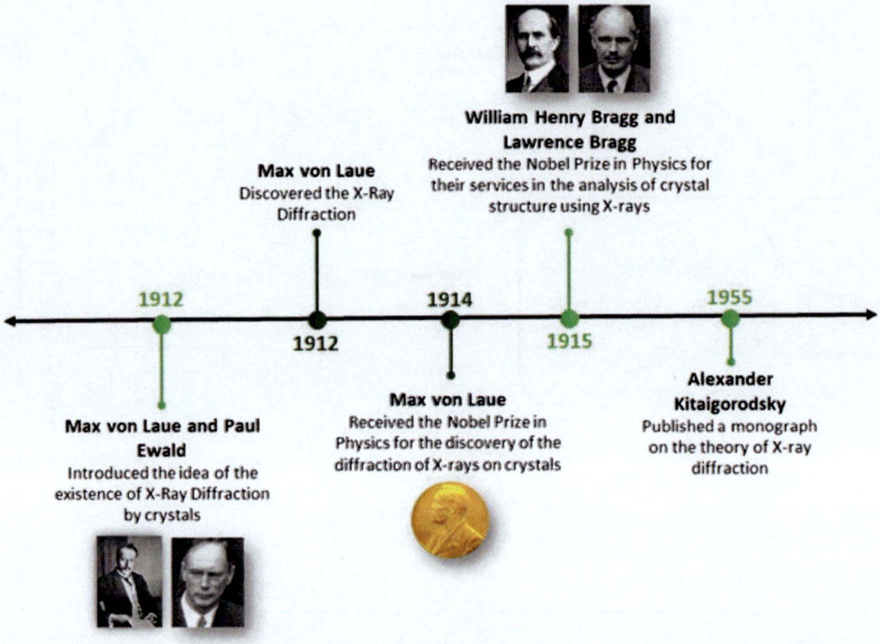

Fig. 4.21 History of XRD

4.6.2 Mechanism of Operation of X-Ray Diffraction

The interaction of wavelengths with periodic structures yields diffraction effects if the wavelength and the periodicity of the crystals are of similar magnitude. X-rays may easily be generated with wavelengths matching the unit cell dimensions of crystals; however, electrons or neutrons of appropriate energy can also be used for diffraction experiments on crystals. Unit cells have dimensions of several Å, demonstrating that crystals with larger sizes of microns composed of billions of cell units repeated in all three dimensions periodically. Since XRD strongly depends on the quality of diffraction effects and the strongly ordered periodicity of atoms, any deviation from the ideal order will show in the X-ray diffraction diagram. Deviations from the ideal order may come from the replacements of atoms by others, slight geometric deviations of atoms from their optimal position due to internal strain, or larger two or three-dimensional aberrations [108].

Considering no diffraction effects, the incidence of a primary X-ray beam onto a sample produces scattering in all directions. Diffraction reallocates intensity from the scattering sphere into different directions. Consequently, intensity peaks hike in specific directions, whereas the intensity decreases considerably in the regions between peaks. The peaks may follow distinct directions. One way of describing these directions is the notion of scattering lattice planes and interference between the

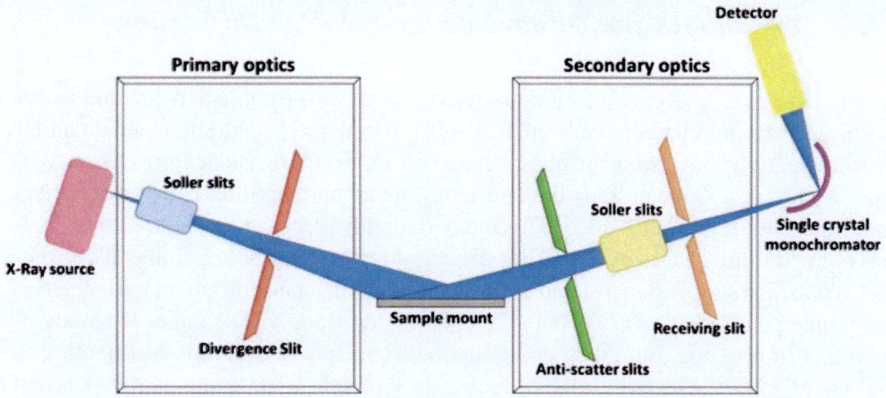

Fig. 4.22 XRD schematic diagram. The IR beam is emitted to the sample surface via primary optics, and the detector receives reflected beams after passing through the secondary optics

wavelets scattered by neighboring lattice planes. Constructive interference and the Bragg reflection are obtained when the path of the wavelet scattered of the lower of the two planes is longer by an integer number of λ than that of the wavelength scattered off the upper plane. A reflection will thus occur following the Bragg equation (Eq. 4.6).

$$n\lambda = 2d \sin \theta \tag{4.6}$$

where λ is the wavelength of the radiation, n is an integer number, θ is the angle between the lattice planes and the incident beam, and d is the distance of the lattice planes from which the peak occurs [109].

The dominant effect of XRD occurs when an incident beam of monochromatic X-rays interacts with a target material. The X-rays are produced by a cathode ray tube, filtered to obtain monochromatic radiation, collimated to concentrate, and directed toward the sample (Fig. 4.22). In materials with regular structures, such as crystals, the scattered X-rays undergo the diffraction process. The interaction of the incident rays with the sample generates constructive interference and a diffracted ray when the conditions satisfy Bragg's law. This law correlates the wavelength of electromagnetic radiation to the diffraction angle and the lattice spacing in a crystalline sample. Finally, these diffracted X-rays are further analyzed and quantified by detectors. Typically, the conversion of the diffraction peaks to d-spacings is achieved by comparing d-spacings with standard reference patterns, which allows the recognition of the minerals since each mineral has a set of unique d-spacings.

4.6.3 Advantages and Disadvantages of X-Ray Diffraction

As an IR-based analytical technique, XRD is a well-established technique for advanced material characterization [110–112]. It generates qualitative and quantitative information concerning the phase composition, structure, and texture of nearly all types of samples, as it can study both pure substances and mixtures in a powder-alike, solid, and liquid state [112–115]. This method mainly offers a precise structural analysis of crystalline phases by acquiring structural features from the sample. Nevertheless, it can also supply data related to the peak position, space group, macro-stresses, and lattice parameters of crystals [110, 114]. In addition, XRD enables the study of the effect of pressure, humidity, and temperature in specimens, thus contributing to the versatility of the technique by widening its application in various study areas and opening new research possibilities [115, 116].

Moreover, the latest advances in the development of computational, microelectronics, and data analysis technologies have enabled a more extensive availability of instruments and have attracted more potential users [110, 116]. Coupled with this, XRD offers a simple and inexpensive preparation of samples and reliable reproducibility of results [112–114]. XRD benefits from collecting bulk properties of long-range order since the technique collects the majority of the intensities scattered from the sample, and hence, the examined characteristics represent an average of the whole sample [111, 112, 117]. However, XRD requires large amounts of samples to achieve this feature (Table 4.6). Additional disadvantages of the technique are the time-consuming process of obtaining the standard diffraction pattern, the varying

Table 4.6 Advantages and disadvantages of XRD

Advantages	Disadvantages
Enables advanced material characterization [110–112]	Requires a large amount of sample [112, 118]
Provides qualitative and quantitative information of powder, solid, and liquid samples [112–115]	
Analyzes pure substances and mixtures [115]	
Provides precise structural analysis of crystal phases [110, 114]	Time-consuming [116, 118, 119]
Collects bulk properties of long-range order [111, 112, 117]	
Low cost [113, 114]	Detection limits depend on the type of equipment [119]
Simple sample preparation [114]	
Reliable reproducibility [112]	
Available in various configurations [110, 116]	Low intensity of diffracted X-rays for materials of low atomic number [118]
Analyzes the effect of pressure, humidity, and temperature in samples [115, 116]	
Versatile [115]	

limits of detection of each type of equipment, and the low intensity of diffracted X-rays for materials of low atomic number [116, 118, 119].

4.6.4 Applications of X-Ray Diffraction

As a nondestructive and high-tech technique, XRD is used to obtain crystallinity, crystal defects, grain size, phase, structure, and other bulk properties of crystalline samples. The interaction between the X-ray radiation and the crystal structure (arrangement of the atoms within the lattice) of the sample provides an XRD diffractogram. The XRD diffractogram is a fingerprint of the atomic distribution of a particular material since the position, given by interplanar spacing d_{hkl} or Bragg angle θ, and intensity of the diffraction peaks are specific to each molecule. Since XRD can readily identify the crystallite size, texture, crystal structure, level of stress, and degree of crystallinity in materials, it is widely used by researchers in forensic science [120], microelectronics [121], corrosion analysis [120], energy generation [117], pharmaceutics [122], geology [120], geopolymers [123], environment [121], glass manufacturing [120], biopolymers [124], catalysis [125], and material science [126].

High glucose concentrations in the human body can manifest as kidney failure, blindness, and cardiovascular diseases. Deficiency in insulin secretion negatively affects the glucose processing in the blood and causes diabetes. Therefore, continuous monitoring of blood glucose levels is essential for timely diagnosis and appropriate treatment of diabetes. Glucose quantification tests are recurrent in clinical diagnostics, meaning that there is a pressing need to develop a highly selective, low-cost, rapid, exceptionally sensitive, and stable system to detect glucose levels in diabetic patients and people prone to develop the disease. The most commonly used techniques for this purpose are colorimetric tests, electrochemical methods, gas chromatography, and high-performance liquid chromatography. Li et al. proposed an enzymatic biofuel cells (EBFCs)-based self-powered glucose biosensor for uninterrupted and noninvasive glucose measurement (Fig. 4.23a). The novelty in this electrochemical system was in the development of the electrodes that encapsulated GO_x and laccase enzymes on cellulose acetate (CA) for the glucose oxidation reaction, and ZIF-8 on CA for the oxygen reduction reaction. Additionally, the authors used multi-walled carbon nanotubes (MWCNTs) and gold nanoparticles (AuNPs) to improve electronic transfer and connection between biomolecules. The authors used XRD (Fig. 4.23b) to evaluate the variations of the crystal structure of each component of the electrode. The results showed the CA was amorphous, and the diffraction peaks of ZIF-8 appeared at 7.8°, 10.9°, 13.1°, 15.1°, 16.8°, 18.4°, 24.8°, and 26.9°. The graphite plane (002) and crystal plane (101) attributed to MWCNTs arose at 26.6° and 43.3°, respectively. Finally, the AuNPs exhibited diffraction peaks at 38.6°, 44.7°, and 78.4° attributed to (111), (200), and (311) crystal planes. The glucose biosensor could record glucose concentrations for 15 continuous hours with a limit of detection of 5.347 μM [127].

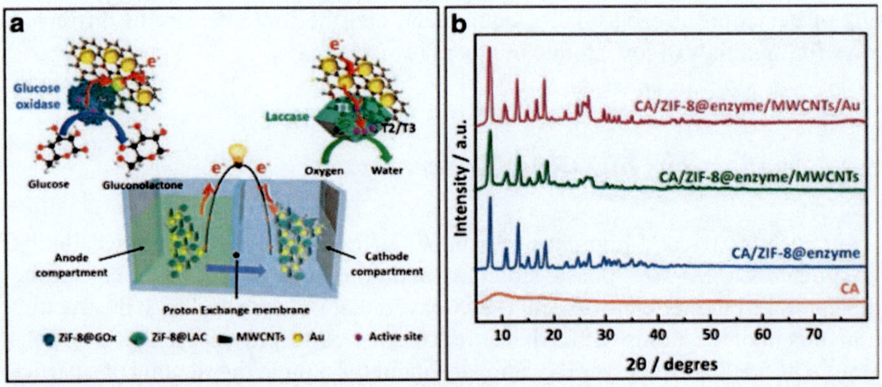

Fig. 4.23 **a** The components and operating principle of the EBFCs-based self-powered glucose detector. **b** The XRD patterns of each element. Reproduced (or adapted) with the permission [127], © 2021, Elsevier

Nonetheless, the instability for long periods, sensitivity to working conditions, and the high cost of enzymes exemplify the disadvantages of using enzymatic electrodes for glucose detection. The use of metal nanoparticles is an attractive alternative to replace enzymes in detection systems. Dayakar et al. proposed a green synthesis for zinc oxide (ZnO) nanoparticles by a biomediated route with *Ocimum tenuiflorum* leaf extract. The ZnO nanoparticles were synthesized by the reaction of a 2 M solution of zinc nitrate hexahydrate ($Zn\ (NO_3)_2 \cdot 6H_2O$) with different volumes of plant extract (5, 10, 15, and 20 ml) and were later used to fabricate a non-enzymatic glucose sensor. The XRD results (Fig. 4.24) showed that the synthesized ZnO nanoparticles exhibited a hexagonal phase with a crystalline wurtzite structure. The obtained data from XRD patterns was helpful to estimate the crystallite size through Scherrer's formula and settled that when the volume of the used plant extract increased, the crystallite size diminished. In addition, the authors concluded that the performed green synthesis reduced the dislocations in the ZnO nanoparticles. Finally, the sensitivity of their non-enzymatic glucose sensor was 631.30 $\mu A\ mM^{-1}\ cm^{-2}$ [128].

Moreover, Şavk et al. fabricated an enzyme-free electrochemical biosensor for glucose identification using a carbon-activated electrode modified with platinum-nickel (Pt-Ni) nanocomposites. The XRD data provided the crystallite size (close to 3.72 $\pm$ 0.51) and crystal structure of the synthesized Pt-Ni nanocomposites. The XRD analysis (Fig. 4.25) showed diffraction peaks at 40.1°, 46.52°, 67.9°, and 81.7°, thus resembling a polycrystalline face-centered cubic (fcc) platinum (Pt) structure. In addition, a diffraction peak arose at 25.4°, attributing it to the activated carbon. Slight shifting in diffraction peaks indicated the lattice parameter decreased for the Pt-Ni composite, and thus, smaller nickel (Ni) atoms replaced a few Pt atoms. The enzyme-free electrochemical biosensor exhibited excellent selectivity, stability, and high sensitivity in analyzing real serum samples, in addition to a low detection limit (0.052 M) [129].

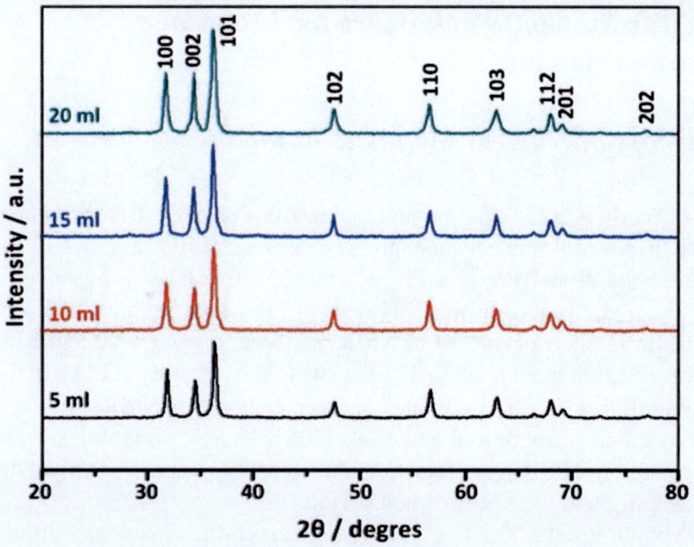

Fig. 4.24 XRD diffractogram of synthesized ZnO nanoparticles through *Ocimum tenuiflorum* leaf extract at different quantities (5, 10, 15, and 20 ml). Reproduced (or adapted) with the permission [128], © 2017, Elsevier

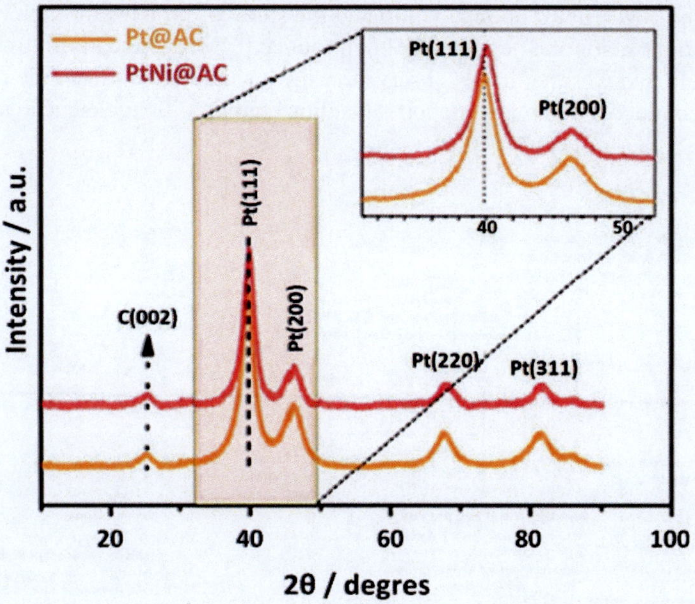

Fig. 4.25 XRD patterns of Platinum@Activated Cabron (Pt@AC) and Platinum-Nickel@Activated Carbon (PtNi@AC) were used as electrodes in the non-enzymatic electrochemical glucose biosensor. Reproduced (or adapted) with the permission [129], © 2020, Elsevier

4.7 Nuclear Magnetic Resonance for Material Characterization

4.7.1 History of Nuclear Magnetic Resonance

By the end of World War II, two prominent scientists, Felix Bloch and Edward Purcell, individually discovered nuclear magnetic resonance (NMR) (Fig. 4.26). Bloch and Purcell shared the Nobel Prize in Physics in 1952 for developing novel methods on nuclear magnetic precision measurement [130]. Since its invention, one of the most significant impacts of the use of NMR has been to make visible the interior of biological objects that were previously invisible to the naked eye [131]. In the 1950s, research laboratories of oil companies showed interest in NMR studies of fluids in rocks, clays, and other porous materials [132]. Hence, a number of intellectual properties related to NMR were registered between 1948 and 1960 in an attempt to adjust the technique to the needs of that period.

In 1949, W. G. Proctor and F. C. Yu first observed the chemical shift effects via NMR. A couple of years later, Arnold et al. confirmed such observation by obtaining a high-resolution NMR spectrum and detecting three lines in the spectrum of ethanol that corresponded to separate signals from methyl, methylene, and hydroxyl protons. In the late 1950s, Jardetzky, Cohn, and Shulman considerably contributed to revolutionizing the NMR field. Jardetzky initiated the study of proteins through NMR by studying amino acids and peptides. Cohn introduced NMR applications to the enzymology field and was the first to observe the signals that led to metabolic research. Shulman conducted paramagnetic ion relaxation studies with nucleic acid solutions

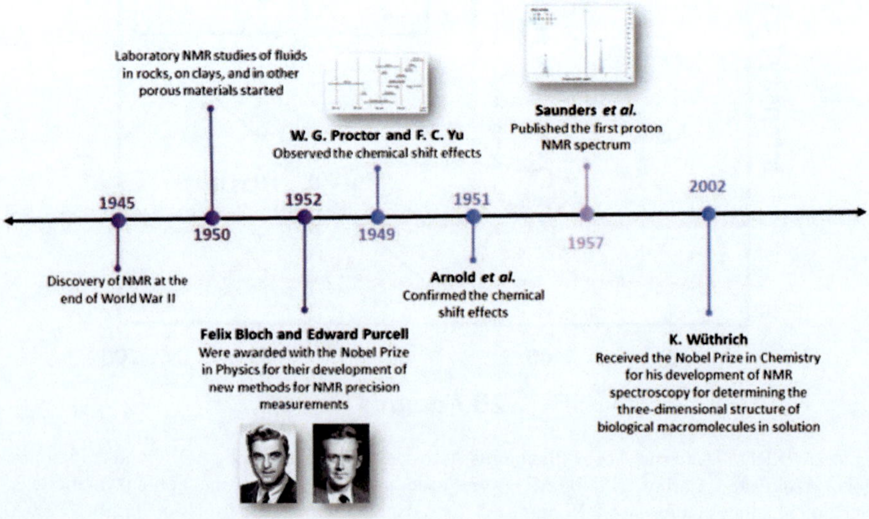

Fig. 4.26 History of NMR

and played an essential role in expanding the scope of NMR in metabolic applications [131]. In 1957, Saunders et al. published the first proton NMR spectrum of a protein.

Richard Ernst was another great enthusiast of the NMR technique who significantly contributed to its development. In 1966, the Belgian physicist, Weston Anderson, demonstrated that pulsed NMR combined with Fourier transformation offered a more elevated sensitivity. Ten years later, Richard Ernst introduced the two-dimensional (2D) NMR, which Kurt Wüthrich used to continue his research on biomolecules. The three-dimensional (3D) NMR was described first on unlabeled proteins and shortly after on a group of triple resonance experiments. Richard Ernst was the laureate of the Nobel Prize in Chemistry in 1991 due to his contributions to the development of the high-resolution NMR spectroscopy methodology. Similarly, in 2002, the Nobel Prize in Chemistry was awarded to Kurt Wüthrich for identifying the three-dimensional structure of biological macromolecules in solution using his developments of NMR spectroscopy [130].

4.7.2 Mechanism of Operation of Nuclear Magnetic Resonance

The nuclei of the atoms in a molecule behave similarly to magnets when such a molecule is placed in a strong magnetic field. Adding a broad radio frequency spectrum to the molecule causes the resonation of the atom's nuclei at a characteristic and exact frequency. In this regard, NMR aims to accurately measure the frequency of a nuclear resonance. The NMR instrument includes a strong and highly stable magnet, normally a superconducting solenoid, surrounded by transmitter and receiver coils. The sample is placed within the magnetic field (Fig. 4.27). The produced

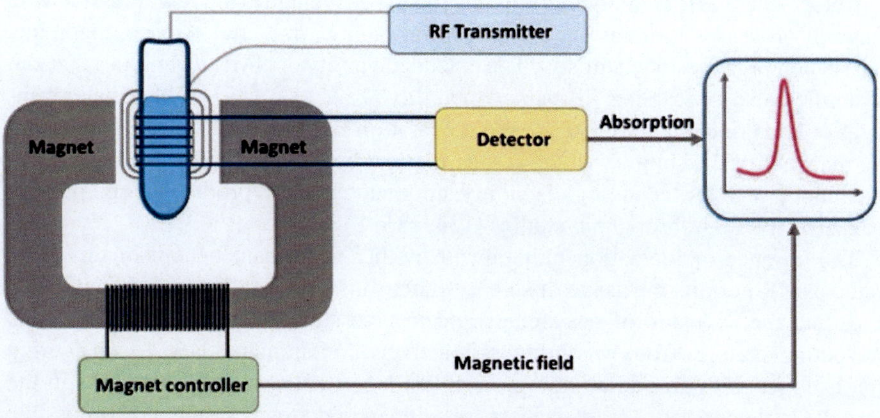

Fig. 4.27 Schematic representation of a conventional NMR spectrometer

signal frequency is not neatly defined since the magnetic field is non-homogeneous throughout the bulk of the sample. Two sets of whim coils encircle the sample to neutralize the field variations and provide a homogeneous field. Consequently, the NMR equipment calculates and transforms the resonant frequencies of the nuclei into an NRM spectrum displaying peaks that represent the measured frequencies. A certain height depicts each peak, and it stands for the number of nuclei that resonate at each specific frequency, known as the intensity of the signal. The intensity is higher when there are more resonating nuclei. The evaluation of each frequency provides information concerning the surroundings of the atom, such as its neighboring atoms and their relative positions. The atoms can even cause each other to resonate when they are near one another. Therefore, scientists can determine the molecule's 3D structure and movements by examining the cross-peaks [130].

In particular, the mechanism of the NMR spectrometer contains a high-frequency generator that drives a transmitter either continuously or with an electronic switch, depending on the type of NMR spectrometer. A high-frequency field surrounds this coil and sample arrangement, and they are submersed in the static magnetic field of a generally superconducting magnet. The digital conversion process involves a computer that reads the obtained signals and processes them to generate NMR graphs.

4.7.3 Advantages and Disadvantages of Nuclear Magnetic Resonance

As a reliable characterization technique, NMR provides rigorous information about the chemical features of substances [133–137]. NMR determines mechanisms and chemical connections at a molecular scale, including the local bonding environment neighboring a particular atom [133]. In addition, the ability to precisely perform the qualitative analysis is an inherent characteristic of NMR [134, 135, 137, 138]. Therefore, the NMR technique is highly versatile as it enables the research of a wide range of processes in fields such as structural biology, material science, medicine, and several branches of chemistry. The extended variety of NMR techniques that can be applied also contributes to such adaptability [133, 137, 138]. Additional advantages of NMR include minimal sample preparation and noninvasiveness, facilitating the recovery of the sample, which results in great benefit for in vivo studies. Moreover, modern NMR technology is highly automatable and reproducible and permits extensive and high throughput studies [134, 135, 138].

The absence of ionization radiation is another significant benefit of this technique. NMR permits the usage of stable isotopes instead of radioactive compounds, impeding the exposure of researchers and in vivo study subjects to radiation and preventing samples from becoming radioactively contaminated [133]. Considering that isotopes with an odd atomic number are NMR active and all elements in the periodic table possess at least one isotope with an odd atomic number, it is possible to analyze most elements utilizing this method. Nonetheless, the principal isotopes

Table 4.7 Advantages and disadvantages of NMR

Advantages	Disadvantages
Provides detailed structural analysis [133–137]	Isotopes with even atomic numbers are NMR inactive [139]
Precisely performs quantitative analysis [134, 135, 137, 138]	
Versatile [133, 137, 138]	Insensitive to certain circumstances [134–138]
Absence of radioactive radiation [133]	
Minimal sample preparation [135, 138]	Limited spatial resolution [136]
Noninvasive [133–138]	
Highly automatable and reproducible [134, 135, 138]	Requires a high magnetic field [133, 134, 136–139]

NMR Nuclear magnetic resonance

of several elements are even atomic number isotopes, which are NMR inactive. Consequently, this represents a considerable constraint of the technique [139].

Another disadvantage of NMR is its relative insensitiveness attributed to the small energy and population differences between the ground and excited spin states. Moreover, several additional components are required to improve the sensitivity of NMR, such as multiple scans, higher magnetic field strength, cryo-cooling and microprobes, and hyperpolarization methods [134–138]. The micrometer-scale of the spatial resolution in NMR also represents a limitation since it even occurs in areas with high magnetic fields [136]. Correspondingly, the fundamental need for a high magnetic field in the surroundings of NMR equipment constitutes a drawback and a danger to users [133, 134, 136–139]. It interferes with the correct operation of nearby electronic devices, including computer-controlled devices and monitors. Furthermore, metal-composed objects are heavily attracted to the magnetic field, leading to potential injuries [133]. Table 4.7 presents a summary of the previously described advantages and disadvantages.

4.7.4 Applications of Nuclear Magnetic Resonance

In essence, NMR is an exact and invaluable tool for providing important information regarding structural elucidation of a sample and performing quantitative analysis. NMR has been greatly applied for the analysis of compounds categories, such as lipids [140], carbohydrates [141], alcohols [142], polyphenols [143], and vitamins [144]. Since NMR can study a large number of molecules, the number of fields in which this technique can be applied is vast as well. The production of green fuels by pyrolysis of renewable biomass is a novel trend to address energy security

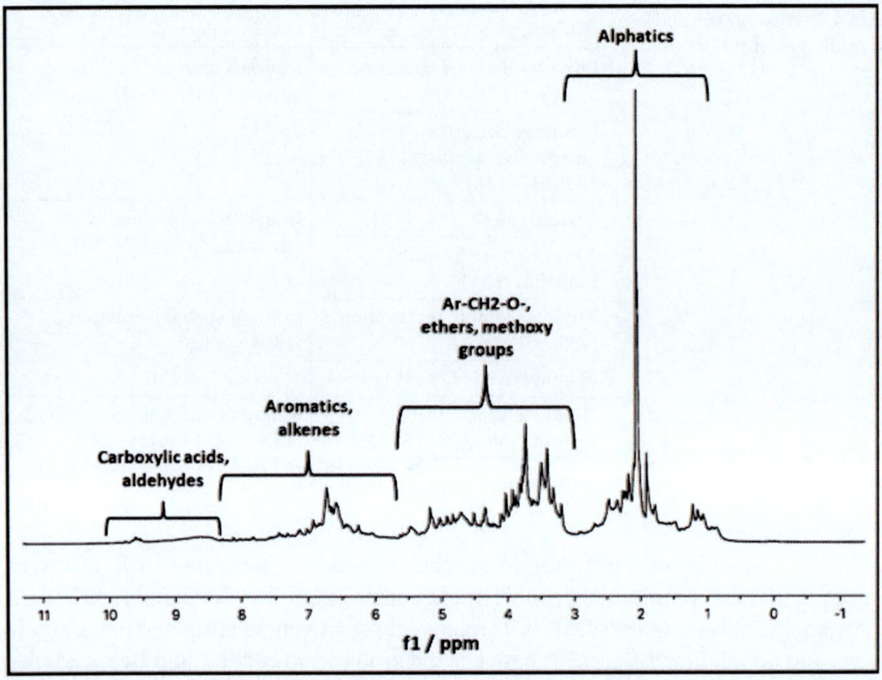

Fig. 4.28 ^{1}H NMR spectrum of a water-insoluble fraction of pyrolysis-derived oil from pine wood. Reproduced (or adapted) with the permission [145], © 2016, American Chemical Society

concerns and lessening environmental matters. However, unprocessed oils cannot blend with petroleum-based transportation fuels, and therefore, pyrolysis-obtained oils still have a limited commercial application [145]. Consequently, researchers have applied diverse characterization techniques for bio-oil studies to understand these challenges, and NMR has proven to be an important spectroscopic characterization method for analyzing bio-oil components [146]. Figure 4.28 presents the standard distributions of a ^{1}H NMR spectrum for bio-oil from pinewood.

In-cell NMR is a novel class of high-resolution NMR spectroscopy that enables the analysis of structural and dynamical features of proteins and other macromolecules in living cells [147]. The first examples of in-cell NMR were reported in *Escherichia coli* cells. Serber et al. employed heteronuclear NMR to show that *E. coli* could overexpress small globular proteins and isotopically label them to identify them from other cellular components [148]. Figure 4.29 summarizes additional approaches for in-cell NMR.

As a matter of fact, profound knowledge of the short- and intermediate-range structure of glasses is vital for the analysis of new compositions. NMR is an excellent and versatile tool for structure elucidation and structural and dynamic identification of macromolecules in organic chemistry and structural biology disciplines.

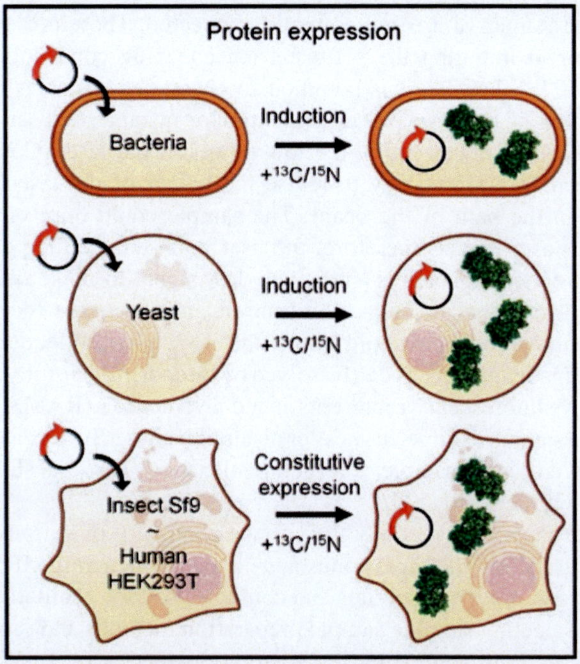

Fig. 4.29 Illustration of different strategies used for in-cell NMR in protein analyses. Reproduced (or adapted) with the permission [147], © 2016, International Union of Crystallography

The NMR has several unique features for metabolic studies, most notably the adaptability in structure elucidation, exploring the dynamics and compartmentation of metabolic networks, isotope-filtered selection of molecules, and analysis of positional isotopomer distributions in complex mixtures. In addition, NMR can spatially select in vivo recording and dynamical analysis of the metabolism of living organism tissues [134]. Furthermore, in biochemistry, more specifically in metabolomics-based studies, the objective is to detect metabolites that take part in several pathways within the metabolic network. For instance, a non-targeted metabolomic or biomarker analysis can determine which NMR or MS spectral characteristics respond to the conditions change, in addition to identifying unknown responsive compounds [149–151].

4.8 Troubleshooting of Chemical and Structural Analyses Techniques

Concerning the technical issues faced while executing RAMAN, it is common that the software used for analysis and data acquisition does not detect the Raman unit. The causes of such a problem are either that the Raman unit is not on or that the

USB cable of the Raman unit is not correctly connected. Therefore, turning on the Raman unit prior to initiating the software and checking connections should fix the difficulty [152]. Likewise, it is common to encounter obstacles related to the incorrect regulation of the laser power intensity. For instance, a laser intensity that is lower than expected is caused by the low power of the laser. Consequently, to solve this problem, it is necessary to revise the power of the laser and check if any obstacle is in the path of the beam. The sample could burn on the contrary scenario of high laser power; therefore, the user must reduce the power intensity [153]. Another frequent concern is a relatively low signal-to-noise ratio, which can be caused by a low value of integration time or an inadequate focus. Moreover, augmenting the integration threshold time value for signal collection may be the solution, and the inadequate focus can be solved by refocusing carefully [152, 153]. In addition, excessive fluorescence represents an inconvenience in RAMAN. A problem would arise if the sample exhibits dryness or needs bleaching. To accordingly fix such inconvenience, wetting the sample or rerunning the analysis with a short integration time is required (Table 4.8) [153].

FTIR spectroscopy aims to study the surface chemistry of various components due to its power to detect chemical bonds and elements. Generally, IR spectroscopy is applicable for a number of materials and could be used for qualitative and quantitative analyses. Selecting suitable sample preparation methods and sample handling techniques are crucial to generating high-resolution spectra [22]. When recording the IR spectrum, it is essential to identify the positions and approximate sizes of all peaks, kind of vibration, and the changes of spectra pattern to obtain valuable and informative discussion [22]. In general, FTIR spectrometers are very easy to operate; however, the user might face unexpected deterrents. For instance, if the user gets poor absorbance accuracy, decreasing the amount of light in the system to improve linearity can be needed to enhance accuracy [154]. Concerning the spectrum, the user may encounter the problem of having no signal but still being able to scan. Usually, the most common causes are related to the nitrogen or the aperture when this takes place. If the cooled detector is not cold, the user can cool it down with liquid nitrogen, and in case the beam is blocked, the user should select a larger aperture size to solve this problem and improve results [154]. There is no other option in the improbable scenario of a dysfunctional moving mirror than to replace it with a new mirror. If the mirror is not working well either, the user must manually move the moving mirror assembly. If unsuccessful, replacement is necessary [154].

In the case of UV–Vis spectroscopy and operational errors due to low signal sensitivity, different solutions related to absorbance or sample preparation can be presented. For instance, signal sensitivity can be enhanced by verifying absorbance values for samples that are among the limits supported by the instrument utilized at all wavelengths or verifying the compatibility of UV–Vis or quartz cuvettes [51]. The acquisition of inaccurate concentration measurements caused by solvent dilatations with temperature is a recurrent phenomenon in UV–Vis implementation. Correcting the factor for the coefficient of thermal expansion constitutes an approach to solve this inconvenience. Additionally, the modification of molecular and electronic structural characteristics due to pH variations leads to absorbance spectra shifts. This problem

Table 4.8 Summarized guide on troubleshooting, problems, possible causes, and solution for RAMAN, FTIR, UV–Vis, XPS, DRIFTS, XRD, and NMR

Technique	Problem	Cause	Solution
RAMAN	The software does not detect Raman unit [152]	The Raman unit is not on [152]	Turn on the Raman unit before initiating the software [152]
		USB cable of Raman unit is not connected properly [152]	Check connections [152]
	Low intensity [153]	The power of the laser is too low [153]	Revise laser power and check if any filter is on the path of the beam [153]
	Sample burns [153]	The power of the laser is too high [153]	Reduce laser power [153]
	Low signal-to-noise ratio [152, 153]	Focus is not adequate [152, 153]	Refocus carefully [152, 153]
		The integration time is set at a too low threshold [152, 153]	Augment integration threshold time value for signal collection [152, 153]
	Excessive fluorescence [153]	Dryness of sample [153]	Wet the sample [153]
		Bleaching [153]	Pre-run with short integration intervals [153]
FTIR	Poor absorbance accuracy [154]	The analog signal is larger than the analog to digital converter range [154]	Decreasing the light intensity in the system can enhance the detector linearity. Turning down the signal gain or increasing the scan speed could further help [154]
	No signal is obtained although scanning is ongoing [154]	A small aperture is selected, the detector is not cold, or the beam is blocked [154]	Cooldown the liquid nitrogen detector or select a larger aperture size [154]
	Frequency-dependent noise [154]	Noise is greater at a higher frequency [154]	Align fixed interferometer mirror or another interferometer component with alignment capability [154]
	Scanning problems [154]	The moving mirror is stuck, or the scanning mirror is broken [154]	Replace the moving mirror if broken. Manually move moving mirror assembly. If it is hard to move, clear by moving back and forth. Inspect interferometer area [154]

(continued)

Table 4.8 (continued)

Technique	Problem	Cause	Solution
	Wavelength inaccuracy [154]	The interferometer is misaligned or laser frequency is incorrect [154]	Realign interferometer mirrors, source, detector, and laser or enter correct laser frequency [154]
UV–Vis	Operator error [51]	Low signal sensitivity [51]	Verify that absorbance values for samples are in the range supported by the instrument at all wavelengths [51]
			Verify compatibility of UV–VIS or quartz cuvettes [51]
	Inaccurate concentration measurements [51]	Solvent dilatation with temperature has occurred [51]	Correct factor for the coefficient of thermal expansion [51]
	Shifts in the absorbance spectra [51]	Modification of molecular and electronic structural properties took place due to pH variations [51]	Adjust into adequate pH and avoid variation [51]
	Low accuracy and reliability [51]	Broader wavelength interval of operation in spectrometers [51]	Narrow wavelength interval from 220 to 780 nm [51]
XPS	Peaks overlap [163]	Insufficient photoelectron intensity [163]	Use different excitation sources or create a reference from the core-levels that offer enough photoelectron intensity [163]
	Fitting spectral data with a weak energy resolution [156]	Low energy resolution [156]	Enough energy resolution should be applied to allow separation of a signal via mathematical fitting [156]
	Peaks are truncating in a narrow scan [156]	Data are recorded over a narrow window hence challenge in fitting the background and component peaks [156]	Find energy acquisition regions with enough width to incorporate the surrounding and the leveled-off background noise [156]
	Inappropriate background for the fit [156]	NS	Select a background suitable for the physical properties of the analyzed materials [156]

(continued)

Table 4.8 (continued)

Technique	Problem	Cause	Solution
	Higher oxidation states are mislabeled [156]	Labels for the C–O and C=O fit components may be reversed, or a carboxyl signal may be mislabeled as a carbonyl [156]	Use reference databases that describe all spectral features for reference materials. Verify the accurate performance of the setup by using standard samples [156]
DRIFTS	Distortion in peak shapes [86, 89, 92]	High absorption indexes of samples [86, 89, 92]	Dilute sample with chemically inert support material or non-absorbing matrix [86]
	Baseline distortions [86, 157, 158]	Sensitivity to temperature and humidity changes of oxide-based catalysts [86, 157, 158]	Establish an appropriate reference spectrum [157]
	Spectra of increased peak widths [86]	Sensitivity to sample morphology [86]	Powder the samples into an appropriate and consistent particle size (<10 μm) [86]
XRD	Graininess [159]	NS	Adjust the divergence slit and beam mask for the largest possible irradiated area [159]
	Smooth, although non-continuous diffraction rings [159]	Orientation of crystallites is not random [159]	Use spinning sample stages [159]
		Orientations are over or under-represented [159]	Press the powder without overworking the surface or use the back-loading sample holder [159]
	Lack of high-resolution XRD unit source [160]	NS	Check the purity of the ingredients and repeatability of the synthesis [160]
	Short cathode life [161]	The cathode has a lifetime of about 2000 h at 50 kV and 200 mA [161]	Reduce the tube current to 100 mA [161]
	Uncertainty about the location of the peaks and their intensity [161]	NS	Make a quick scan with a large step size and a short time per step over a wide angular range [161]
NMR	Shims are not optimal [162]	NS	Load an acceptable shim parameter file for each solvent prior to shimming [162]

(continued)

Table 4.8 (continued)

Technique	Problem	Cause	Solution
	Clipping [162]	NS	Set the receiver gain one or two dB lower from the receiver gain settings [162]
	Not sufficient time for the complete FID to be obtained [162]	NS	Allow more signals till they thoroughly decay [162]
	Inhomogeneity [162]	NS	Use a higher concentration of sample with less volume [162]
	NMR time is limited [162]	NS	Tune, match the probe, and perform careful pulse calibration [162]
	Low sensitivity [162]	NS	Solvent signals can be reduced by solvent pre-saturation [162]
	Peaks are folded in [162]	NS	Increase spectral width to the highest value and slightly move the transmitter frequency [162]

DRIFTS diffuse reflectance infrared fourier transform spectroscopy; *FTIR* Fourier transform infrared spectroscopy; *NS* Not specific; *NMR* nuclear magnetic resonance; *RAMAN* Raman spectroscopy; *UV–Vis* ultraviolet–visible spectroscopy; *XRD* X-ray diffraction; *XPS* X-ray photoelectron spectroscopy; *FID* free induction decay

can be prevented by avoiding pH variations and adjusting pH accordingly during sample preparation [51]. Another common obstruction in UV–Vis is low accuracy and reliability. Although the UV–Vis light spectrum is between 100 and 800 nm, UV–Vis spectrometers can regularly operate in an extended range of wavelengths from 200 to 1100 nm. However, operating in the extremes of such wavelength ranges often leads to results with compromised quality. Consequently, to narrow the wavelength interval of operation from 220 to 780 nm is suggested [51].

In XPS, the successful acquisition of high-quality photoemission depends on several operating parameters. Obtaining unfitted data was found to be among the most reported problems (more than 65% of the time) when using XPS [155], in addition to misidentified peaks and inadequate data collection. Many examples of XPS analysis and peak fitting guides address important considerations of the physical and chemical nature for obtaining acceptable spectra [156]. A few of the errors are rather easy to resolve and do not affect the conclusions; however, some principally divert the data interpretation and change the course of discussion. Table 4.8 lists the errors in XPS peak fitting that are mainly observed in the scientific literature and suggests measures to avoid them.

The most common problem when applying DRIFTS is the distortion of peak shapes due to the high absorption rates of the samples. Such distortion can be reduced by diluting the sample with chemically inert support material including alumina or silica, or a non-absorbing matrix, such as KCl and KBr. Another common drawback encountered is baseline distortions caused by the sensitivity to temperature and humidity changes of the oxide-based catalysts often used with this technique. Establishing a suitable reference spectrum is essential to avoid baseline distortions. Since sensitivity to sample morphology is an intrinsic feature of DRIFTS, it is common to find spectra with increased peak widths when the particle size is inappropriate and inconsistent. In such cases, it is suggested to pulverize the sample to an approximate size below 10 μm (Table 4.8) [86, 89, 92, 157, 158].

Users of XRD may encounter obstacles that are intrinsic to the sample or the sample preparation. For instance, avoiding graininess involves a response from the user to adjust the divergence slit and beam mask for obtaining the most extensive possible irradiated area or use spinning sample stages [159]. The obtained diffraction rings may be non-continuous due to non-random orientation crystallites and over or under-represented orientations. To correct this problem, the user can press the powder without overworking the surface or use the back-loading sample holder if the user wants to avoid orientation at the sample's surface [159]. Moreover, if the user is not careful enough while preparing the sample, the qualitative data can be different from sample to sample [160].

The cathode of the equipment has a lifetime of about 2000 h at 50 kV and 200 mA. If the user wants to increase the cathode lifetime by 50%, the user should reduce the tube current to 100 mA. In addition, it is crucial to consider that turning the XRD system on and off causes stress to the filaments. Therefore, it is recommended to keep it on as much as possible. If it is not possible, it is recommended to reduce the power to 20 kV and 10 mA when it is not used [161]. Finally, it is important to remember that the scan velocity and step size depend on the sample size and the quality of the sample. In this regard, the equipment measures a small sample of poor crystalline quality in a longer time than a large sample of good crystalline quality. In order to observe the location of the peaks and their intensity, it is recommended to quickly scan with a large step size and a short time per step over a wide angular range and secondly to adjust the slits if necessary and continue measuring (Table 4.8) [161].

Distorted peaks are one of the most encountered problems in NMR; however, the shims are also a source of error while implementing this technique. The shims have limitations, and distorted NMR tubes should not be used. It is advised to load a suitable shim parameter file for a given probe and solvent before shimming since the user should consider that the previous user could have run a sample with a different solvent. Moreover, assuming that the receiver gain is set too high, the initial most intense part of the free induction decay (FID) truncates as it is beyond the capability of the analog to digital converter. The solution to this limitation is to set the receiver gain one or two dB lower from the receiver gain settings [162].

In multidimensional datasets, there is usually not enough time for the full FID to be obtained. In this case, the acquisition time can be longer to accommodate more signals. However, extending the acquisition time in diluted samples could provoke a harmful increase in the proportion of S/N included in the FID. Since NMR sensitivity is proportional to the quantity of sample in the detection region of the probe, the use of a shorter sample would introduce inhomogeneity caused by the liquid and air interface, leading to shimming problems and distorted lines. A higher concentration of sample with less volume should be used in order to avoid such a problem. NMR usage time is commonly limited, and thus, optimizing the experiments to obtain the highest sensitivity each time is needed. It is necessary to tune and match the probe and perform accurate pulse calibration to optimize the equipment successfully [162]. Furthermore, in spectrometers with no oversampling and digital filtering, an insufficient spectral width may lead to the peaks outside the spectral range being "folded in". To check for the presence of folded peaks, the user should increase spectral width (SW) to the maximum value and move the transmitter frequency. If the positions in the spectrum remain in the same relative position when the user changes the transmitter frequency, it is a fold in peak (Table 4.8) [162].

References

1. R.S. Krishnan, R.K. Shankar, Raman effect: history of the discovery. J. Raman Spectrosc. **10**(1), 1–8 (1981). https://doi.org/10.1002/jrs.1250100103
2. D.A. Long, Early history of the Raman effect. Int. Rev. Phys. Chem. **7**(4), 317–349 (1988). https://doi.org/10.1080/01442358809353216
3. E. Smith, G. Dent, *Modern Raman Spectroscopy* (Wiley Online Library, 2019)
4. J.R. Beattie, J.J. McGarvey, A.W. Stitt, Raman spectroscopy for the detection of AGEs/ALEs, in *Methods in Molecular Biology*, vol. 965, ed. by L. Galluzzi, I. Vitale, O. Kepp, G. Kroemer (Springer, 2013), pp. 297–312
5. K. Eberhardt, C. Stiebing, C. Matthäus, M. Schmitt, J. Popp, Advantages and limitations of Raman spectroscopy for molecular diagnostics: an update. Expert Rev. Mol. Diagn. **15**(6), 773–787 (2015). https://doi.org/10.1586/14737159.2015.1036744
6. E.E. Lawson, H.G.M. Edwards, A.C. Williams, B.W. Barry, Applications of Raman spectroscopy to skin research. Ski. Res. Technol. **3**(3), 147–153 (1997). https://doi.org/10.1111/j.1600-0846.1997.tb00179.x
7. D. Chenery, H. Bowring, Infrared and Raman spectroscopic imaging in biosciences. Spectrosc. Eur. **15**(4), 8–14 (2003) [Online]. Available: https://www.spectroscopyasia.com/system/files/pdf/IR_15_4.pdf
8. A. Saha et al., Precision of Raman spectroscopy measurements in detection of microcalcifications in breast needle biopsies. Anal. Chem. **84**(15), 6715–6722 (2012). https://doi.org/10.1021/ac3011439
9. C. Zheng et al., Pursuing shell-isolated nanoparticle-enhanced Raman spectroscopy (SHINERS) for concomitant detection of breast lesions and microcalcifications. Nanoscale **7**(40), 16960–16968 (2015). https://doi.org/10.1039/C5NR05319F
10. R.E. Kast, S.C. Tucker, K. Killian, M. Trexler, K.V. Honn, G.W. Auner, Emerging technology: applications of Raman spectroscopy for prostate cancer. Cancer Metastasis Rev. **33**(2–3), 673–693 (2014). https://doi.org/10.1007/s10555-013-9489-6
11. K. Kong, C. Kendall, N. Stone, I. Notingher, Raman spectroscopy for medical diagnostics—from in-vitro biofluid assays to in-vivo cancer detection. Adv. Drug Deliv. Rev. **89**, 121–134 (2015). https://doi.org/10.1016/j.addr.2015.03.009
12. M.F. Kircher et al., A brain tumor molecular imaging strategy using a new triple-modality MRI-photoacoustic-Raman nanoparticle. Nat. Med. **18**(5), 829–834 (2012). https://doi.org/10.1038/nm.2721
13. P. Gao et al., The clinical application of Raman spectroscopy for breast cancer detection. J. Spectrosc. **2017**, 1–10 (2017). https://doi.org/10.1155/2017/5383948

14. B. Li, N.M.S. Sirimuthu, B.H. Ray, A.G. Ryder, Using surface-enhanced Raman scattering (SERS) and fluorescence spectroscopy for screening yeast extracts, a complex component of cell culture media. J. Raman Spectrosc. **43**(8), 1074–1082 (2012). https://doi.org/10.1002/jrs.3141
15. B. Li, B.H. Ray, K.J. Leister, A.G. Ryder, Performance monitoring of a mammalian cell based bioprocess using Raman spectroscopy. Anal. Chim. Acta **796**, 84–91 (2013). https://doi.org/10.1016/j.aca.2013.07.058
16. Z. Wen, Raman spectroscopy of protein pharmaceuticals. J. Pharm. Sci. **96**(11), 2861–2878 (2007). https://doi.org/10.1002/jps.20895
17. B. Li, P.W. Ryan, B.H. Ray, K.J. Leister, N.M.S. Sirimuthu, A.G. Ryder, Rapid characterization and quality control of complex cell culture media solutions using Raman spectroscopy and chemometrics. Biotechnol. Bioeng. **107**(2), 290–301 (2010). https://doi.org/10.1002/bit.22813
18. A. Kamińska, A. Kowalska, P. Albrycht, E. Witkowska, J. Waluk, ABO blood groups' antigen–antibody interactions studied using SERS spectroscopy: towards blood typing. Anal. Methods **8**(7), 1463–1472 (2016). https://doi.org/10.1039/C5AY02658J
19. Y. Zhang, W. Zhou, Y. Xue, J. Yang, D. Liu, Multiplexed imaging of trace residues in a single latent fingerprint. Anal. Chem. **88**(24), 12502–12507 (2016). https://doi.org/10.1021/acs.analchem.6b04077
20. M.-M. Blum, H. John, Historical perspective and modern applications of attenuated total reflectance—Fourier transform infrared spectroscopy (ATR-FTIR). Drug Test. Anal. **4**(3–4), 298–302 (2012). https://doi.org/10.1002/dta.374
21. S. Tanaka, Fourier transform infrared spectroscopy. J. Japan Oil Chem. Soc. **32**(10), 586–592 (1983). https://doi.org/10.5650/jos1956.32.586
22. M. Mohamed, J. Jaafar, A. Ismail, M.H.D. Othman, M.A. Rahman, Fourier transform infrared (FTIR) spectroscopy, in *Membrane Characterization*, ed. by N. Hilal, A.F. Ismail, T. Matsuura, D. Oatley-Radcliffe (Elsevier, 2017), pp. 3–29
23. J.S. Gaffney, N.A. Marley, D.E. Jones, Fourier transform infrared (FTIR) spectroscopy, in *Characterization of Materials* (Wiley, Hoboken, NJ, USA, 2012)
24. J.A. Larrabee, S. Choi, Fourier transform infrared spectroscopy. Methods Enzymol. **226**, 289–305 (1993). https://doi.org/10.1016/0076-6879(93)26014-Z
25. B. Smith, Introduction to infrared spectroscopy, in *Fundamentals of Fourier Transform Infrared Spectroscopy*, 2nd edn. (CRC Press, 2011), pp. 1–17
26. S.F. Sim, M.X.L. Chai, A.L. Jeffrey Kimura, Prediction of lard in palm olein oil using simple linear regression (SLR), multiple linear regression (MLR), and partial least squares regression (PLSR) based on Fourier-transform infrared (FTIR). J. Chem. **2018**, 1–8, (2018). https://doi.org/10.1155/2018/7182801
27. A. Dutta, Fourier transform infrared spectroscopy, in *Spectroscopic Methods for Nanomaterials Characterization*, vol. 2 (Elsevier, 2017), pp. 73–93
28. M. Mecozzi, L. Nisini, The differentiation of biodegradable and non-biodegradable polyethylene terephthalate (PET) samples by FTIR spectroscopy: a potential support for the structural differentiation of PET in environmental analysis. Infrared Phys. Technol. **101**, 119–126 (2019). https://doi.org/10.1016/j.infrared.2019.06.008
29. H. Barndõk, N. Merayo, L. Blanco, D. Hermosilla, Á. Blanco, Application of on-line FTIR methodology to study the mechanisms of heterogeneous advanced oxidation processes. Appl. Catal. B Environ. **185**, 344–352 (2016). https://doi.org/10.1016/j.apcatb.2015.12.036
30. J. Zięba-Palus, B. Trzcińska, A. Wesełucha-Birczyńska, P. Moskal, J. Sacharz, The sequence of changes observed during degradation process of paper by the use of UV/VIS and FTIR spectrometry with application of the PCA and 2D correlation method for forensic purposes. J. Mol. Struct. **1205**, 127651 (2020). https://doi.org/10.1016/j.molstruc.2019.127651
31. H. Lin, Y. Zhang, Q. Wang, B. Li, P. Huang, Z. Wang, Estimation of the age of human bloodstains under the simulated indoor and outdoor crime scene conditions by ATR-FTIR spectroscopy. Sci. Rep. **7**(1), 13254 (2017). https://doi.org/10.1038/s41598-017-13725-1

32. N.N. Atia, M.A. Marzouq, A.I. Hassan, W.E. Eltoukhi, A rapid FTIR spectroscopic assay for quantitative determination of memantine hydrochloride and amisulpride in human plasma and pharmaceutical formulations. Spectrochim. Acta Part A Mol. Biomol. Spectrosc. **236**, 118377 (2020). https://doi.org/10.1016/j.saa.2020.118377
33. H. Tiernan, B. Byrne, S.G. Kazarian, ATR-FTIR spectroscopy and spectroscopic imaging for the analysis of biopharmaceuticals. Spectrochim. Acta Part A Mol. Biomol. Spectrosc. **241**, 118636 (2020). https://doi.org/10.1016/j.saa.2020.118636
34. C. Agudelo-Cuartas, D. Granda-Restrepo, P.J.A. Sobral, W. Castro, Determination of mechanical properties of whey protein films during accelerated aging: application of FTIR profiles and chemometric tools. J. Food Process. Eng. **44**(5) (2021). https://doi.org/10.1111/jfpe.13477
35. O.F. Vázquez-Vuelvas, F.A. Chávez-Camacho, J.A. Meza-Velázquez, E. Mendez-Merino, M.M. Ríos-Licea, J.C. Contreras-Esquivel, A comparative FTIR study for supplemented agavin as functional food. Food Hydrocoll. **103**, 105642 (2020). https://doi.org/10.1016/j.foodhyd.2020.105642
36. R. Kumar, S. Sharma, D. Pathak, N. Dhiman, N. Arora, Ionic conductivity, FTIR and thermal studies of nano-composite plasticized proton conducting polymer electrolytes. Solid State Ionics **305**, 57–62 (2017). https://doi.org/10.1016/j.ssi.2017.04.020
37. A. Pawlicka et al., Dielectric behavior and FTIR studies of xanthan gum-based solid polymer electrolytes. Electrochim. Acta **305**, 232–239 (2019). https://doi.org/10.1016/j.electacta.2019.03.055
38. M. Mecozzi, M. Pietroletti, Y.B. Monakhova, FTIR spectroscopy supported by statistical techniques for the structural characterization of plastic debris in the marine environment: application to monitoring studies. Mar. Pollut. Bull. **106**(1–2), 155–161 (2016). https://doi.org/10.1016/j.marpolbul.2016.03.012
39. A.P.M. Michel, A.E. Morrison, V.L. Preston, C.T. Marx, B.C. Colson, H.K. White, Rapid identification of marine plastic debris via spectroscopic techniques and machine learning classifiers. Environ. Sci. Technol. **54**(17), 10630–10637 (2020). https://doi.org/10.1021/acs.est.0c02099
40. O. Anjos, A.J.A. Santos, L.M. Estevinho, I. Caldeira, FTIR-ATR spectroscopy applied to quality control of grape-derived spirits. Food Chem. **205**, 28–35 (2016). https://doi.org/10.1016/j.foodchem.2016.02.128
41. J. Bell, P. Nel, B. Stuart, Non-invasive identification of polymers in cultural heritage collections: evaluation, optimisation and application of portable FTIR (ATR and external reflectance) spectroscopy to three-dimensional polymer-based objects. Herit. Sci. **7**(1), 1–18 (2019). https://doi.org/10.1186/s40494-019-0336-0
42. T. Rymsza, E.A. Ribeiro, L.F. das Chagas e Silva de Carvalho, T. Bhattacharjee, R. de Azevedo Canevari, Human papillomavirus detection using PCR and ATR-FTIR for cervical cancer screening. Spectrochim. Acta Part A Mol. Biomol. Spectrosc. **196**, 238–246 (2018). https://doi.org/10.1016/j.saa.2018.02.004
43. K. Chrabaszcz et al., Label-free FTIR spectroscopy detects and visualizes the early stage of pulmonary micrometastasis seeded from breast carcinoma. Biochim. Biophys. Acta Mol. Basis Dis. **1864**(11), 3574–3584 (2018). https://doi.org/10.1016/j.bbadis.2018.08.022
44. V.E. Sitnikova, M.A. Kotkova, T.N. Nosenko, T.N. Kotkova, D.M. Martynova, M.V. Uspenskaya, Breast cancer detection by ATR-FTIR spectroscopy of blood serum and multivariate data-analysis. Talanta **214**(Oct 2019), 120857 (2020). https://doi.org/10.1016/j.talanta.2020.120857
45. E. Gray et al., Health economic evaluation of a serum-based blood test for brain tumour diagnosis: exploration of two clinical scenarios. BMJ Open **8**(5), e017593 (2018). https://doi.org/10.1136/bmjopen-2017-017593
46. D. Kumar et al., Results in chemistry detection of flonicamid insecticide in vegetable samples by UV–visible spectrophotometer and FTIR. Results Chem. **2**, 10059 (2020). https://doi.org/10.1016/j.rechem.2020.100059
47. W. Schmidt, Introduction to optical spectroscopy, in *Optical Spectroscopy in Chemistry and Life Sciences: An Introduction* (Wiley-VCH Verlag, Weinheim, Germany, 2005), pp. 1–12

48. N.C. Thomas, The early history of spectroscopy. J. Chem. Educ. **68**(8), 631 (1991). https://doi.org/10.1021/ed068p631
49. S.S. Nielsen, *Food Analysis* (Springer US, Boston, MA, USA, 2010)
50. H. Wang, P.K. Chu, Surface characterization of biomaterials, in *Characterization of Biomaterials* (Elsevier, 2013), pp. 105–174
51. F.S. Rocha, A.J. Gomes, C.N. Lunardi, S. Kaliaguine, G.S. Patience, Experimental methods in chemical engineering: ultraviolet visible spectroscopy-UV–Vis. Can. J. Chem. Eng. **96**(12), 2512–2517 (2018). https://doi.org/10.1002/cjce.23344
52. E. Tomaszewska et al., Detection limits of DLS and UV–Vis spectroscopy in characterization of polydisperse nanoparticles colloids. J. Nanomater. **2013**, 1–10 (2013). https://doi.org/10.1155/2013/313081
53. J. Roberts, A. Power, J. Chapman, S. Chandra, D. Cozzolino, The use of UV–Vis spectroscopy in bioprocess and fermentation monitoring. Fermentation **4**(1), 18 (2018). https://doi.org/10.3390/fermentation4010018
54. J. Zeng et al., Au/AgI dimeric nanoparticles for highly selective and sensitive colorimetric detection of hydrogen sulfide. Adv. Funct. Mater. **1800515**, 1–10 (2018). https://doi.org/10.1002/adfm.201800515.
55. F.A. Saif, S.A. Yaseen, A.S. Alameen, S.B. Mane, P.B. Undre, Identification and characterization of aspergillus species of fruit rot fungi using microscopy, FT-IR, Raman and UV–Vis spectroscopy. Spectrochim. Acta Part A Mol. Biomol. Spectrosc., 119010 (2020). https://doi.org/10.1016/j.saa.2020.119010
56. M.U.M. Patel, R. Dominko, Application of in operando UV/Vis spectroscopy in lithium-sulfur batteries. Chemsuschem **7**(8), 2167–2175 (2014). https://doi.org/10.1002/cssc.201402215
57. B. Li et al., A self-designed versatile and portable sensing device based on smart phone for colorimetric detection. Anal. Bioanal. Chem. **413**(2), 533–541 (2021). https://doi.org/10.1007/s00216-020-03024-6
58. P. Liu et al., Fluorescence-enhanced bio-detection platforms obtained through controlled 'step-by-step' clustering of silver nanoparticles. Nanoscale **10**(2), 848–855 (2018). https://doi.org/10.1039/C7NR07486G
59. F. Mohammadparast, R. Teja Addanki Tirumala, S. Bhardwaj Ramakrishnan, A.P. Dadgar, M. Andiappan, Operando UV–Vis spectroscopy as potential in-line PAT system for size determination of functioning metal nanocatalysts. Chem. Eng. Sci. **225**, 115821 (2020). https://doi.org/10.1016/j.ces.2020.115821
60. R. Begum et al., Applications of UV/Vis spectroscopy in characterization and catalytic activity of noble metal nanoparticles fabricated in responsive polymer microgels: a review. Crit. Rev. Anal. Chem. **0**(0), 1–14 (2018). https://doi.org/10.1080/10408347.2018.1451299
61. K.-H. Shin, C.-S. Jin, J.-Y. So, S.-K. Park, D.-H. Kim, S.-H. Yeon, Real-time monitoring of the state of charge (SOC) in vanadium redox-flow batteries using UV–Vis spectroscopy in operando mode. J. Energy Storage **27**(Nov 2019), 101066 (2020). https://doi.org/10.1016/j.est.2019.101066
62. J.M. Antosiewicz, D. Shugar, UV–Vis spectroscopy of tyrosine side-groups in studies of protein structure. Part 2: selected applications. Biophys. Rev. **8**(2), 163–177 (2016). https://doi.org/10.1007/s12551-016-0197-7
63. Y. Li, Q. Luo, R. Hu, Z. Chen, P. Qiu, A sensitive and rapid UV–vis spectrophotometry for organophosphorus pesticides detection based on Ytterbium (Yb^{3+}) functionalized gold nanoparticle. Chinese Chem. Lett. **29**(12), 1845–1848 (2018). https://doi.org/10.1016/j.cclet.2018.11.016
64. H.V. Tran et al., Silver nanoparticles-decorated reduced graphene oxide: a novel peroxidase-like activity nanomaterial for development of a colorimetric glucose biosensor. Arab. J. Chem. **13**(7), 6084–6091 (2020). https://doi.org/10.1016/j.arabjc.2020.05.008
65. T.K. Patle, K. Shrivas, R. Kurrey, S. Upadhyay, R. Jangde, R. Chauhan, Phytochemical screening and determination of phenolics and flavonoids in Dillenia pentagyna using UV–vis and FTIR spectroscopy. Spectrochim. Acta Part A Mol. Biomol. Spectrosc. **242**, 118717 (2020). https://doi.org/10.1016/j.saa.2020.118717

66. J.M. Hollander, D.A. Shirley, The 1981 nobel prize in physics. Science (80–) **214**(4521), 629–631 (1981). https://doi.org/10.1126/science.214.4521.629
67. N.V. Alov, Fifty years of X-ray photoelectron spectroscopy. J. Anal. Chem. **60**(3), 297–300 (2005). https://doi.org/10.1007/s10809-005-0087-9
68. T. Susi, T. Pichler, P. Ayala, X-ray photoelectron spectroscopy of graphitic carbon nanomaterials doped with heteroatoms. Beilstein J. Nanotechnol. **6**, 177–192 (2015). https://doi.org/10.3762/bjnano.6.17
69. G. Greczynski, L. Hultman, X-ray photoelectron spectroscopy: towards reliable binding energy referencing. Prog. Mater. Sci. **107**, 100591 (2020). https://doi.org/10.1016/j.pmatsci.2019.100591
70. J.D. Andrade, X-ray photoelectron spectroscopy (XPS), in *Surface and Interfacial Aspects of Biomedical Polymers* (Springer, Boston, MA, 1985), pp. 105–195
71. T.A. Carlson, Basic assumptions and recent developments in quantitative XPS. Surf. Interface Anal. **4**(4), 125–134 (1982). https://doi.org/10.1002/sia.740040402
72. L. Yahia, L.K. Mireles, X-ray photoelectron spectroscopy (XPS) and time-of-flight secondary ion mass spectrometry (ToF SIMS), in *Characterization of Polymeric Biomaterials* (Elsevier, 2017), pp. 83–97
73. E. Korin, N. Froumin, S. Cohen, Surface analysis of nanocomplexes by X-ray photoelectron spectroscopy (XPS). ACS Biomater. Sci. Eng. **3**(6), 882–889 (2017). https://doi.org/10.1021/acsbiomaterials.7b00040
74. S. Tougaard, Improved XPS analysis by visual inspection of the survey spectrum. Surf. Interface Anal. **50**(6), 657–666 (2018). https://doi.org/10.1002/sia.6456
75. D. Briggs, Industrial applications of XPS: recent developments in catalysis and polymer adhesion studies. Appl. Surf. Sci. **6**(3–4), 188–203 (1980). https://doi.org/10.1016/0378-5963(80)90012-4
76. A.V. Boryakov, S.I. Surodin, R.N. Kryukov, D.E. Nikolichev, S.Y. Zubkov, Spectral fit refinement in XPS analysis technique and its practical applications. J. Electron Spectros. Relat. Phenom. **229**(2018), 132–140 (2018). https://doi.org/10.1016/j.elspec.2017.11.004
77. C. Fu et al. Ultrasensitive sandwich-like electrochemical biosensor based on core-shell Pt@CeO_2 as signal tags and double molecular recognition for cerebral dopamine detection. Talanta **223** (July 2020), 121719 (2021). https://doi.org/10.1016/j.talanta.2020.121719
78. S.A. Al-Bataineh, L.G. Britcher, H.J. Griesser, XPS characterization of the surface immobilization of antibacterial furanones. Surf. Sci. **600**(4), 952–962 (2006). https://doi.org/10.1016/j.susc.2005.12.028
79. S.L. McArthur, Applications of XPS in bioengineering. Surf. Interface Anal. **38**(11), 1380–1385 (2006). https://doi.org/10.1002/sia.2498
80. B. Putz et al., Combined TEM and XPS studies of metal-polymer interfaces for space applications. Surf. Coatings Technol. **332**, 368–375 (2017). https://doi.org/10.1016/j.surfcoat.2017.07.079
81. F. Reniers, C. Tewell, New improvements in energy and spatial (*x*, *y*, *z*) resolution in AES and XPS applications. J. Electron Spectros. Relat. Phenom. **142**(1), 1–25 (2005). https://doi.org/10.1016/j.elspec.2004.07.004
82. V. Shutthanandan et al., Applications of XPS in the characterization of battery materials. J. Electron Spectros. Relat. Phenom. **231**(Nov 2017), 2–10 (2019). https://doi.org/10.1016/j.elspec.2018.05.005
83. V.S. Smentkowski, Applications of XPS in biology and biointerface analysis. Surf. Anal. Tech. Biol., 1–326 (2014). https://doi.org/10.1007/978-3-319-01360-2
84. N. Vandencasteele, F. Reniers, Plasma-modified polymer surfaces: characterization using XPS. J. Electron Spectros. Relat. Phenom. **178–179**(C), 394–408 (2010). https://doi.org/10.1016/j.elspec.2009.12.003
85. A.V. Pushkarev et al., Data on characterization of glass biochips and validation of the label-free biosensor for detection of autoantibodies in human serum. Data Br. **30**, 105648 (2020). https://doi.org/10.1016/j.dib.2020.105648

86. M.B. Mitchell, Fundamentals and applications of diffuse reflectance infrared Fourier transform (DRIFT) spectroscopy, in *Structure-Property Relations in Polymers* (American Chemical Society, 1993), pp. 351–375
87. P. Larkin, *Infrared and Raman Spectroscopy* (Elsevier, 2018)
88. G. Anbalagan, G. Sivakumar, A.R. Prabakaran, S. Gunasekaran, Spectroscopic characterization of natural chrysotile. Vib. Spectrosc. **52**(2), 122–127 (2010). https://doi.org/10.1016/j.vibspec.2009.11.007
89. G. Accardo, R. Cioffi, F. Colangelo, R. D'Angelo, L. De Stefano, F. Paglietti, Diffuse reflectance infrared Fourier transform spectroscopy for the determination of asbestos species in bulk building materials. Materials (Basel) **7**(1), 457–470 (2014). https://doi.org/10.3390/ma7010457
90. R.T. So, N.E. Blair, A.L. Masterson, Carbonate mineral identification and quantification in sediment matrices using diffuse reflectance infrared Fourier transform spectroscopy. Environ. Chem. Lett. **18**(5), 1725–1730 (2020). https://doi.org/10.1007/s10311-020-01027-4
91. M. Manfredi, E. Barberis, A. Rava, E. Robotti, F. Gosetti, E. Marengo, Portable diffuse reflectance infrared Fourier transform (DRIFT) technique for the non-invasive identification of canvas ground: IR spectra reference collection. Anal. Methods **7**(6), 2313–2322 (2015). https://doi.org/10.1039/c4ay02006e
92. I. Arrizabalaga et al., Diffuse reflectance FTIR database for the interpretation of the spectra obtained with a handheld device on built heritage materials. Anal. Methods **7**(3), 1061–1070 (2015). https://doi.org/10.1039/c4ay02189d
93. Z.M. Gibbs, A. Lalonde, G.J. Snyder, Optical band gap and the Burstein-Moss effect in iodine doped PbTe using diffuse reflectance infrared Fourier transform spectroscopy. New J. Phys. **15**, 0–18 (2013). https://doi.org/10.1088/1367-2630/15/7/075020
94. J. Niemeyer, Y. Chen, J.-M. Bollag, Characterization of humic acids, composts, and peat by diffuse reflectance Fourier-transform infrared spectroscopy. Soil Sci. Soc. Am. J. **56**(1), 135–140 (1992). https://doi.org/10.2136/sssaj1992.03615995005600010021x
95. T. Armaroli, T. Bécue, S. Gautier, Diffuse reflection infrared spectroscopy (DRIFTS): application to the in situ analysis of catalysts. Oil Gas Sci. Technol. **59**(2), 215–237 (2004). https://doi.org/10.2516/ogst:2004016
96. L. Tremblay, J.P. Gagné, Fast quantification of humic substances and organic matter by direct analysis of sediments using DRIFT spectroscopy. Anal. Chem. **74**(13), 2985–2993 (2002). https://doi.org/10.1021/ac011043g
97. P.E. Fanning, M.A. Vannice, A DRIFTS study of the formation of surface groups on carbon by oxidation. Carbon N. Y. **31**(5), 721–730 (1993). https://doi.org/10.1016/0008-6223(93)90009-Y
98. M.P. Fuller, P.R. Griffiths, Diffuse reflectance measurements by infrared Fourier transform spectrometry. Anal. Chem. **50**(13), 1906–1910 (1978). https://doi.org/10.1021/ac50035a045
99. A.A. Kamnev, A.V. Tugarova, A.G. Shchelochkov, K. Kovács, E. Kuzmann, Diffuse reflectance infrared Fourier transform (DRIFT) and Mössbauer spectroscopic study of Azospirillum brasilense Sp7: evidence for intracellular iron (II) oxidation in bacterial biomass upon lyophilisation. Spectrochim. Acta Part A Mol. Biomol. Spectrosc. **229** (2020). https://doi.org/10.1016/j.saa.2019.117970
100. L. Granado et al., Non-destructive DRIFT spectroscopy measurement of the degree of curing of industrial epoxy/silica composite buildup layers. Polym. Test. **70**(July), 188–191 (2018). https://doi.org/10.1016/j.polymertesting.2018.07.007
101. B.A. Bhongade, S.R. Dhaneshwar, Diffuse reflectance infrared Fourier transform spectroscopy (DRIFTS) application in pharmaceutical analysis: method for quantification of ranolazine in tablet dosage form. Vib. Spectrosc. **93**(95635), 52–56 (2017). https://doi.org/10.1016/j.vibspec.2017.10.001
102. S. Steger, H. Stege, S. Bretz, O. Hahn, Capabilities and limitations of handheld diffuse reflectance infrared Fourier transform spectroscopy (DRIFTS) for the analysis of colourants and binders in 20th-century reverse paintings on glass. Spectrochim. Acta Part A Mol. Biomol. Spectrosc. **195**, 103–112 (2018). https://doi.org/10.1016/j.saa.2018.01.057

103. K. Ushizawa, Y. Sato, T. Mitsumori, T. Machinami, T. Ueda, T. Ando, Covalent immobilization of DNA on diamond and its verification by diffuse reflectance infrared spectroscopy. Chem. Phys. Lett. **351**(1–2), 105–108 (2002). https://doi.org/10.1016/S0009-2614(01)01362-8
104. F. Herold et al., Methodology for the identification of carbonyl absorption maxima of carbon surface oxides in DRIFT spectra. Carbon Trends **3**, 100020 (2021). https://doi.org/10.1016/j.cartre.2020.100020
105. M. Eckert, Max von Laue and the discovery of X-ray diffraction in 1912. Ann. Phys. **524**(5), 83–85 (2012). https://doi.org/10.1002/andp.201200724
106. S.V. Borisov, N.V. Podberezskaya, X-ray diffraction analysis: A brief history and achievements of the first century. J. Struct. Chem. **53**, 1–3 (2012). https://doi.org/10.1134/S0022476612070013
107. G. Hildebrandt, The discovery of the diffraction of X-rays in crystals—a historical review. Cryst. Res. Technol. **28**(6), 747–766 (1993). https://doi.org/10.1002/crat.2170280602
108. H. Stanjek, W. Häusler, Basics of X-ray diffraction. Hyperfine Interact. **154**(1–4), 107–119 (2004). https://doi.org/10.1023/B:HYPE.0000032028.60546.38
109. M. Lee, *X-Ray Diffraction for Materials Research: From Fundamentals to Applications* (Apple Academic Press, 2016)
110. J. Epp, X-ray diffraction (XRD) techniques for materials characterization. Mater. Charact. Using Nondestruct. Eval. Methods, 81–124 (2016). https://doi.org/10.1016/B978-0-08-100040-3.00004-3
111. A.K. Keshari, A.C. Pandey, Size and distribution: a comparison of XRD, SAXS and SANS study of II–VI semiconductor nanocrystals. J. Nanosci. Nanotechnol. **8**(3), 1221–1227 (2008). https://doi.org/10.1166/jnn.2008.370
112. L. Lu, V. Sahajwalla, C. Kong, D. Harris, Quantitative X-ray diffraction analysis and its application to various coals. Carbon N. Y. **39**(12), 1821–1833 (2001). https://doi.org/10.1016/S0008-6223(00)00318-3
113. M.R. Gauna, M.S. Conconi, S. Gomez, G. Suárez, E.F. Aglietti, N.M. Rendtorff, Monoclinic-tetragonal zirconia quantification of commercial nanopowder mixtures by XRD and DTA. Ceram. Silikaty **50**(4), 318–325 (2015)
114. T. Theivasanthi, M. Alagar, X-ray diffraction studies of copper nanopowder. Arch. Phys. Res. **1**(2), 112–117 (2010)
115. M. Ermrich, D. Opper, *XRD for the Analyst: Getting Acquainted with the Principles* (PANalytical, 2011)
116. R. Svoboda, Reaction/crystallization kinetics studied via in situ XRD: experimental conditions versus methods of kinetic analysis. Philos. Mag. **99**(23), 2941–2956 (2019). https://doi.org/10.1080/14786435.2019.1648899
117. C. Ghanty et al. Li^+-ion extraction/insertion of Ni-rich Li_{1+x} $(Ni_yCo_zMn_z)_wO_2$ ($0.005 < x < 0.03$; $y{:}z = 8{:}1$, $w \approx 1$) electrodes: in situ XRD and Raman spectroscopy study. ChemElectroChem **2**(10), 1479–1486 (2015). https://doi.org/10.1002/celc.201500160
118. N. Raval, R. Maheshwari, D. Kalyane, S.R. Youngren-Ortiz, M.B. Chougule, R.K. Tekade, Importance of physicochemical characterization of nanoparticles in pharmaceutical product development. Basic Fundam. Drug Deliv. 369–400 (2018). https://doi.org/10.1016/B978-0-12-817909-3.00010-8
119. G. Koorösy, K. Tomolya, D. Janovszky, J. Sólyom, Evaluation of XRD analysis of amorphous alloys. Mater. Sci. Forum **729**, 419–423 (2013). https://doi.org/10.4028/www.scientific.net/MSF.729.419
120. A.A. Bunaciu, E.G. Udriştioiu, H.Y. Aboul-Enein, X-ray diffraction: instrumentation and applications. Crit. Rev. Anal. Chem. **45**(4), 289–299 (2015). https://doi.org/10.1080/10408347.2014.949616
121. H. Khan, A.S. Yerramilli, A. D'Oliveira, T.L. Alford, D.C. Boffito, G.S. Patience, Experimental methods in chemical engineering: X-ray diffraction spectroscopy—XRD. Can. J. Chem. Eng. **98**(6), 1255–1266 (2020). https://doi.org/10.1002/cjce.23747
122. S. Tariq et al., Synthesis and structural analysis of novel indole derivatives by XRD, spectroscopic and DFT studies. J. Mol. Struct. **1203**, 127438 (2020). https://doi.org/10.1016/j.molstruc.2019.127438

123. C.A. Rosas-Casarez et al., Experimental study of XRD, FTIR and TGA techniques in geopolymeric materials. Int. J. Adv. Comput. Sci. Its Appl. **4**(4), 25–30 (2014)
124. S.H. Kim, C.M. Lee, K. Kafle, Characterization of crystalline cellulose in biomass: basic principles, applications, and limitations of XRD, NMR, IR, Raman, and SFG. Korean J. Chem. Eng. **30**(12), 2127–2141 (2013). https://doi.org/10.1007/s11814-013-0162-0
125. A. Patlolla, E.V. Carino, S.N. Ehrlich, E. Stavitski, A.I. Frenkel, Application of operando XAS, XRD, and Raman spectroscopy for phase speciation in water gas shift reaction catalysts. ACS Catal. **2**(11), 2216–2223 (2012). https://doi.org/10.1021/cs300414c
126. L. Zhao et al. Molecular structure characterization of lignite treated with ionic liquid via FTIR and XRD spectroscopy. Fuel **272**, 117705 (2020). https://doi.org/10.1016/j.fuel.2020.117705
127. X. Li et al., Encapsulating enzyme into metal-organic framework during in-situ growth on cellulose acetate nanofibers as self-powered glucose biosensor. Biosens. Bioelectron. **171**(July 2020), 112690 (2021). https://doi.org/10.1016/j.bios.2020.112690
128. T. Dayakar, K. Venkateswara Rao, K. Bikshalu, V. Rajendar, S.H. Park, Novel synthesis and structural analysis of zinc oxide nanoparticles for the non enzymatic glucose biosensor. Mater. Sci. Eng. C **75**, 1472–1479 (2017). https://doi.org/10.1016/j.msec.2017.02.032
129. A. Şavk, H. Aydın, K. Cellat, F. Şen, A novel high performance non-enzymatic electrochemical glucose biosensor based on activated carbon-supported Pt-Ni nanocomposite. J. Mol. Liq. **300** (2020). https://doi.org/10.1016/j.molliq.2019.112355
130. D. Marion, An introduction to biological NMR spectroscopy. Mol. Cell. Proteom. **12**(11), 3006–3025 (2013). https://doi.org/10.1074/mcp.O113.030239
131. J.S. Cohen, J.W. Jaroszewski, O. Kaplan, J. Ruiz-Cabello, S.W. Collier, A history of biological applications of NMR spectroscopy. Prog. Nucl. Magn. Reson. Spectrosc. **28**(1), 53–85 (1995). https://doi.org/10.1016/0079-6565(95)01020-3
132. R.L. Kleinberg, J.A. Jackson, An introduction to the history of NMR well logging. Concepts Magn. Reson. **13**(6), 340–342 (2001). https://doi.org/10.1002/cmr.1018
133. K. Zia, T. Siddiqui, S. Ali, I. Farooq, M.S. Zafar, Z. Khurshid, Nuclear magnetic resonance spectroscopy for medical and dental applications: a comprehensive review. Eur. J. Dent. **13**(1), 124–128 (2019). https://doi.org/10.1055/s-0039-1688654
134. T.W.M. Fan, A.N. Lane, Applications of NMR spectroscopy to systems biochemistry. Prog. Nucl. Magn. Reson. Spectrosc. **92–93**, 18–53 (2016). https://doi.org/10.1016/j.pnmrs.2016.01.005
135. J.L. Markley et al., The future of NMR-based metabolomics. Curr. Opin. Biotechnol. **43**, 34–40 (2017). https://doi.org/10.1016/j.copbio.2016.08.001
136. S.J. Devience et al., Nanoscale NMR spectroscopy and imaging of multiple nuclear species. Nat. Nanotechnol. **10**(2), 129–134 (2015). https://doi.org/10.1038/nnano.2014.313
137. Atta-ur-Rahman, M.I. Choudhary, Atia-tul-Wahab, Sensitivity enhancement, in *Solving Problems with NMR Spectroscopy*, 2nd edn. (Elsevier, 2016), pp. 99–132
138. A.H. Emwas et al., NMR spectroscopy for metabolomics research. Metabolites **9**(7) (2019). https://doi.org/10.3390/metabo9070123
139. Atta-ur-Rahman, M.I. Choudhary, Atia-tul-Wahab, The basics of modern NMR spectroscopy, in *Solving Problems with NMR Spectroscopy*, 2nd edn. (Elsevier Inc, 2016), pp. 1–34
140. A. Spyros, P. Dais, *NMR Spectroscopy in Food Analysis* (Royal Society of Chemistry, 2012)
141. D.W.H. Merkx, G.T.S. Hong, A. Ermacora, J.P.M. Van Duynhoven, Rapid quantitative profiling of lipid oxidation products in a food emulsion by ^1H NMR. Anal. Chem. **90**(7), 4863–4870 (2018). https://doi.org/10.1021/acs.analchem.8b00380
142. E. Hatzakis, E. Archavlis, P. Dais, Determination of glycerol in wines using 31P-NMR spectroscopy. JAOCS J. Am. Oil Chem. Soc. **84**(7), 615–619 (2007). https://doi.org/10.1007/s11746-007-1099-4
143. P. Charisiadis, V. Exarchou, A.N. Troganis, I.P. Gerothanassis, Exploring the 'forgotten'–OH NMR spectral region in natural products. Chem. Commun. **46**(20), 3589 (2010). https://doi.org/10.1039/b927256a
144. S.M. Ackermann et al., Automated multicomponent analysis of soft drinks using 1D ^1H and 2D ^1H–^1H J-resolved NMR spectroscopy. Food Anal. Methods **10**(3), 827–836 (2017). https://doi.org/10.1007/s12161-016-0643-y

145. N. Hao, H. Ben, C.G. Yoo, S. Adhikari, A.J. Ragauskas, Review of NMR characterization of pyrolysis oils. Energy Fuels **30**(9), 6863–6880 (2016). https://doi.org/10.1021/acs.energyfuels.6b01002
146. C.A. Mullen, G.D. Strahan, A.A. Boateng, Characterization of various fast-pyrolysis bio-oils by NMR spectroscopy. Energy Fuels **23**(5), 2707–2718 (2009). https://doi.org/10.1021/ef801048b
147. E. Luchinat, L. Banci, In-cell NMR: a topical review. Int. Union Crystallogr. J. **4**(2), 108–118 (2017). https://doi.org/10.1107/S2052252516020625
148. Z. Serber, A.T. Keatinge-Clay, R. Ledwidge, A.E. Kelly, S.M. Miller, V. Dötsch, High-resolution macromolecular NMR spectroscopy inside living cells. J. Am. Chem. Soc. **123**(10), 2446–2447 (2001). https://doi.org/10.1021/ja0057528
149. C. Chen et al., Exploring metabolic profile differences between colorectal polyp patients and controls using seemingly unrelated regression. J. Proteome Res. **14**(6), 2492–2499 (2015). https://doi.org/10.1021/acs.jproteome.5b00059
150. A.H. Emwas et al., Standardizing the experimental conditions for using urine in NMR-based metabolomic studies with a particular focus on diagnostic studies: a review. Metabolomics **11**(4), 872–894 (2015). https://doi.org/10.1007/s11306-014-0746-7
151. R. Kaddurah-Daouk, R. Weinshilboum, Metabolomic signatures for drug response phenotypes: pharmacometabolomics enables precision medicine. Clin. Pharmacol. Ther. **98**(1), 71–75 (2015). https://doi.org/10.1002/cpt.134
152. N.E. Leadbeater, J.R. Schmink, Use of Raman spectroscopy as a tool for in situ monitoring of microwave-promoted reactions. Nat. Protoc. **3**(1), 1–7 (2008). https://doi.org/10.1038/nprot.2007.453
153. N. Gierlinger, T. Keplinger, M. Harrington, Imaging of plant cell walls by confocal Raman microscopy. Nat. Protoc. **7**(9), 1694–1708 (2012). https://doi.org/10.1038/nprot.2012.092
154. Madison Instruments, FTIR Troubleshooting Guide
155. G.H. Major et al., Assessment of the frequency and nature of erroneous X-ray photoelectron spectroscopy analyses in the scientific literature. J. Vac. Sci. Technol. A **38**(6), 061204 (2020). https://doi.org/10.1116/6.0000685
156. G.H. Major et al., Practical guide for curve fitting in X-ray photoelectron spectroscopy. J. Vac. Sci. Technol. A **38**(6), 061203 (2020). https://doi.org/10.1116/6.0000377
157. F.C. Meunier, Pitfalls and benefits of: In situ and operando diffuse reflectance FT-IR spectroscopy (DRIFTS) applied to catalytic reactions. React. Chem. Eng. **1**(2), 134–141 (2016). https://doi.org/10.1039/c5re00018a
158. A. Paredes-Nunez, I. Jbir, D. Bianchi, F.C. Meunier, Spectrum baseline artefacts and correction of gas-phase species signal during diffuse reflectance FT-IR analyses of catalysts at variable temperatures. Appl. Catal. A Gen. **495**, 17–22 (2015). https://doi.org/10.1016/j.apcata.2015.01.042
159. N. Döbelin, Lesson 3 Sample Preparation and Problems (2013)
160. V. Ramaswamy, Problems encountered in powder X-ray diffraction analysis of zeolites and molecular sieves. Stud. Surf. Sci. Catal. **113**, 683–688 (1998). https://doi.org/10.1016/s0167-2991(98)80346-7
161. S. How, X. How, X. General, XRD-System Instruction Manual
162. A.M. Torres, W.S. Price, Common problems and artifacts encountered in solution-state NMR experiments. Concepts Magn. Reson. Part A Bridg. Educ. Res. **45A**(2) (2016). https://doi.org/10.1002/cmr.a.21387
163. P.R. Davies, D.J. Morgan, Practical guide for X-ray photoelectron spectroscopy: applications to the study of catalysts. J. Vac. Sci. Technol. A **38**(3), 033204 (2020). https://doi.org/10.1116/1.5140747

Chapter 5
Characterization Techniques for Thermal Analysis

Hamed Hosseinian, Euth Ortiz Ortega, Ingrid Berenice Aguilar Meza, Andrea Rodríguez Vera, María José Rosales López, and Samira Hosseini

Abbreviations

AC	Activated carbon
Al_2O_3	Aluminum oxide
CNF	Cellulosic nanofibers
DASM	Differential adiabatic scanning microcalorimeter
DSC	Differential scanning calorimetry
DSC–TM	Simultaneous differential scanning calorimetry thermomicroscopy
DTA	Differential thermal analysis
IR	Infrared
IR-heated DSC	Infrared-heated differential scanning calorimetry
MDSC	Modulated differential scanning calorimetry
MTMA	Ethyltrimethylammonium chloride
NAC	Nanocellulose and activated carbon
NMR	Nuclear magnetic resonance
PMMA	Polymethylmethacrylate
PnDSC	Parallel nanodifferential scanning calorimetry
SMDSC	Supercooling modulated differential scanning calorimetry
T_c	Crystallization temperature
T_d	Degradation temperature
T_g	Glass transition temperature
TGA	Thermogravimetric analysis
TG–DSC	Thermogravimetry differential scanning calorimetry
$T_{l\text{-}l}$	Liquid–liquid transitions
T_m	Melting temperature
$T_{s\text{-}s}$	Solid–solid transitions

E. O. Ortega et al., *Material Characterization Techniques and Applications*,
Progress in Optical Science and Photonics 19,
https://doi.org/10.1007/978-981-16-9569-8_5

5.1 Thermogravimetric Analysis for Material Characterization

5.1.1 History of Thermogravimetric Analysis

The process of functioning furnaces at a laboratory scale has been described since the studies of alchemy. In terms of thermogravimetric analysis, a need surged for a controlled process of heating furnaces that can be carefully adjusted to precise temperatures and provide a temperature-time profile to maintain the conditions. Many pre-established tests have evolved into modern physicochemical characterization methods. Thermometers and temperature scales were first established in the eighteenth century by Celsius in 1742 (Fig. 5.1). At this time, the thermal expansion of materials indicated the temperature, while measurements of the change in length of materials as in correlation with temperature, dilatometry, had already been accomplished [1].

Measuring heat quantities was introduced in the nineteenth century after the distinction between temperature and enthalpy had been explained using thermodynamic principles. In 1887, Le Chatelier implemented the first actual thermoanalytical measurements [2]. Le Chatelier placed a thermocouple in a sample of clay and heated the sample in a furnace. The heating curve was reported on a photographic plate with a mirror galvanometer. Consequently, Roberts-Austen in 1899 significantly enhanced the sensitivity and meaningfulness of these measures via the presentation of two differential thermocouples connected in one opposition [3]. This allowed the temperature variation between the sample and an inert reference to be

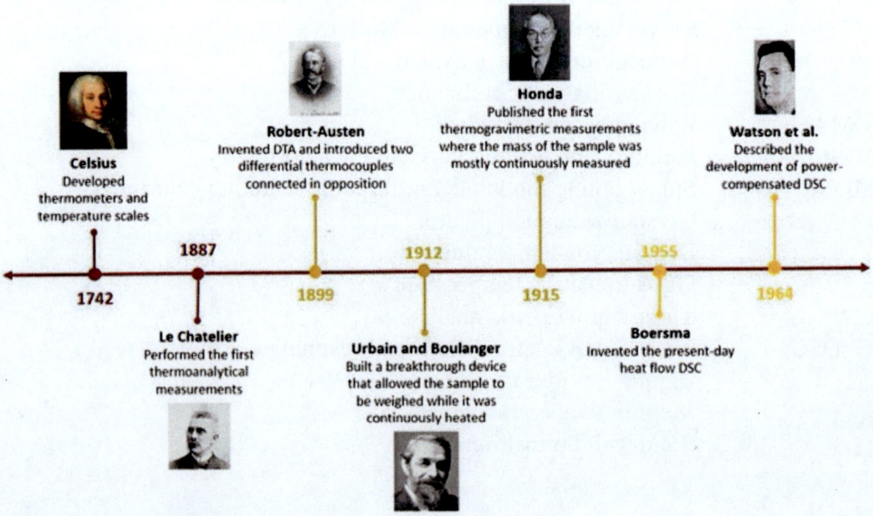

Fig. 5.1 History of the evolution of TGA

estimated, which later granted Roberts-Austen the title of inventor of differential thermal analysis (Fig. 5.1).

In 1912, Urbain and Boulanger developed an invention that encompassed a regular balance modified for null point electromagnetic compensation [4]. The sample was hanging from a balance arm into an electrically heated furnace in a gas-tight environment. The design allowed evolved gases to be inserted as an input for the supply of absorbent materials. In this arrangement, the sample was continuously heated and could, therefore, be weighed. Honda in 1915 published the first thermogravimetric measurements of its own kind where the mass of the sample was nearly continuously measured [5]. Prior to Honda's method, the only way to achieve this was by measuring the mass difference after a thermal experiment by back-weighing. In 1955, Boersma invented the present-day heat-flow differential scanning calorimetry with the idea of placing the thermal resistance outside of the crucible [6]. A decade later, the development of power-compensated differential scanning calorimetry (DSC) was mentioned for the first time in a publication by Watson et al. [7]. Dynamic mechanical measurements with a constant selectable frequency have been recently discussed. Before this, similar measurements were conducted using a torsion pendulum at resonance frequencies. Robert Mackenzie investigated the history of thermal analysis to a great extent and contributed to the field the following year [8].

In recent years, the accessibility of useful computer hardware and software greatly benefited thermal analysis and heavily influenced the improvement of thermoanalytical analyses. Before 1980, measurement curves were manually designated and evaluated with linear pen recorders [9]. Selecting the wrong measurement range would lead to repeated computation using a more suitable range which was time-consuming and tedious. Nowadays, due to new technologies, the curve can be saved, then displayed optimally for interpretation, and, if necessary, automatically evaluated [9]. Furthermore, modern TGA defines the amount and the frequency of the weight variation of the samples against temperature and time in a restricted environment (e.g., the purging of nitrogen gas). TGA can also be used principally to examine the strength of the material at a given temperature (thermal stability), the oxygen absorption rate on the material (oxidative stabilities), as well as the compositional properties (e.g., fillers, polymer resin, solvents) of the specimens [4].

5.1.2 Mechanism of Operation of Thermogravimetric Analysis

Thermogravimetric analysis (TGA) is an analytical characterization method where the mass of a substance is marked as a function of the temperature and time under a controlled environment. As an appropriate upper limit for polymer applications, the temperature range for commercial TGA can typically rise to 1000 °C or more. The flowing of purge gas can create an atmosphere that is inert (nitrogen, argon, or helium), oxidizing (air or oxygen), or reducing (forming gas) [10]. The moisture content of the purge gas varies between dry and saturated. Measured substances usually demonstrate mass loss, while mass gain may be recognized as well, before degradation at slow heating rates in an oxidizing atmosphere. Mass loss can be considered as volatile elements including absorbed moisture, residual solvents, or low-molecular-mass additives or oligomers [10]. These normally evaporate within ambient temperature and 300 °C. Other components involved are reaction products, such as water and formaldehyde from the cure of phenolic and amino resins, which surge amid 100 and 250 °C. Moreover, the degradation of volatile products occurs at chain scissions between 200 and 800 °C. TGA can be used to characterize these mass loss procedures in order to understand the composition, degree of cure, and thermal stability of the analyzed compounds [10].

There are three different designs for thermobalances that are represented schematically in Fig. 5.2: top-loading, hang down, and horizontal arrangement. According to these types of balances, the sample position in the furnace continues the same when the mass changes [11]. One of the main disadvantages of the horizontal arrangement is that samples that move horizontally during heating cause an apparent change in mass. This can be solved by using a parallel-guided system. Main measurements should be included between the balance and the furnace to preserve the balance against the influences of heat radiation and the ingress of corrosive decomposition products. In most cases, the balance area is purged with a protective gas. Some thermobalances include an "external" furnace whereby the furnace is not in contact with the atmosphere which can be beneficial if measurements are made in pure hydrogen atmospheres [11]. According to the resolution provided, balances are categorized as semimicro- (10 μg), micro- (1 μg), or ultra-micro- (0.1 μg) balances. Another vital factor to take into account is the maximum capacity of the balance, especially when measuring inhomogeneous materials since, for instance, a few milligrams would not be representative when a larger sample mass is desired [11].

Thermogravimetric instruments are made up of several elements to accommodate flexibility for specific analytical data. These include a balance for measuring

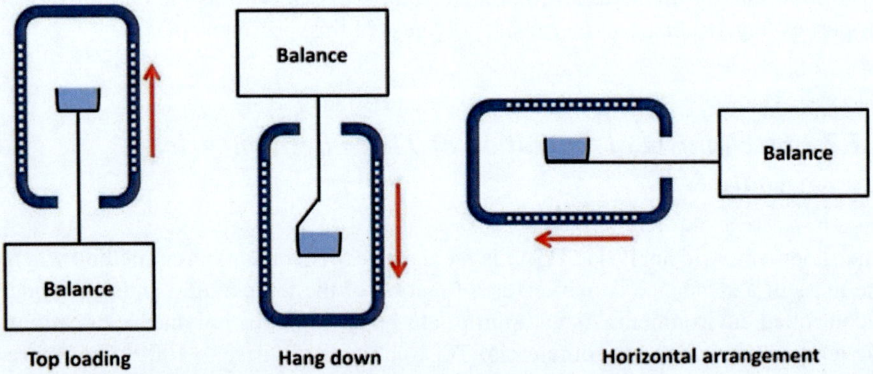

Fig. 5.2 Representation of three thermobalance structure schematic designs including top-loading, hang down, and horizontal arrangements

the weight of the sample, a furnace for providing linear heating, a temperature programmer for temperature measurement and control, and a recorder for observing the unit for weight and temperature changes [12]. As the sample weight varies, a balance beam favors deviating from its normal position. There exist two varieties of balances that may be employed: null or deflection balance. The former is used when a sensor is used to detect the deviation of the balance beam from the position of the null point. Subsequently, a reset force is utilized to balance its null point. In some systems, the conducted beam deflection is measured by transforming it into a recognizable weight by means of different techniques such as photographic record, electrical signals, or utilizing an electrochemical device. There remain multiple determinants that can influence the sensitivity of the device and as a result change the shape of the thermogravimetric curves. By examining these factors, the tools must be calibrated prior to carrying out the new experiments for improved results [13–16]. Some of these key elements are included below:

- Known sample: The chemical fingerprint which includes the source, purity, pretreatments, and particle size needs to be clearly defined.
- Sample holder: The size of the holder must be adequate to contain the complete sample to bypass loss of signal as a result of spurting and creeping. To collect precise data or signal, the heat transfer connecting the furnace and sample holder should be active, which is why a metal holder is a more desirable choice.
- Crucible: A suggested class of crucible is made of platinum, which is tubular and has enough height. Other varieties of approved crucible are gauze crucible, polyplate crucible, and labyrinth crucible.
- Atmosphere: Selection of various atmospheric environments in correlation with the type of gas purged in can impact the outcomes as it can be inert or oxidative to the specimen. The chemical differences as a result of the oxidative conditions can be measured by the analyzer, and it can offer valuable data regarding sample characterization.
- Heating rate: The decomposition rate is a function of the heating rate. At greater heating rates, a greater decomposition temperature would be observed as well. Regularly, a suggested heating rate is 3.5 °C/min for reproducible TGA.

5.1.3 Advantages and Disadvantages of Thermogravimetric Analysis

TGA is a method in which the mass of a specimen is measured as a function of temperature and time, whereas a thermogravimetric analyzer is a tool applied to perform such measures [17]. TGA results are plotted with mass against temperature and time, displaying a TGA curve [17, 18]. Nowadays, there exist three types of TGA: dynamic, isothermal, and quasistatic. These are distinguished by the heating rate that is applied to the sample. In a dynamic TGA, a constant linear rise of the temperature occurs with respect to time. For an isothermal TGA, the sample weight

Table 5.1 Advantages and disadvantages of TGA

Advantages	Disadvantages
Minimal sample preparation requirements [18, 19]	The sample volume is limited [18]
High exactitude of atmospheric conditions and temperature controlling system [18]	The interpretation of the results is rather complex [17, 18]
Easy control of the sample [18, 20]	The curve can be easily affected by sample characteristics or instrumental factors [18]
The heating rate can be maintained with good resolution [18, 20]	Operation requires high control over temperature
Low construction requirements [18, 19]	

is variable, while the temperature is kept constant. Finally, for a quasistatic TGA, the sample is heated to a constant weight over a series of increasing temperatures [18, 19].

A thermogravimetric analyzer has the advantage of easy control of the sample, operation, weight loss, and temperature. These are precisely recorded and controlled which allows uniformity of the procedure. Additionally, it is a time-saving technique due to its minimum procedure in sample preparation and easy implementation [18–20]. On the other hand, thermogravimetric experiments are dependent on procedural details such as furnace heating rate, the atmosphere inside the furnace, sample size, or weight [17–19]. Also, the sample volume is limited by the capacities of the analyzer. Table 5.1 reviews the advantages and disadvantages of the TGA technique.

5.1.4 Applications of Thermogravimetric Analysis

One key part of biosensor fabrication is the development of smart bioreceptors for enzyme immobilization which can be achieved by surface modification of the material [21]. TGA can characterize materials through the assessment of the impact of a corrosive and reactive atmosphere on polymers, as well as its decomposition kinetics and oxidative stabilities. It can also be utilized to analyze the thermal stability of polymers [18, 19]. Thus, it is particularly beneficial for the investigation of polymeric materials when injection, molding, or coating is involved in their application as electrical and electronic segments, pharmaceuticals, among others [9]. When using TGA, it is important to choose materials for specific end-use applications, anticipate product performance, and enhance product quality [4, 22].

A method to modify materials is adding fillers that enhance the properties of the products. The TGA technique can determine the quantity of a filler in a polymer [4]. An example is a study carried out by Sobhan et al. in which nanocellulose and activated carbon (NAC) films were developed with different cellulosic nanofibers (CNF) and activated carbon (AC) contents [23]. The thermal properties were analyzed, and Fig. 5.3 displays the thermal degradation curves of the films. Films containing 15%,

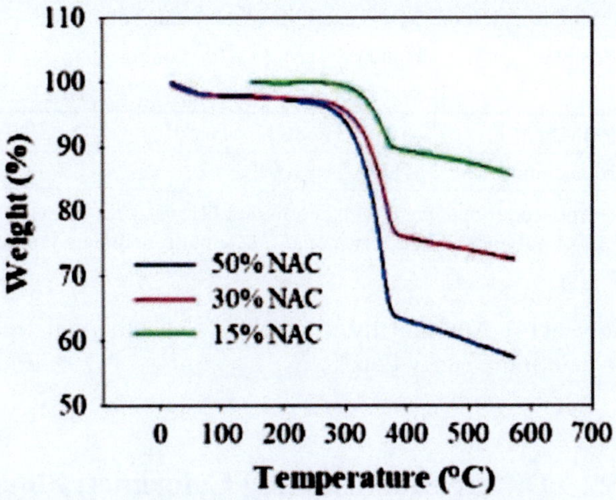

Fig. 5.3 TGA curves of NAC films. Reproduced (or adapted) with permission [23], ©2019, Elsevier

30%, and 50% of CNF with corresponding 85%, 70%, and 50% content of AC were analyzed. TGA was performed under a nitrogen atmosphere from 20° to 400 °C with a heating rate of 10 °C/min. The curve shows the series of thermal degradation, divided into three stages. In the first stage, dehydration occurs between 100 and 270 °C where water molecules were evaporated from the film. The following stage occurs between 270 and 390 °C which is the transition region, where a weight change is observed by the loss of 60% of the component due to the structural degradation of NAC film composites. The third step was recorded between 390 and 590 °C. The change that occurred in this stage is due to the carbonization of residual organic matter. Therefore, it was concluded that the content of cellulosic nanofibers significantly affected the physicochemical properties. It is important to note that all films were thermally stable until 270 °C [23].

The thermal stability of a substance is described as the ability of a material to maintain constant characteristics when exposed to heat, which is relevant for various industrial sectors to determine the temperature range in which a material can be used [19]. This is exemplified in a study by Emre et al. based on analyzing the development of polymethylmethacrylate (PMMA)-clay nanocomposite as a bioreceptor for biomolecule deposition [21]. TGA was performed for a temperature range of 25–800 °C at a rate of 10 °C/min under a nitrogen atmosphere [21]. The TGA data obtained for the composite is presented in Table 5.2. It is shown that the PMMA-clay nanocomposite had a higher decomposition temperature than that of the pure polymer, thus the inclusion of additives into the polymer matrix resulted in a more thermally stable material [21].

In summary, the TGA technique facilitates the evaluation of the compositional study of a polymer mixture. It is utilized to estimate the additive content and measure low levels of volatile matter present in polymer matrix, including water or residual

Table 5.2 TGA of PMMA and PMMA-clay nanocomposite

Sample	Modifier	$M_w \times 10^3$ (g/mol)	W		
			At 270 °C	At 350 °C	At 600 °C
PMMA	–	126	76	40	0
Nanocomposite	MTMA	109	85	47	4

Reproduced (or adapted) with permission [21], ©2015, Elsevier.
PMMA polymethylmethacrylate; *MTMA* ethyltrimethylammonium chloride.

solvent [4]. Additionally, it serves to predict thermal decomposition mechanisms and to determine purity [24].

5.2 Differential Scanning Calorimetry for Material Characterization

5.2.1 History of Differential Scanning Calorimetry

Differential scanning calorimetry (DSC) is a method that correlates the molar heat capacity of samples with temperature. The origins of this technique date back to the nineteenth century, when heat quantity measurement became possible. DSC has been optimized since then with new and demanding applications involving technological advances in hardware and software [25].

In 1887, Le Chatelier, a French chemist, performed the first thermoanalytical measurements by inserting a thermocouple into a sample of clay to be further heated in a furnace (Fig. 5.4). The thermocouple output was fed to a mirror galvanometer to record the heating curve on a photographic plate [25, 26]. Subsequently, in 1899, William Robert-Austen enhanced the sensitivity of the measurements by developing the first classical differential thermal analysis through the connection of two differential thermocouples in opposition [25, 26]. The present-day flow DSC was invented by Boersma in 1955 by introducing a design with the thermocouple outside of the crucible. Since the thermocouple does not touch the specimen directly, the reproducibility of the differential thermal analysis (DTA) curve is greatly improved [25, 26]. The differential adiabatic scanning microcalorimeter (DASM) was introduced in 1963. It suppressed heat losses and provided consecutive recordings of heat capacity as a function of a set heating rate [27]. Similarly, in 1964, O'Neil and Watson invented the first differential scanning calorimeter to measure the heat flow of solid materials [25]. For its operation, specific shields, vacuum, or water jackets were utilized by DSCs to preserve temperature feedback loops in the environment [17, 18].

The capabilities of thermal analysis of materials have dramatically improved due to the availability of powerful computer hardware and software [25]. Moreover, several DSC models became feasible based on their application [27]. One example is the hyperDSC which allows high scanning rates between 400° and 500 °C. Another

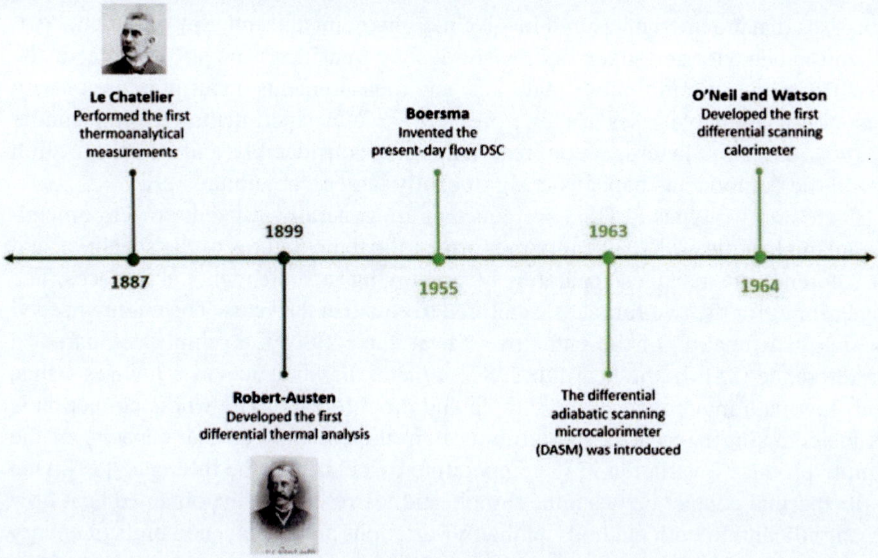

Fig. 5.4 Evolution of DSC, from the first thermoanalytical measurements in 1887 to the development of DSC in 1964

program functions as a modulated DSC which can separate reversible and non-reversible heat flows. Moreover, nanoDSC operates with very small quantities of sample per trial (130μL or 100 μg) while sustaining the precision of larger volume calorimeters [27].

5.2.2 *Mechanism of Operation of Differential Scanning Calorimetry*

As a thermal analysis method, DSC is applied to define the heat flows associated to transitions in materials in correlation with time and temperature. These measurements present qualitative and quantitative data about chemical and physical changes regarding heat capacity variations or endothermic and exothermic processes [28]. Modern DSC methods have a wide range of temperature capabilities with values from -60° to more than 1500 °C. A change in the sample temperature may occur due to different phenomena, including loss of solvent, phase transitions, melting point, crystallization temperature, and degradation temperature of the plastic sample [28].

The technical features of DSC allow users to identify endothermic and exothermic effects, measure transition and reaction enthalpies (peak areas), and determine specific heat capacity [9]. In terms of quality, physical transitions and chemical reactions can also be obtained through this powerful analysis. Certain characteristics and processes that are often measured involve melting point and enthalpy of fusion, crystallization behavior and supercooling, solid–solid transitions and polymorphism, the glass transitions of amorphous materials, and measurements under pressure or with poisonous/flammable gases in a high-pressure DSC. In experiments performed under pressure, the rate of heterogeneous reactions raises considerably, and the vaporization of volatile components happens at significantly larger temperatures [9].

There are two types of DSC systems that are generally utilized: power compensation and heat flux DSCs [28]. In the former, the temperatures of the specimen and the reference are managed separately by employing separate identical furnaces. The power input for the two furnaces is tailored to equalize the heats. The energy needed for this is a function of the enthalpy or heat capacity of the sample compared to the reference [28]. In the heat flux DSC, a metal disk that acts as a low-resistance heat-flow path interconnects the sample and the reference. The whole collection is set inside a single furnace. The differences in the enthalpy or heat capacity of the sample produce a variation in its temperature associated to the reference [28]. Due to the thermal contact between the sample and the reference, the obtained heat flow is insignificant. In both methods, enthalpy variations are displayed in units of energy per unit of mass. In the following sections, both categories are discussed, and the instrumentation for each type is further explained.

5.2.2.1 Power Compensation Differential Scanning Calorimetry

There exist two small identical furnaces at the core of the power compensation DSC apparatus: one for the reference (usually an empty pan) and one for the sample (Fig. 5.5) [29]. These are both heated at a pre-programmed heating (or cooling) rate maintained by power compensations utilized in either furnace as needed. In the output DSC trace, the difference in energy flowing into the sample furnace is compared to the inert reference and considered as a function of temperature or time [29]. The underlying equation of DSC measures the flow of energy directly in mW or J/s as is presented below (Eq. 5.1)

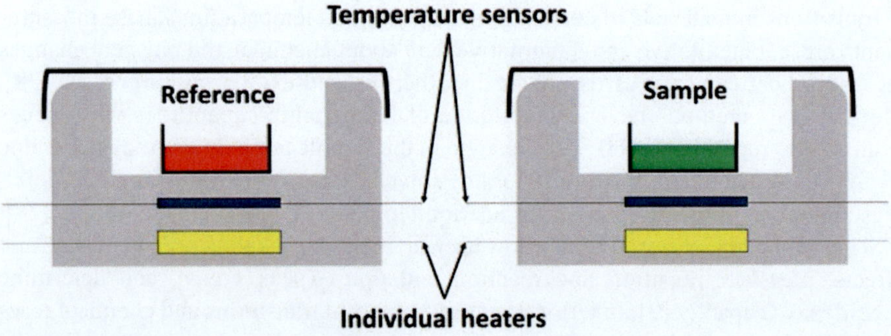

Fig. 5.5 Schematic illustration of a power-compensated DSC

$$\text{DSC signal(W/g)} = \text{Heat Capacity(J/(Kg))} \times \text{Scanning Rate(K/s)}$$
$$\mathrm{d}H/\mathrm{d}t = \mathrm{d}H/\mathrm{d}T \times \mathrm{d}T/\mathrm{d}t \tag{5.1}$$

Accordingly, the raw heat-flow signal could be regarded as a form of heat capacity, when in essence, it displays the changes occurring in heat capacity. The absolute value is calculated when the method in question takes into account the role of the empty pans and reference jointly with the scan rate [29]. The small furnaces of this method can be heated or cooled from very low rates to very high rates and are excellent for a wide range of different procedures, especially fast scan DSC. Power-compensated DSC further allows true isothermal process, considering that, under steady temperature conditions, both the specimen and furnace are held isothermally. The temperature series is selected from that of liquid nitrogen (about −190 °C) to about 730 °C [29].

5.2.2.2 Heat Flux Differential Scanning Calorimetry

Another class of DSC is the heat flux DSC which has an individual furnace design with a temperature sensor or added sensors for each of the specimen and reference pans within the same furnace (Fig. 5.6) [29]. After positioning the sample and reference pans, the furnace is heated at the pre-programmed heating (or cooling) rate. A temperature variation is initiated between sample and reference when some transitions are encountered in the sample [29]. As the heating continues, the variance in

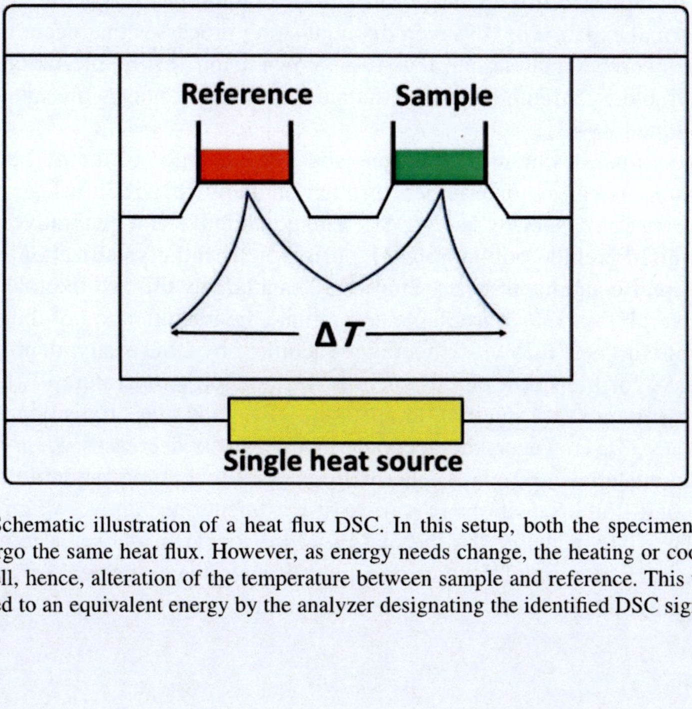

Fig. 5.6 Schematic illustration of a heat flux DSC. In this setup, both the specimen and reference undergo the same heat flux. However, as energy needs change, the heating or cooling effect vary as well, hence, alteration of the temperature between sample and reference. This variation is transformed to an equivalent energy by the analyzer designating the identified DSC signal

temperature lowers, while the system approaches equilibrium with time. These are the changes in temperature or ΔT signal, which is the basic parameter measured. Modern analyzers are carefully calibrated in order to transform the ΔT signal into a heat-flow equivalent and then represent it as a function of temperature or time [29]. Considering the melting point as a reason for the difference in temperature makes the system practical to use. Melting a single crystal produces a mixture of solid and liquid which stabilizes at the melting point until the melting is completed. This causes the sample temperature to fall behind that of the reference. Standard heat flux DSC analyzers operate between the liquid nitrogen temperature to a maximum ~700 °C, similar to the power-compensated DSC [29]. However, advanced high-temperature DTA analyzers commonly offer a calibrated DTA (heat flow) signal, providing a measurement obtained from the heat flux to considerably higher temperatures [29].

5.2.3 Advantages and Disadvantages of Differential Scanning Calorimetry

Enthalpy variations modify the physical and chemical characteristics of a sample. DSC can define qualitatively and quantitatively how temperature affects these properties in the range of −90° to 550 °C. For instance, DSC can analyze the phase transitions in polymeric materials, allowing the understanding of how the material structure is related to their properties. DSC is a quick and sensitive thermal analysis method for identifying chemical reactions and thermodynamic properties, including isothermal compressibility, glass transition temperature, heat capacity, and coefficient of thermal expansion. However, distinguishing processes that occur at similar temperatures can be challenging, thus it is known as one of the limitations of this technique. Table 5.3 highlights the advantages and disadvantages presented by the DSC technique [30–33].

A great number of chemical reactions and a broad range of thermal properties in solid materials can be determined through enthalpy change in a fast, inexpensive, and versatile manner by using DSC. The quantitative and qualitative data that can be acquired include polymorphism, purity, melt, and crystallization enthalpy, as well as precise and latent heats. Since DSC is a largely utilized thermal analysis method, several libraries are available to facilitate the identification of the samples. The thermogram (heat flow vs. temperature) acquired by DSC easily identifies well-defined peaks for transition phases such as degradation temperature (T_d), crystallization temperature (T_c), melting temperature (T_m), and glass transition temperature (T_g) [30–33]. However, the accuracy of the results decreases when trying to analyze nanomaterials or complex thermal processes (e.g., liquid–liquid transitions ($T_{l\text{-}l}$) and solid–solid transitions ($T_{s\text{-}s}$)). As it was mentioned above, when samples have similar components or the thermal processes occur in the same temperature ranges, the interpretation of the data becomes difficult due to the peaks overlapping in the thermogram. In some cases, the temperature range at which DSC operates

Table 5.3 Advantages and disadvantages of DSC

Advantages	Disadvantages
Versatility [30–32]	The accuracy decreases for complex thermal processes [30, 32]
Reduced impact of solvents and other external elements on thermal measurement [31]	
A broad range of temperature (−90° to 550 °C) [30]	The resulting thermograms might be sensitive to sample preparation protocols [33]
Ease of qualitative and quantitative data acquisition [30, 31, 33]	
Simple sample preparation [31]	
Relatively quick [31, 33]	Phase transitions overlap [30, 31]
Low cost [33]	
The versatility of the modules (MDSC, SMDSC, PNDSC, IR-heated DSC, DSC–TM, TG–DSC, etc.) [30–32]	

DSC–TM simultaneous differential scanning calorimetry thermomicroscopy; *IR-heated DSC* infrared-heated differential scanning calorimetry; *MDSC* modulated differential scanning calorimetry; *PnDSC* parallel nanodifferential scanning calorimetry; *SMDSC* supercooling modulated differential scanning calorimetry; *TG–DSC* thermogravimetry differential scanning calorimetry.

does not allow the identification of the last transition phase in polymeric materials. The resulting thermograms are sensitive to sample preparation protocols since a low amount of the sample can produce low-resolution peaks. Some of the limitations of DSC can be avoided using extended modules, for example, modulated differential scanning calorimetry (MDSC) and supercooling modulated differential scanning calorimetry (SMDSC) [30–33].

5.2.4 Applications of Differential Scanning Calorimetry

Identifying compounds or analyzing chemical reactions due to a temperature change is possible in a simple way using a thermodynamic technique such as DSC. The basic operation principle consists of recording the heat capacity, while a temperature scan occurs between the sample and a reference. This thermal technique allows the user to acquire data of conformational changes and phase transitions in the specimen as a dependence of the temperature; for instance, thermal transition midpoints (T_m) describe the macromolecule stability. In the biopharmaceutical industry, DSC is a precise means for analyzing the thermotropic properties of proteins such as thermal stability, domain folding integrity, and structural integrity [34]. Since the extracts and biological macromolecules (proteins and lipids) show very reproducible phase transitions, the DSC is a sensitive and versatile tool to determine purity, stability under various environmental requirements, and the type of binding interaction for biochemical and pharmaceutical applications [27]. The oxidative stability and autoxidation process in food components such as fats, lipids, and edible oils are rather simple to follow by DSC due to kinetic parameters being determined faster and consuming less amount of solvent than other methods as infrared (IR) or nuclear magnetic resonance (NMR) [35].

Determining the energetic and physical properties, such as glass transition temperature (T_g), recrystallization exotherm temperature (T_c), as well as melting endotherm temperature (T_m), is vital in drug development. In this sense, DSC obtains this data accurately, easily, and vastly in comparison to other techniques [36]. Biochemistry and biotechnology industries have a high interest in understanding how the biomolecules interact with one another, as well as how their structure impacts their thermal stability. DSC can analyze the conformational transitions of biomolecules through parameters depending on temperature, for instance, the unfolded and folded state in DNA. Furthermore, other thermal transitions in biological systems where the DSC is an accurate and useful technique are biomolecular recognition, protein–ligand, and drug-DNA interactions [37]. Since the DSC offers a precise qualitative and quantitative study of transition phases in materials, the relation between their structure and properties can be comprehended. Different applications utilize this thermal technology such as food and drug industries [30], polymer science [31], drug delivery nanosystems [33], biochemical [38], oils and fat research [39], skin collagen hydrolysate for the degraded leather [40], structure of proteins [41], and multilamellar vesicles [42].

Invasive grass, *Pennisetum purpureum*, could be utilized as biomass to produce bioenergy, significantly reducing greenhouse emissions [43]. Its composition presents high quantity of volatile matter and low humidity content; thus, gasification and pyrolysis are suitable methods to transform invasive biomass into clean energy. Reza et al. applied the DSC technique to analyze the enthalpy change pattern of *Pennisetum purpureum* [43]. From the thermogram, the reaction route could be obtained from the invasive biomass. The DSC curve (Fig. 5.7) presented three processes: first, the evaporation of water molecules due to an endothermic reaction indicated by a peak at 89 °C; second, an exothermic reaction of the degradation of cellulose and hemicellulose appeared at 315 °C; and finally, at 564 °C, an endothermic reaction occurred again indicating the decomposition of lignin [43].

In the pharmaceutical application, the co-crystal, a novel solid form of active pharmaceutical ingredients (APIs), is of high interest due to its properties that can greatly impact the dissolution of medicines in water [44]. In order to do so, the preparation of co-crystals should be based on benzodiazepines. DSC is a quick screening tool used to recognize the formation of co-crystals from the physical blend of co-formers and drugs. Saganowska et al. studied co-crystals based on benzodiazepines using non-toxic and soluble co-formers [44]. The oxazepam mixed with fumaric acid showed temperature peaks considerably different from the melting temperatures presented by individual components (Fig. 5.8). This result suggested that a new structure could be formed due to the physical interaction between oxazepam and fumaric acid [44]. For the possible new co-crystal, an endothermic process appeared

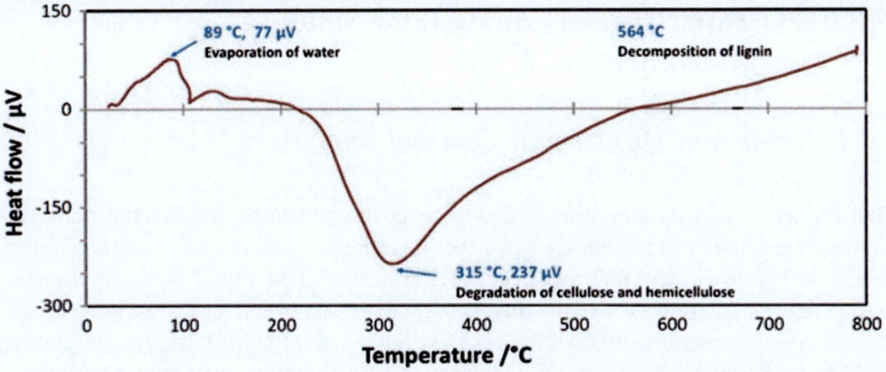

Fig. 5.7 DSC thermogram for *Pennisetum purpureum*. Temperature ranges from 25 to 780 °C at a heating rate of 10 °C min^{-1} Reproduced (or adapted) with permission [43], © 2020, Elsevier

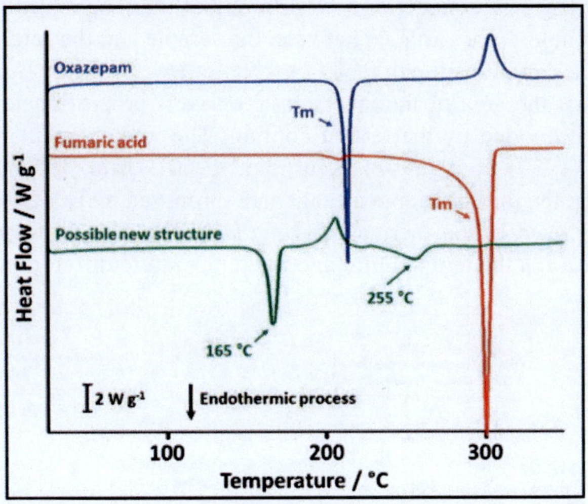

Fig. 5.8 DSC thermogram for oxazepam, fumaric acid, and their physical mixture. Temperature ranges between 25° and 300 °C at a heating speed of 10 °C min^{-1}. Reproduced (or adapted) with permission [44], © 2018, Springer

at 165 °C below the melting temperature (Tm) of oxazepam. Another endothermic process was identified close to 255 °C, between the Tm of fumaric acid and oxazepam. From the DSC results, the authors assumed that the new endothermic peaks did not overlap with the endothermic process of the separate components, suggesting possible co-crystallization [44].

5.3 Differential Thermal Analysis for Material Characterization

5.3.1 *History of Differential Thermal Analysis*

Thermometers and temperature scales were developed in the eighteenth century, but heat quantities did not become possible to measure until the nineteenth century (Fig. 5.9) [5]. It was in 1887 that Le Chatelier accomplished the first thermal analysis experiments. It was also in this year that the microbalance was introduced [2]. The development of this device was a stimulus to follow the weight variations that happened on heating materials. Nernst and Riesenfeld were notable scientists among those who attempted this line of work [3]. They announced a design of a quartz beam balance implemented with an electric oven. In the late nineteenth century, Robert-Austen developed the first classical DTA by improving the sensitivity of the measurements by connecting two differential thermocouples in opposition, measuring the temperature variance between the sample and the reference material [5, 15, 18]. Further modification of the DTA technique was done by Burgess in 1909, who assumed that the area of the inverse-rate curve is proportionate to the extent of produced heat divided by the rate of cooling. The improvement was accompanied by Norton (1939), Kerr (1948), Kauffman (1950), Grim (1951), and Foldvári (1958). However, the thermocouple arrangement proposed by Robert-Austen is still extensively employed in the thermal analysis [17]. Nowadays, modern apparatus automatically records the temperature and the temperature difference by using two

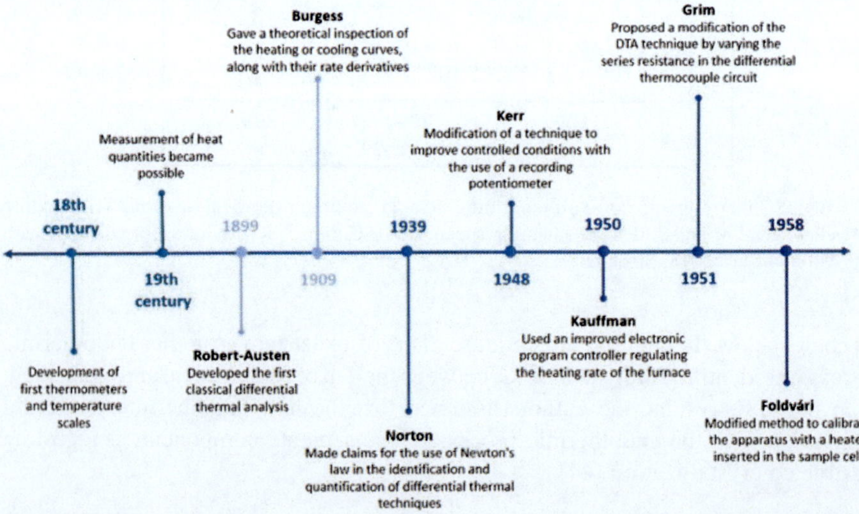

Fig. 5.9 Timeline history of DTA, including the advancements of the device from its invention in 1899 by Robert-Austen until nowadays where fully automated apparatuses are available

galvanometers and a photographic plate with prisms. Nevertheless, fully automated devices only became commercially available at the end of the fifties.

5.3.2 *Mechanism of Operation of Differential Thermal Analysis*

DTA measures the difference among the observed temperatures of the specimen and a thermally inert material known as the reference. This difference is then recorded while the sample and the reference are heated by the equivalent heat source as well as undergoing the same temperature raise profile. Subsequently, a plot (curve or thermogram) is obtained by correlating the differential temperature against time or temperature [18]. Since the sample and the reference present different heat capacities, there is a constant temperature difference (ΔT). However, when the specimen experiences the endothermic or exothermic changes, ΔT displays a difference due to the absorption or evolution of heat. This basic principle of measuring temperature change is correlated with physical or chemical alterations of the specimen throughout the controlled heating procedure [18].

5.3.2.1 Instrumentation

DTA is composed of eight fundamental components (Fig. 5.10) [18]. Furnace assembly serves as a temperature programmer; it varies depending on the sample material and the rate of heating. There are two holders: one for the specimen and for the reference. The furnace and the holders are enclosed by an insulating material. A

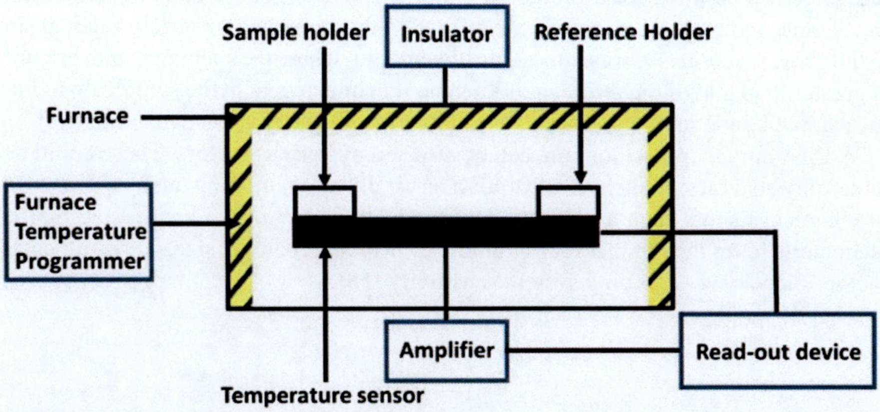

Fig. 5.10 Graphic representation of fundamental components of DTA. The sample and reference are held within an insulator, and heat signals are represented as a thermogram

temperature detector connects both holders and measures the difference in temperature that is growing at a steady linear rate. Also, there is an electronic temperature regulator to ensure a constant rate of heating, as well as thermocouples for temperature regulation which are usually inserted into holders. An amplifier has the function of converting the heat signal into an electrical signal. The results are displayed in the frame of a thermogram by a read-out device [18].

5.3.2.2 Operation

The sample is first loaded into a container and placed onto the sample pan. An equal quantity of inert material is placed in another container to place onto the reference pan. The dimensions of the pans should be nearly identical in order to heat the pans at a uniform rate. Additionally, the specimen and reference weights must be balanced, thermally adjusted, and arranged symmetrically with the furnace. Two sets of thermocouples are usually employed, where one is in contact with the sample and the other one with the reference. Next, the thermocouples are attached to an amplifier which will send the differential thermocouple outcome to the computer device to display in the form of a DTA curve or thermogram. The signal is only produced if a temperature deviation is observed; otherwise, even when the actual temperatures of the sample and reference are hiking, no signal will be generated [18].

5.3.2.3 Thermogram

A thermogram is recorded in the form of a DTA curve obtained, while the sample is analyzed. It is a plot between differential temperature (ΔT) and temperature of reference (T). A downward plot is an endothermic DTA curve demonstrating that the sample temperature is smaller than that of the reference material, whereas an exothermic reaction is represented as an upward plot where the specimen temperature is greater than that of the reference. When no reaction occurs in the sample material, the sample temperature continues the same as that of the reference material [18].

A DTA curve, or thermogram, can be affected by various factors. These could be either physical factors such as adsorption, crystallization, melting, and vaporization or chemical factors such as oxidation and reduction. It can also be affected by the characteristics of the sample, such as amount, shrinkage, particle size, or instrumental factors such as the recording system sensitivity [18].

5.3.3 Advantages and Disadvantages of Differential Thermal Analysis

Some thermophysical properties that can be studied from a sample DTA are recrystallization, solidification, evaporation, and melting temperatures. If the degree of contamination between the sample holder and the crucible is unspecified, DTA is recommended over DSC since it utilizes fewer sensitive materials (e.g., aluminum oxide (Al_2O_3)) as sample holders. DTA is a simple way to determine material transformations due to the released or absorbed heat in the sample, for instance, dehydration. The enthalpy change can be determined through the mass difference baseline method. Moreover, DTA is considered a cheap technique since most of its components are made of ceramic; therefore, it is possible to study reactive and volatile samples. Also, the operating temperature range used in the DTA is from −150 to 2400 °C [45]. DTA is a versatile technique; thus, it can simultaneously determine thermal processes, heat capacity, reaction kinetics, and heat transitions supported by complementary technologies such as TGA [46]. Another advantage is that low amounts of samples (app. tenths of a gram) are required to perform a DTA analysis. This benefit allows the temperature to be homogeneous in the whole sample, avoiding a temperature difference along the sample that may affect results [47]. In summary, this thermal technique is simple, easy to handle, and versatile.

The most important disadvantage of this technique is low sensitivity. On some occasions, the absorbed or released heat during the measurements is not enough (too low) to be recognized; thus, the peaks associated with these thermal processes are not appreciated or are missed in the DTA curve. When high loads of samples are employed, the responding time increases considerably; hence, the interpretation of the curve becomes more challenging. Similarly, an excess of sample evaporation can affect the interpretation of the results (incorrect temperatures) since the sample mass reduces and the chemical composition can shift [45]. Moreover, DTA cannot analyze metals due to their high specific surface tension promoting oxidation processes in the sample during the study. Identifying the temperatures of phase transitions experimentally is not trivial; therefore, in some cases, computational software, such as JMatPro, is required. However, this software employs simplifications that cause the calculated results to vary from the real values [47]. Table 5.4 reviews the advantages and disadvantages presented by the DTA technique.

5.3.4 Applications of Differential Thermal Analysis

The most straightforward method of thermal analysis technique is DTA where the materials undergo changes during heating and can be observed by thermodynamic processes (endothermic and exothermic). It is considered a direct approach to determine the activation energy in chemical reactions. In thermograms, the reaction kinetics can be recorded since the characteristics of the DTA peaks (e.g., slope,

Table 5.4 Advantages and disadvantages of DTA

Advantages	Disadvantages
Provides qualitative and quantitative analyses [45]	Limited sensitivity [45]
Ease of use [45, 46]	
Allows the creation of high pressure or vacuum conditions [45]	Complicated determination of phase transformation temperatures [47]
Small sample size (tenths of a gram order) [47]	
Enables mass loss of samples during analysis [45]	The longer acquisition time for heavy loads [45]
Possibility of analyzing reactive and volatile systems [45]	
Simultaneous measurements [45, 46]	Difficulty in the interpretation of heavy loads [45]
Adaptability [46]	

height, area) vary with thermal processes. Different rates of heating are used to obtain the DTA curves. Additionally, it is possible to obtain quantitative measurements with the proper operating conditions [46, 48]. Overall, DTA is an easy-handle, low-cost, and rapid thermal technique and can be employed in the determination of moisture content in food [49], composition analysis of carbon electrodes [50], characterization of solid divalent transition metal compounds [51], analysis of water and fuel absorption [52], and degradation of different starches [53], among others.

Clay minerals are widely used in the design and construction of buildings; therefore, identifying their composition is vital. A low-cost and rapid method to determine the composition of clays is through DTA. This technique exhibits the effect of endothermic and exothermic reactions and does not provide the composition with certainty. However, the data acquired offers basic information about the behavior of the minerals. Arab et al. used kaolinite (KAO) and bentonite (BEN) at different proportions to evaluate the feasibility of the DTA technique as an identification tool for clay minerals [54]. The DTA curves identified the main endothermic and exothermic peaks that correspond to KAO and BEN (Fig. 5.11). In the case of the 100% BEN, three endothermic peaks appeared at 175 °C, 580 °C, and 820 °C. These were attributed to the lack of water, the lack of structural hydroxyl, and the destruction of crystalline structure, respectively (Fig. 5.11a). In the case of the 50% KAO and 100% BEN, the typical endothermic peaks of each mineral were diminishing (Fig. 5.11b). Lastly, two endothermic processes were identified in 100% KAO. The loss of adsorbed water appeared at 130 °C, followed by hydroxylation at 575 °C. The metakaolinite is formed at 990 °C through an exothermic process due to the nucleation of the mullite (Fig. 5.11c). The authors concluded that the DTA coupled with TGA was an excellent and inexpensive combination for the indirect detection of clay minerals such as KAO and BEN [54].

In the food industry, a cereal sample that presents amorphous states after undergoing starch gelatinization can acquire more ordered states in its structure. This phenomenon is known as retrogradation and alters the properties of the cereal (e.g., the solubility decreases). The data obtained from the DTA provides information

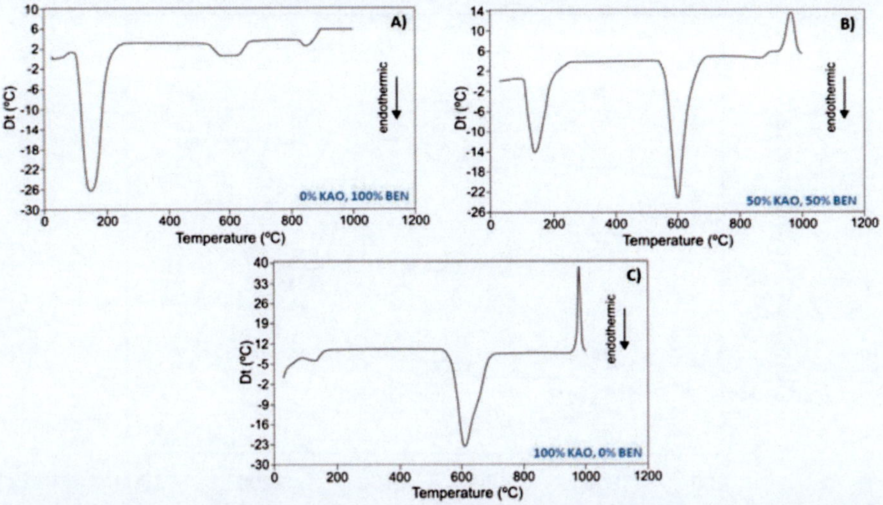

Fig. 5.11 DTA curves for **a** 0% kaolinite—100% bentonite, **b** 50% kaolinite—50% bentonite, and **c** 100% kaolinite—0% bentonite. Reproduced (or adapted) with permission [54], © 2015, Elsevier

about physical and chemical alterations including crystallization. For this reason, DTA is suitable for analyzing the degree of retrogradation of rice starch. Tian et al. demonstrated using the DTA technique that the onset (To), peak (Tp), and conclusion (Tc) temperatures are decreased for retrograded rice starch after 35 days of storage (Fig. 5.12) [55]. The authors associated this behavior to the basis that the initial recrystallization of amylose decreased, affecting the number of perfect crystal nuclei and favoring the formation of imperfect crystal nuclei [55].

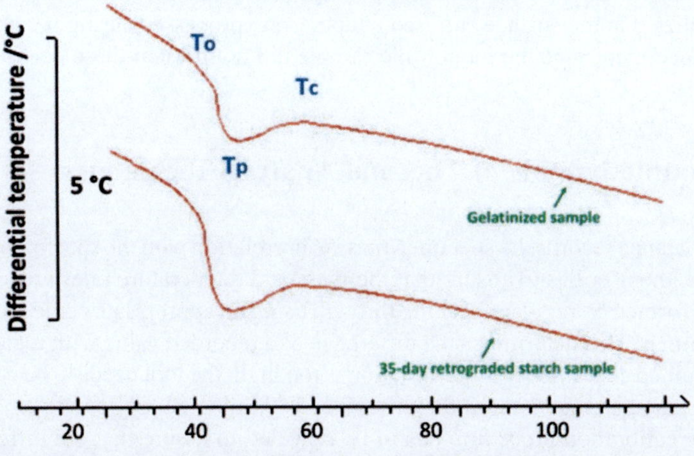

Fig. 5.12 DTA curves for the gelatinized and the 35-day retrograded starch specimens made from rice starch with low-level amylose. Reproduced (or adapted) with permission [55], © 2011, Elsevier

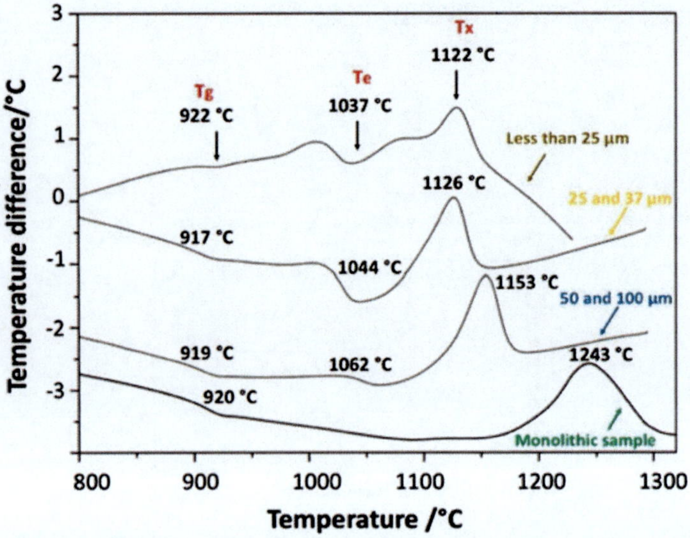

Fig. 5.13 DTA curves for monolithic and various YAS glass powder fractions. Glass transition temperature (T_g), endothermic peak temperature (T_E), and crystallization temperature (T_X) are indicated. Reproduced (or adapted) with permission [56], © 2013, Elsevier

During sintering, the crystallization kinetics in glasses are analyzed by the DTA technique. For this purpose, Lago et al. prepared yttrium aluminosilicate (YAS) glass samples and followed the crystallization process evaluating the effect of particle size [56]. The DTA curves demonstrated that the crystallization peaks in YAS powder specimens appeared at more moderate temperatures than the monolithic sample (Fig. 5.13). Also, endothermic peaks appeared in YAS samples, and this behavior was better appreciated in the smallest particle sizes. The presence of endothermic peaks implies that the surface suffered endothermic processes during the heating of the YAS specimen since the monolithic sample did not present these peaks [56].

5.4 Troubleshooting of Thermal Analysis Techniques

A thermobalance records the specimen mass in correlation with the specimen temperature; thus, the user has to ensure that the mass and temperature rates are adequate. This is performed by regular checking through using an appropriate choice of a reference specimen. This determines the difference of a recorded value with regard to the pre-determined exact value. If the variation is too high, the tool needs to be calibrated by varying the parameters until the difference is eliminated. Following any modification, the calibration procedure has to be repeated to assure that the difference is now within the permitted range. The balance is monitored and regulated utilizing reference masses, while certain TGAs are automatically adjusted with built-in reference masses when turned on and throughout extended periods of inactivity. While performing the TGA process, there are possible problems that may arise. In Table 5.5, the most common impediments for obtaining excellent results, causes, and solutions to solve these problems are presented. For instance, at the start of every operating day, the gas supplies and purge rates for the instrument must be checked. If the instrument is using bottled gas, recheck for how full the tank is. In case the tank shows less than 300 psi, replacing the tank should be considered. It is ideal to change the purge gas before the tank turns dry. On the other hand, if the instrument is using house gas, it is recommended to view installing a filter, a dryer, or both in the gas supply line. If the instrument has been moved or is inactive for a lengthy period of time, then the calibration should be revised [10]. In some cases, there are specimens that produce heavy black decomposition products that collect in the furnace tube, which must be cleaned, and the dirty furnace tube must be scrubbed carefully. To cleanse the furnace tube, mild soap, water, and a bottle brush to scrub the interior are the best choices. If this does not eliminate the residue, then a suitable solvent must be used. While applying solvents, the safety data sheet must always be read, and if possible, a laboratory safety specialist should be consulted for precaution. Once cleaned, the furnace tube should be rinsed with water and left to air-dry [29]. In the case of fully decomposed samples, the chosen temperature is greater than the approximate temperature of the material's transition. To solve this problem, a survey scan must be taken at 20 °C per minute starting and ending at 100 °C under and over the transition of interest. The survey scan can be further reviewed, and the starting and ending temperatures can be, therefore, modified accordingly [29].

While DSC is the most employed thermal analytical technique by the research community due to its low price and efficiency of handling, several general failures may occur during the operation. Troubleshooting of DSC is summarized in Table 5.5 which lists possible causes of error and solutions. The most common challenge when dealing with the DSC technique is the interpretation of the data when phase transitions occur at similar temperatures. Since the DSC technique is sensitive to the heat-flow changes and the instrument calculates them as an average of multiple read-outs, using low heating rates or large quantities of samples could prevent this issue. The thermal shock produced by loading the sample affects the baseline; however, this phenomenon could be minimized by a preliminary scan of temperatures prior to the transition phase. The sample encapsulation process is crucial since the resolution of the results depends on the heat transfer. Therefore, when the sample is not well compressed, the heat transfer becomes slow and, expectedly, a poor resolution is obtained. Furthermore, it is recommended that the sample environment is purged with nitrogen, argon, or helium to avoid any oxidation process, condensation, or contamination of volatiles affecting the quality of the results [30, 33, 57]. If samples subjected to temperature change experience different physical or chemical transformations including phase transitions and chemical reactions, or nucleation and crystal growth, they can pose a challenge in understanding the read-outs as well. Overall, while the DTA technique is an economical, fast, and simple method of thermal analysis, the problems encountered during the measurements could be

Table 5.5 Guide to problems when working with the thermal analytical techniques, causes of errors, and solutions for the TGA, DSC, and DTA techniques

Technique	Problem	Cause	Solution
TGA	The instrument is not working properly [10]	Running out of gas supplies [10]	If using bottled gas, it should contain above 300 psi capacity. If using house gas, install a filter, dryer, or both in the gas supply line [10]
	No exact measurements can be made [10]	The instrument has moved or was inactive for a long period of time [10]	Perform a calibration process [10]
	The furnace tube is affecting the analysis [29]	Samples generate heavy black decomposition products [29]	TGA furnace tube should be thoroughly cleaned [29]
	Samples are immediately fully decomposed [29]	The selected temperature is higher than the approximate temperature of the material's transition [29]	Perform a survey scan and change the starting and ending temperatures [29]
DSC	Phase transition overlap occurs [30]	Heat flow is obtained by the instrument as the mean of diverse heat-flow signals [30]	Augment the sample quantity [30]
			Reduce the heating rate [30]
	Baseline artifact is observed [33, 57]	Thermal shock might have happened during the loading of reference cells or samples [33]	Run the analysis for collecting prior single or multiple scans of samples or standard materials at temperatures before the start of phase transitions [33, 57]
	Significant thermal lag happens [57]	Energy originated from the furnace takes time to pierce the sample [57]	Calibrate the system for each scan rate required and for each essential method modification [57]
		The separation between the temperature sensor and the sample might have taken place [57]	
	Altered transition peaks are observed [57]	Sample collapse might have happened [57]	Ensure adequate thermal contact of the sample with the pan [57]
		The instrument might have contamination [57]	Review for sample leakage out of the pan and reduce sample size [57]
	Broad transitions and poor resolution are obtained [57]	Thermal gradients are built over the sample [57]	Perform improved sample encapsulation to prevent slow heat transfer [57]

(continued)

Table 5.5 (continued)

Technique	Problem	Cause	Solution
			Reduce the sample size [57]
			Slow down the heating rate (≈ 10–20 °C min^{-1}) [57]
	Inhibition of reaction processes takes place [57]	Fast scan rates may not allow kinetic events to occur [57]	Measure at two scan rates separated by one order of magnitude to capture more kinetic events [57]
DTA	Low-quality thermograms are obtained [58]	Heating rate, the composition of the crucible, distribution of the particle size, and pre-treatments of the sample might be the cause of the error [58]	Control the operating conditions [58]
			Use the adequate material, shape, and size of the crucible [46]
	The shape of the thermogram is irreproducible [58]	Sample contamination might have occurred [58]	Perform adequate pre-treatment [58]
	The equipment is unable to operate satisfactorily at desired temperature ranges [59]	Condensation of atmospheric moisture could be the reason [59]	Control the environmental conditions [59]
	Displacement of the baseline takes place	Significant differences might exist between the thermal characteristics of the sample and the reference material [59]	Employ an electrical compensation [59]
		An asymmetry might exist in the manufacturing of sample and reference holders [46, 59]	
	Peak area is not reproducible [6]	Sample volume, sample conductivity, thermocouple interference, sample shrinkage, or loss of material may be the possible causes [6]	Change the sample holder or employ a new method of diagram interpretation [6]
			Use the same pretreatment of the sample in all measurements [46]
			Weight the sample after each measurement [46]
	The results demonstrate low sensitivity of the equipment [46]	The device might be out of calibration [46]	Use samples with separated phase transitions to calibrate the temperature [46]

DSC Differential scanning calorimetry; *TGA* thermogravimetric analysis.

mainly attributed to the sample (storage, pretreatment, nature, and collection), the crucible (material, shape, and size), the rate of heating (too fast or slow), and the environment [6, 46, 58, 59].

References

1. C.J. Keattch, D. Dollimore, Studies in the history and development of thermogravimetry. I. Early development. J. Therm. Anal. **37**, 2089–2102 (1991)
2. H.L. le Chatelier, A general statement of the laws of chemical equilibrium. Comptes Rendus **99**, 786–789 (1884)
3. W.C. Roberts-Austen, Historical Paper.
4. H.M. Ng, N.M. Saidi, F.S. Omar, K. Ramesh, S. Ramesh, S. Bashir, Thermogravimetric analysis of polymers, *Encyclopedia of Polymer Science and Technology* (2002), pp. 1–29
5. K.H.-T.S.R. of the T. I. University and undefined 1915, On a thermobalance. *ci.nii.ac.jp*
6. S.L. Boersma, A theory of differential thermal analysis and new methods of measurement and interpretation. J. Am. Ceram. Soc. **38**(8), 281–284 (1955). https://doi.org/10.1111/j.1151-2916.1955.tb14945.x
7. E.S. Watson, M.J. O'Neill, J. Justin, N. Brenner, A differential scanning calorimeter for quantitative differential thermal analysis. Anal. Chem. **36**(7), 1233–1238 (1964). https://doi.org/10.1021/ac60213a019
8. R.C. Mackenzie, Thermochimica ACTA, Thermochim. Acta **73**(3), 1984 (1984)
9. M. Wagner, *Thermal Analysis in Practice: Fundamental Aspects* (Carl Hanser Verlag GmbH Co KG, 2017)
10. J.D. Menczel, R.B. Prime, *Thermal Analysis of Polymers: Fundamentals and Applications* (Wiley, 2009)
11. R. Bottom, Thermogravimetric analysis. Principles Appl. Thermal Anal. **3**, 87–118 (2008)
12. P.J. Haines, *Thermal Methods of Analysis: Principles, Applications and Problems* (Springer Science & Business Media, 2012)
13. R.S. Mikhail, E. Robens, *Microstructure and Thermal Analysis of Solid Surfaces* (1983)
14. L.E. Nielsen, *Mechanical Properties of Polymers, Reinhold Pub* (Corp, NY, 1962)
15. O.T. Sørensen, J. Rouquerol, in *Sample Controlled Thermal Analysis: Origin, Goals, Multiple Forms, Applications and Future*, vol. 3 (Springer Science & Business Media, 2003)
16. S. Materazzi, S. Vecchio, Evolved gas analysis by infrared spectroscopy. Appl. Spectrosc. Rev. **45**(4), 241–273 (2010)
17. M. Wagner, Thermogravimetric analysis, in *Thermal Analysis in Practice: Fundamental Aspects*, 2018, pp. 162–186. https://doi.org/10.3139/9781569906446.010
18. M.S. Hamid Akash, K. Rehman, Thermo gravimetric analysis, in *Essentials of Pharmaceutical Analysis*, 2020, pp. 215–222
19. O.O. Olatunji, S.A. Akinlabi, M.P. Mashinini, S.O. Fatoba, O.O. Ajayi, Thermo-gravimetric characterization of biomass properties: a review, in *IOP Conference Series: Materials Science and Engineering*, vol. 423 (2018). https://doi.org/10.1088/1757-899X/423/1/012175
20. Q.-V. Bach, W.-H. Chen, Pyrolysis characteristics and kinetics of microalgae via thermogravimetric analysis (TGA): a state-of-the-art review. Biores. Technol. **246**, 88–100 (2017)
21. F.B. Emre et al., A benzimidazole-based conducting polymer and a PMMA-clay nanocomposite containing biosensor platform for glucose sensing. Synth. Met. **207**, 102–109 (2015). https://doi.org/10.1016/j.synthmet.2015.06.015
22. E. Pindelska, A. Sokal, W. Kolodziejski, Pharmaceutical cocrystals, salts and polymorphs: advanced characterization techniques. Adv. Drug Deliv. Rev. **117**, 11–146 (2017)

23. A. Sobhan, K. Muthukumarappan, Z. Cen, L. Wei, Characterization of nanocellulose and activated carbon nanocomposite films' biosensing properties for smart packaging. Carbohydr. Polym. **225**, 115189 (2019)
24. E.M. Cahill et al., Metallic microneedles with interconnected porosity: a scalable platform for biosensing and drug delivery. Acta Biomater. **80**, 401–411 (2018)
25. A. Fortunato, in *DSC: history, instruments and devices* (Woodhead Publishing Limited, 2013).https://doi.org/10.1533/9781908818348.169
26. M. Wagner, A brief history of thermal analysis, in *Thermal Analysis in Practice: Fundamental Aspects* (2018), pp. 16–18. https://doi.org/10.3139/9781569906446.002
27. M.H. Chiu, E.J. Prenner, Differential scanning calorimetry: an invaluable tool for a detailed thermodynamic characterization of macromolecules and their interactions. J. Pharm. Bioall Sci. **3**(1), 39–59 (2011). https://doi.org/10.4103/0975-7406.76463
28. S. Ebnesajjad, C. Ebnesajjad, in *Surface Treatment of Materials for Adhesive Bonding* (William Andrew, 2013)
29. P. Gabbott, in *Principles and Applications of Thermal Analysis* (Wiley, 2008)
30. C. Leyva-Porras et al., Application of differential scanning calorimetry (DSC) and modulated differential scanning calorimetry (MDSC) in food and drug industries. Polymers **12**(1), 5 (2020)
31. J. Drzeżdżon, D. Jacewicz, A. Sielicka, L. Chmurzyński, Characterization of polymers based on differential scanning calorimetry based techniques. TrAC, Trends Anal. Chem. **110**, 51–56 (2019)
32. G.P. Ashton, E.L. Charsley, L.P. Harding, G.M.B. Parkes, Applications of a simultaneous differential scanning calorimetry–thermomicroscopy system. J. Therm. Anal. Calorim. 1–9 (2021)
33. M. Chountoulesi, N. Naziris, N. Pippa, S. Pispas, C. Demetzos, Differential scanning calorimetry (DSC): An invaluable tool for the thermal evaluation of advanced chimeric liposomal drug delivery nanosystems, in *Thermodynamics and Biophysics of Biomedical Nanosystems* (Springer, 2019), pp. 297–337
34. J. Wen, K. Arthur, L. Chemmalil, S. Muzammil, J. Gabrielson, Y. Jiang, Applications of differential scanning calorimetry for thermal stability analysis of proteins: qualification of DSC. J. Pharm. Sci. **101**(3), 955–964 (2012)
35. P. Gill, T.T. Moghadam, B. Ranjbar, Differential scanning calorimetry techniques: applications in biology and nanoscience. J Biomol. Tech. JBT **21**(4), 167 (2010)
36. S.-D. Clas, C.R. Dalton, B.C. Hancock, Differential scanning calorimetry: applications in drug development. Pharm. Sci. Technol. Today **2**(8), 311–320 (1999)
37. G. Bruylants, J. Wouters, C. Michaux, Differential scanning calorimetry in life science: thermodynamics, stability, molecular recognition and application in drug design. Curr. Med. Chem. **12**(17), 2011–2020 (2005)
38. J.M. Sturtevant, Biochemical applications of differential scanning calorimetry. Annu. Rev. Phys. Chem. **38**(1), 463–488 (1987)
39. R.I.M. Almoselhy, Applications of differential scanning calorimetry (DSC) in oils and fats research. a review. Am. Res. J. Agric. **6**(1), 1–9 (2020)
40. Y. Zhang, Z. Chen, X. Liu, J. Shi, H. Chen, Y. Gong, SEM, FTIR and DSC investigation of collagen hydrolysate treated degraded leather. J. Cult. Herit. **48**, 205–210 (2021)
41. R. Vaskoska, et al., Thermal denaturation of proteins in the muscle fibre and connective tissue from bovine muscles composed of type I (masseter) or type II (cutaneous trunci) fibres: DSC and FTIR microspectroscopy study. Food Chem. **343**, 128544 (2021)
42. K.A. Okotrub, I.V. Zaytseva, S.V. Adichtchev, N.V. Surovtsev, Raman spectroscopy and DSC assay of the phase coexistence in binary DMPC/cholesterol multilamellar vesicles. Biochimica et Biophysica Acta (BBA)-Biomembranes 1863(2) 183514 (2021)
43. M.S. Reza, S.N. Islam, S. Afroze, M.S.A. Bakar, J. Taweekun, A.K. Azad, Data on FTIR, TGA–DTG, DSC of invasive Pennisetum purpureum grass. Data Brief **30**, 105536 (2020)
44. P. Saganowska, M. Wesolowski, DSC as a screening tool for rapid co-crystal detection in binary mixtures of benzodiazepines with co-formers. J. Therm. Anal. Calorim. **133**(1), 785–795 (2018)

45. G. Klančnik, J. Medved, P. Mrvar, Differential thermal analysis (DTA) and differential scanning calorimetry (DSC) as a method of material investigation Diferenčna termična analiza (DTA) in diferenčna vrstična kalorimetrija (DSC) kot metoda za raziskavo materialov, RMZ–Mater. Geoenvironment **57**(1), 127–142 (2010)
46. P.J. Haines, M. Reading, F.W. Wilburn, Differential thermal analysis and differential scanning calorimetry, in *Handbook of Thermal Analysis and Calorimetry*, vol. 1 (Elsevier, 1998), pp. 279–361
47. M. Kawuloková et al., Study of equilibrium and nonequilibrium phase transformations temperatures of steel by thermal analysis methods. J. Therm. Anal. Calorim. **127**(1), 423–429 (2017)
48. G.O. Piloyan, I.D. Ryabchikov, O.S. Novikova, Determination of activation energies of chemical reactions by differential thermal analysis. Nature **212**(5067), 1229 (1966)
49. G. Vuataz, V. Meunier, J.C. Andrieux, TG–DTA approach for designing reference methods for moisture content determination in food powders. Food Chem. **122**(2), 436–442 (2010)
50. N. Raje, D.A. Aacherekar, A.V.R. Reddy, Compositional characterization of carbon electrode material: a study using simultaneous TG–DTA–FTIR. Thermochim. Acta **496**(1–2), 143–150 (2009)
51. J.A. Teixeira, W.D.G. Nunes, R.P. Fernandes, A. do Nascimento, F.J. Caires, M. Ionashiro, Thermal behavior in oxidative and pyrolysis conditions and characterization of some metal p-aminobenzoate compounds using TG–DTA, EGA and DSC-photovisual system. J. Anal. Appl. Pyroly. **128**, 261–267 (2017)
52. C. Ayrault, J.S. Chang, D. Ewing, J.S. Cotton, I.E. Gerges, J. Burgers, Differential thermal analysis, thermal gravimetric analysis, and solid phase micro-extraction gas chromatography analysis of water and fuel absorption in diesel soot. J. Aerosol Sci. **41**(2), 237–241 (2010)
53. M. Pigłowska, B. Kurc, Ł Rymaniak, P. Lijewski, P. Fuć, Kinetics and thermodynamics of thermal degradation of different starches and estimation the OH group and H_2O content on the surface by TG/DTG-DTA. Polymers **12**(2), 357 (2020)
54. P.B. Arab, T.P. Araújo, O.J. Pejon, Identification of clay minerals in mixtures subjected to differential thermal and thermogravimetry analyses and methylene blue adsorption tests. Appl. Clay Sci. **114**, 133–140 (2015)
55. Y. Tian, X. Xu, Z. Xie, J. Zhao, Z. Jin, Starch retrogradation determined by differential thermal analysis (DTA). Food Hydrocolloids **25**(6), 1637–1639 (2011)
56. D.C. Lago, M.O. Prado, Dehydroxilation and crystallization of glasses: a DTA study. J. Non-Cryst. Solids **381**, 12–16 (2013)
57. P. Gabbot, T. Mann, Differential scanning calorimetry, in *Principles of Thermal Analysis and Calorimetry* (2016), pp. 67–103
58. A.M. Zambrano Arévalo, G.C. Castellar Ortega, W.A. Vallejo Lozada, I.E. Piñeres Ariza, M.M. Cely Bautista, J.S. Valencia Ríos, Conceptual approach to thermal analysis and its main applications. Prospectiva **15**(2), 117–125 (2017)
59. M.E. Brown, Differential thermal analysis (DTA) and differential scanning calorimetry (DSC), in *Introduction to THERMAL Analysis* (Springer, 1988), pp. 23–49

Chapter 6
Characterization Techniques for Wettability Analysis

Ingrid Berenice Aguilar Meza, Euth Ortiz Ortega, Hamed Hosseinian, Andrea Rodríguez Vera, María José Rosales López, and Samira Hosseini

Abbreviations

NS	Not specific
OTS	Octadecyl trichlorosilane
SERS	Surface-enhanced Raman scattering
UV	Ultraviolet
WCA	Water contact angle

6.1 Water Contact Angle Analysis for Material Characterization

6.1.1 History of Water Contact Angle Analysis

Water contact angle (WCA) is the main data obtained from wettability studies, identifying the wetting degree in the interaction among a solid and a liquid. The angle is formed at the intersection of the liquid of interest and the particular solid surface, thus projecting the properties and structure of the material's surface that is in contact with the droplet. The theory and implementation of contact angles have undergone various developments since the 1800s (Fig. 6.1) [1, 2].

Thomas Young, a British polymath and physician, was the first scholar to describe a specific contact angle in a three-phase system analysis in 1805. By examining the adhesion of a liquid to a solid, the correlation between the contact angle and the surface tension for a stable solid was established. This was applied to liquid/vapor, liquid/liquid, and solid/liquid interactions. During the next years, the behavior of liquids on substrate surfaces continued to be researched and has presented new directions to understanding contact angles on the surfaces of various materials [3].

E. O. Ortega et al., *Material Characterization Techniques and Applications*,
Progress in Optical Science and Photonics 19,
https://doi.org/10.1007/978-981-16-9569-8_6

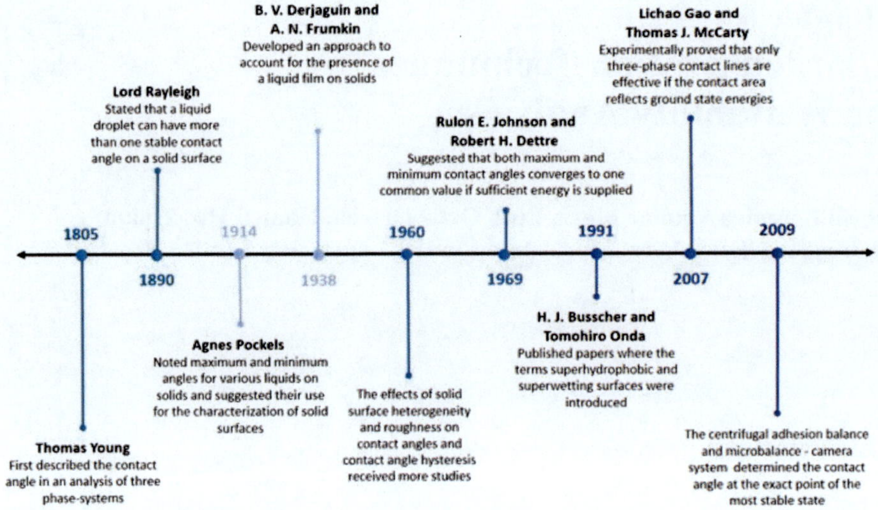

Fig. 6.1 Timeline of the development of the science and applications of contact angles since its proposed mathematical equation

Some months after the publication of Young's essay about fluid cohesion in 1805, the capillary phenomenon was observed and explored by numerous academics. Nonetheless, at the time there was no reference to contact angle (commonly referred to as angle or angle of contact in early research). Thereafter, Lord Rayleigh in 1890 identified that a liquid droplet can present multiple constant contact angles on a solid surface, however, further analysis on this observation was not provided [4].

Agnes Pockels, one of the collaborators of Lord Rayleigh, gained a deep understanding of the presence of maximum and minimum contact angles, namely hydrophobic and hydrophilic components. In 1914, Pockels published an article with the reported maximum and minimum angles for numerous liquids on solids, discussed their viable applications in the analysis of solid surfaces, and proposed some extensions to the work [1].

In the late 1930s, the effect of liquid film on a solid surface and the impact on the deviation of surface energy was recognized and discussed by Donald H. Bangham and R. I. Razouk, A. N. Frumkin, and B. V. Derjaguin. The work of Frumkin and Derjaguin suggested that a "microscopic" contact angle is made by the molecular forces of the interfaces, and macroscopic contact angles are additionally dependent on some external forces. These scientists also developed the wetting theory which explains the disjoining pressure in a liquid film and the repercussions on the contact angle. In the late nineteenth century, the detection of contact angle hysteresis and related highest and lowest contact angles initiated a pursuit for variables affecting contact angle measurements. Surface roughness and heterogeneity were the foremost aspects for describing contact angle hysteresis in solid interactions [1].

Wenzel [5] in 1936 and later Cassie [6] in 1944 proposed equations to explain the equilibrium contact angle for a liquid on a rough surface, depending on the magnitude of the liquid/solid contact area below the drop. On the one hand, the Wenzel equation is utilized when the drop size can completely penetrate the roughness grooves. However, Cassie's equation describes when a liquid interacts with a heterogeneous solid surface, while the Cassie–Baxter equation applies to apparent contact angles for rough and porous surfaces with trapped air underneath [1, 6]. However, in 2007, the publication of Lichao Gao and Thomas J. McCarty titled "How Wenzel and Cassie were wrong" analytically demonstrated that contact areas beneath the droplet were independent of the contact angle, but three-phase contact lines are efficient in determining wettability. This opened a discussion regarding the importance of contact lines in contact angle analysis on heterogeneous and irregular surfaces [7, 8].

From the beginning of the twentieth century, experimental advancements of contact angles were investigated for samples of different surface quality. It was, nevertheless, not before the 1960s and 1970s that the impacts of solid surface heterogeneity and roughness on contact angles and contact angle hysteresis gained methodical attention. Rulon E. Johnson Jr. and Robert H. Dettre (1964), and subsequently, A. W. Neumann and R. J. Good (1972), pioneered the modeling of contact angles on heterogeneous and rough surfaces, an advancement that remained rather limited to symmetrical structures. In the 1960s, the theoretical analysis of energetic states of liquids on heterogeneous and rough surfaces begun and continued through the rest of the twentieth century, which provided valuable insights into the knowledge of contact angles. However, certain misleading judgments were further drawn such as advancing contact angles representing or resembling equilibrium contact angles, when they actually differ significantly [1].

In 1969, Johnson and Dettre suggested that if adequate energy was provided for a liquid to subdue the energy blocks while spreading and running amid the metastable states, then both maximum and minimum contact angles may converge to a constant angle. This proposal was briefly investigated until the 1990s and 2000s [1]. It was concluded that the relaxation of liquid on a solid surface to a more stable state at about the weakest system energy is due to external energy through mechanical or acoustic oscillations throughout contact angle measurements [1].

The use of words such as superhydrophobicity, superhydrophilicity, and superwetting dates back to 1991 and 1996 when H. J. Busscher et al. and Tomohiro Onda et al. published the findings on ion-etched Teflon surfaces and wettability of fractal (rough) surfaces, sequentially [1].

In 2009, the centrifugal adhesion balance introduced some applications of centrifugal applications and gravitational forces to cause various regular and lateral force combinations for direct adhesion recording among a fluid drop and a solid surface [1, 9]. Moreover, a microbalance equipped with software was developed to automatically record the interplays of the liquid drop with the surface as a function of the surface position [1].

6.1.2 *Mechanism of Operation of Water Contact Angle Analysis*

6.1.2.1 Surface Tension and Contact Angle

Geometrically, the contact angle is defined by using a tangent line from the contact position along with the liquid–vapor interface in the droplet profile. In another word, WCA is defined as the angle made by the intersection of the liquid–solid interface and the liquid–vapor interface. The term "Three-phase contact line" is commonly used when solid, liquid, and vapor coexist (Fig. 6.2) [10]. For that reason, this technique is often referred to as a water-in-air contact angle.

A large contact angle is recognized when the liquid beads on the surface and a small contact angle is recognized when the liquid spreads over the surface. When a contact angle is smaller than 90°, it indicates that the surface is hydrophilic, and the fluid spreads across a larger area of the surface (Fig. 6.3). However, when contact angles are greater than 90°, it suggests that the surface is hydrophobic, and therefore, the fluid minimizes its contact with the surface and forms a compact liquid droplet. When

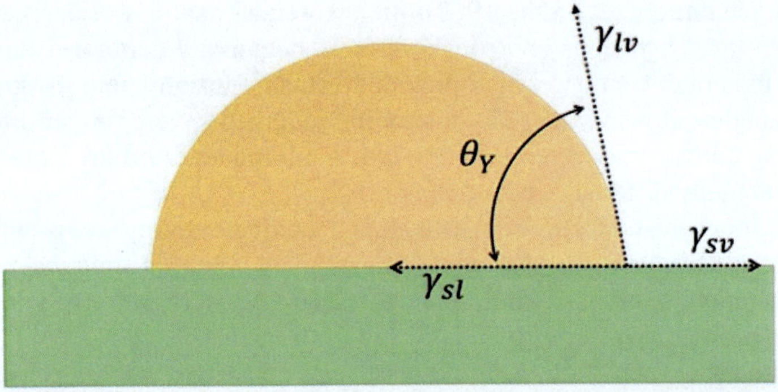

Fig. 6.2 Illustration of the three-phase contact line, where γ_{lv}, γ_{sv}, and γ_{sl} represent the liquid–vapor, solid–vapor, and solid–liquid intermolecular tensions, and θ_Y is the contact angle

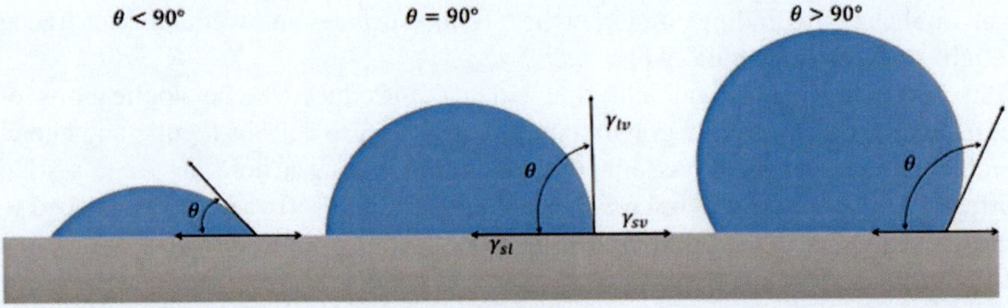

Fig. 6.3 Schematic representation of different measured contact angles where less than 90° surface is hydrophilic and more than 90° surface is hydrophobic

the droplet turns into a flat puddle, a complete wetting happens, and therefore, the contact angle is ~ 0° which marks a superhydrophilic surface. For superhydrophobic surfaces, however, the water contact angles are regularly larger than 150° representing minimal contact among the surface and the liquid, which explains the lotus effect. The lotus effect is a self-cleaning phenomenon in living beings where rolling water drops on the surface collect particles due to the interplay between the water droplet and the surface nanoscale architecture [11, 12].

Generally, the surface tension of the liquid will take the form of a liquid droplet. Due to the spherical form of small water droplets, the least surface area for a fixed volume is acquired. This intermolecular force to contract the liquid surface into the minimum surface area possible is called surface tension which determines the shape of the liquid droplets. In essence, some external forces including gravity can affect the shape and distort the droplet [3]. Consequently, the contact angle is defined by a mixture of surface tension and external forces including gravity. Thomas Young, for the first time, theoretically described that the contact angle of a liquid drop on an optimal solid surface is defined by the mechanical equilibrium of the drop beneath the action of three interfacial tensions (Eq. 6.1) [3]:

$$\gamma_{lv} \cos \cos \theta_Y = \gamma_{sv} - \gamma_{sl} \tag{6.1}$$

where γ_{lv}, γ_{sv}, and γ_{sl} denote the liquid–vapor, solid–vapor, and solid–liquid intermolecular tensions, sequentially, and θ_Y is the contact angle. Equation 6.1 is related to Young's equation, and θ_Y is Young's contact angle [3].

6.1.2.2 Contact Angle Measurement

The sessile drop method is the renowned method for analyzing the contact angle and directly measuring the tangent angle at a three-phase equilibrium interfacial position. By viewing the drop profile on flat surfaces, the wetting property can be defined by direct recording of contact angle. In a study presented in 1946, W. C. Bigelow et al. claimed that a telescope-goniometer was able to observe the liquid drop profile positioned across the smooth surface and calculate the angle between the three interfacial tensions [10]. Nowadays, an image of the adhered bubble could be transmitted onto a screen, the edges outlined, and the angle calculated via a computer program. The graphical illustration of the sessile drop method is depicted in Fig. 6.4. For measuring the contact angle, the stage for the sample needs to be flattened, so that the droplet does not move during deposition. Subsequently, a droplet of liquid is dispensed onto the surface. A source of light illuminates the droplet from behind, and therefore, an image is projected onto a screen and recorded by a camera for further analysis using program software. The contact angle analysis is conducted by the software, while the images and/or recordings of the phenomenon are stored on the connected computer. Using comparatively high magnifications improve the accuracy of the method and allows the detailed exploration of the intersection between the solid surface and the drop profile [13]. The technique is simple and straightforward

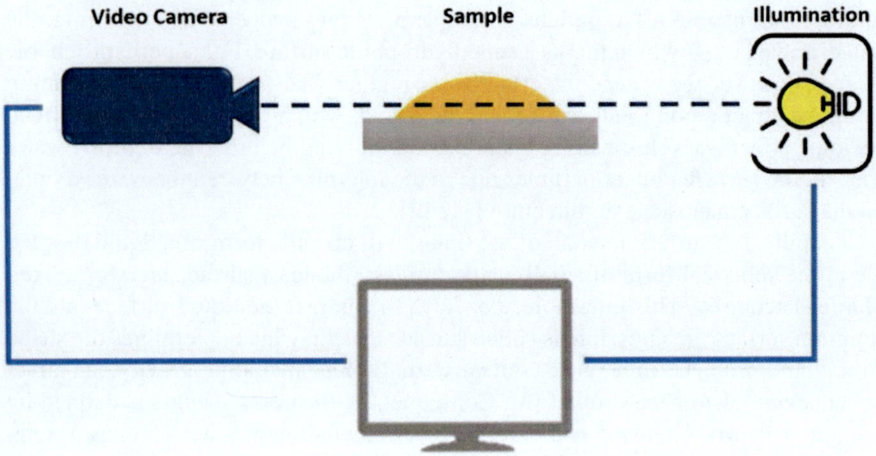

Fig. 6.4 Graphical illustration of the sessile drop technique. The camera takes images of the droplet, while the light is in the background and the contact angle is assessed by the software

in execution and requires a small surface area of substrates and small quantities of liquid to perform the analysis. However, impurities and irregularities on the surface can largely influence the analysis outcomes [10].

The accuracy and reproducibility of the contact angle measurements are mostly affected by the designation of the tangent line and the appropriateness of the operator. Thus, it is vital to define instructions for operators to follow. Additionally, it is recommended that the telescope be shifted slightly down (1–2 degrees) off the horizon so that the proximal edge of the stage is out of focus and a portion of the profile reflected by the sample's surface is focused. This limits the production of a hazy liquid substrate contact line in the profile. A light source in the background is regularly utilized to support the view, while a distinct light source is selected to inhibit the unfavorable heating of the liquid or specimen. To verify an exact contact angle, it is advised to gradually increase the sessile drop to a diameter of nearly 5 mm employing a micrometer syringe with narrow-gauge stainless steel. In order to avoid distortion of the drop profile, the needle size must be small. Since the drop may be asymmetrical, it is recommended to estimate the contact angles on both corners of the liquid drop and to use the average. In the case of a comparatively large substrate, contact angles should be measured at different points to give an average value that is demonstrative of the complete surface. A microscope detects the interaction between the drop profile and the sheath surface, while the goniometer analyzes the contact angle [14]. Typically, the contact angle of the covering layer is defined by the sessile drop methods in its dry state. Nevertheless, the contact angle of the surface might vary with the neighboring environment, and therefore, the contact angle of a dry membrane might be different from the contact angle of a wet membrane [10].

With the evolution of the theories of wetting procedures, a variety of methods have surged to record contact angles. Some of these techniques are summarized in Table

Table 6.1 Methods for measuring the contact angle

Category	Method	Description
Direct [15]	Sessile drop [10, 12, 15]	The contact angle is obtained from the tangent angle at the tri-phase contact point [10, 15]
	Captive bubble [10, 12, 15]	A gas bubble is confined under the solid surface, contact angle formed by the air bubble in the testing liquid [10, 15]
	Tilting plate [12, 15]	A solid rectangle plate is covered in the liquid so that a meniscus is shaped on both sides of the plate. Subsequently, the plate is rotated toward the liquid surface until one of the meniscuses displays horizontally. In this method, the angle within the solid surface and the liquid surface is the corresponding contact angle [12, 15]
Indirect [15]	Wilhelmy balance [10, 12, 15]	The contact angle is recorded by calculating the force change when a narrow, smooth, vertical, and solid plate is exposed to a probe liquid [10, 15]
	Capillary rise [10, 12, 15]	A capillary tube connects with a probe liquid. The meniscus enclosed in the tube serves to calculate the contact angle [10, 15]

6.1. In the first group, the direct group, measurements are based on optical imaging techniques, whereas in the second group, the indirect group, the contact angles are determined through force balance [10, 12, 15].

6.1.3 *Advantages and Disadvantages of Water Contact Angle Analysis*

As previously mentioned, measuring the tangent angle at the three-phase contact point is the commonly applied approach for calculating contact angle. This method is advantageous due to its simplicity of operation. Moreover, it offers information with respect to surface uniformity, porosity, and chemistry without the need for a large sample size. However, considering the quantity of the liquid and substrate, there is a comparatively high risk for error and the intervention of impurities in measurements [10, 16, 17]. Additionally, for materials with a great area, computations at various points are needed, therefore, tedious [17]. Also, the measurement accuracy and reproducibility depend on the reliability of the operator when determining the tangent line which could cause notable error and divergence among various users. Thus, specific guidelines for operators are necessary [10, 17]. Some of the advantages and disadvantages of WCA are detailed in Table 6.2.

Table 6.2 Advantages and disadvantages of utilizing WCA

Advantages	Disadvantages
Supplies data on uniformity, hydrophobicity, roughness, and chemistry of the surfaces [16–18]	Strict protocol must be followed to prevent operator errors [17, 18]
Simple operation [16–18]	Time-consuming analysis of large surface areas due to measurements at multiple locations [17]
Highly reproducible and cost-effective [19]	Impurities have a high influence on experimental error [17, 18]
Small amount of liquid is required [16–18]	The surface constitution may cause air trapping and drop penetration [16, 19]
The small substrate surface can be analyzed [17, 18]	Requires multiple measurements at different locations of the sample for a reliable mean value [17]
A wide range of materials can be analyzed [16, 17]	The surrounding atmosphere can impact the measurement over time [17]

6.1.4 Applications of Water Contact Angle Analysis

Surface properties dictate the performance of the materials in their intended applications. For the determination of the hydrophilicity or hydrophobicity of a surface, water contact angle measurement is the customary technique employed to analyze the wettability of a solid material [12]. In some emerging areas of biomedical application, advanced materials with unique wettability properties are needed to evoke desired bioactivities. Therefore, controlling and measuring the surface wettability of the designed materials are of significant importance [20].

An example of a material using specific wettability are platforms patterned with extreme differences in wettability that achieve the selective and sensitive detection of biomolecules by combining different approaches such as surface-enhanced Raman scattering (SERS), fluorescence, colorimetric, electrochemical, and mass spectrometry [21]. This is exemplified in the work undertaken by Sawsan Almohammed et al. where a platform for SERS-based sensing was developed [22]. In order to create wettability gradients and align peptide nanotubes, UV/ozone exposure served as the pattering technique. Silver nanoparticles were incorporated into the peptide nanotubes substrate to test the eligibility of the device to detect an analyte molecule through SERS [22].

Another approach for the detection of biomolecules is the fluorescence detection technique. Fluorescence is a discharge event from a molecule following the beginning electron excitation in a light-absorption procedure [21]. An example is a study carried out by Li Jiang et al. in which different surfaces were modified to develop an appropriate substrate for detecting toxoplasmosis, rubella virus, cytomegalovirus, and herpes infections [23]. Hydrophilic microwells were used to confine analytes, while hydrophobic substrates acted as a barrier for the restrictive spread of microdroplets. Contact angle measurements served to analyze the modified surfaces and understand their nature upon surface treatment (Fig. 6.5). Based on the measurements, it was concluded that agarose-modified slides presented a higher potential for binding to proteins. The developed platform enhances fluorescence sensing and achieves an effective biorecognition procedure. With this research, signal amplification assay with a cyanine dye labeled biotin-streptavidin and nanogold labeling was assessed, revealing a higher sensitivity of the method (Fig. 6.5) [23].

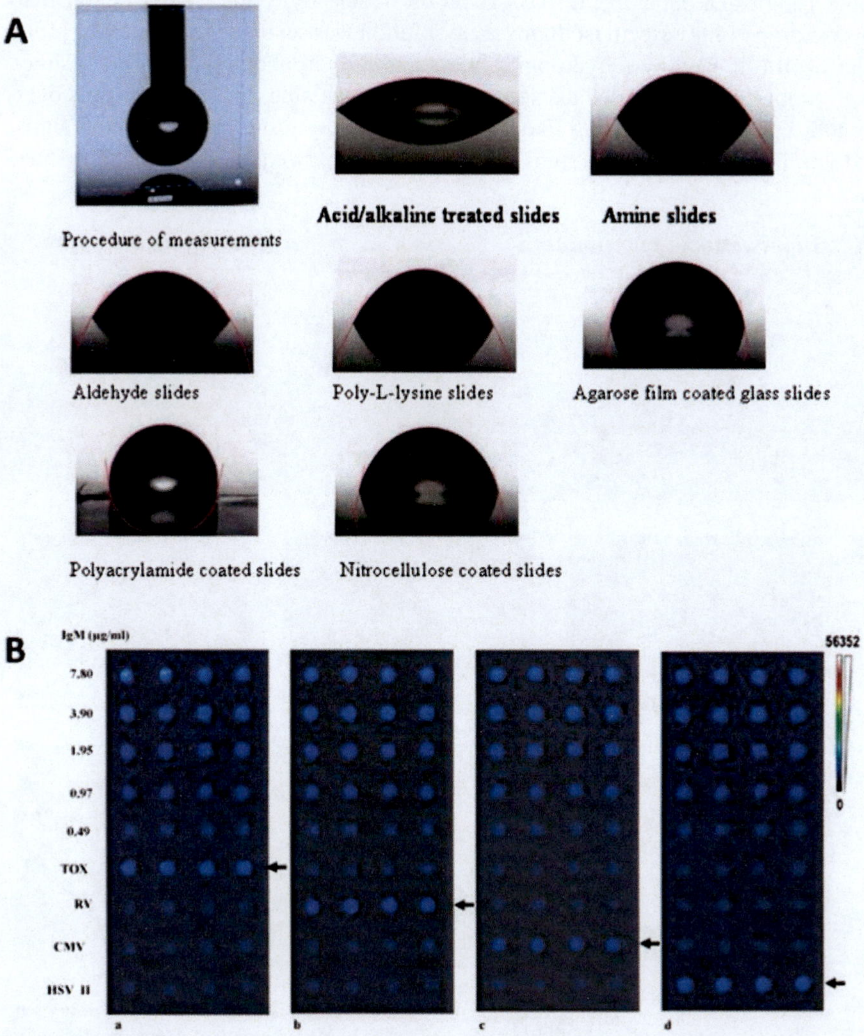

Fig. 6.5 Wettable fluorescence detection. **a** Measurements of contact angles on modified surfaces. **b** Fluorescence detection enhancement of immunoglobulin M antibodies in subjects with potential infections. The arrows indicate the positive samples. Reproduced (or Adapted) with permission [23], ©2008, Elsevier

In order to obtain effective biochips, the microarray quality and spot homogeneity are fundamental elements. Yanxia Chen et al. developed a chip with superhydrophilic microwells on a superhydrophobic substrate [24]. A platform of n-octadecyltrichlorosilane (OTS)-modified nanodendritic SiO_2-shell coating with low surface energy and a permeable nanostructure, used as a superhydrophobic surface, had high water contact angles of 157.5 ± 1.8°. However, the superhydrophilic microwells, formed after a UV exposure of the substrate to degrade the OTS, showed a contact angle of around 0° (Fig. 6.6). To probe the sensitivity and accuracy of the platform, a drop of fluorescein isothiocyanate solution was set onto the superhydrophilic microwell followed by a hydrophilic glass and a hydrophobic glass. This resulted in a homogeneous fluorescence spot on the superwettable micropattern, thus overcoming a phenomenon of ring-like morphology which causes inhomogeneous signal and unreliable readout. The study confirms that superwettable micropatterns have

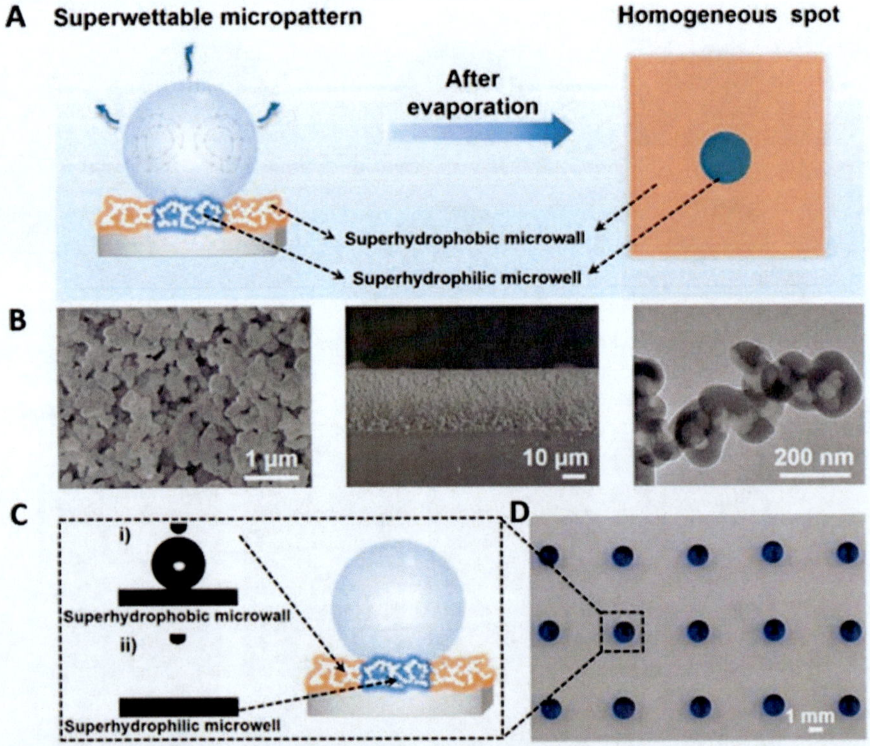

Fig. 6.6 Superwettable micropattern developed from superhydrophobic substrates patterning with superhydrophilic microwell arrays. **a** Illustrative image of homogeneous spot deposition following droplet evaporation. **b** SEM and TEM images of micropattern coating. **c** Water contact angles for microwall (i) and microwell substrate (ii). **d** The methylene blue trihydrate microarray droplets in a superwettable micropattern. Reproduced (or Adapted) with permission [24], ©2018, Elsevier

excellent potentials as biosensing platforms for biomarker detection while controlling the spot homogeneity [24].

Zi-Xia Zhao et al. also demonstrated the importance of the wettability on the surface modification of a material [25]. It is important to note that once the wettability of a material is altered, its bioactivity is changed as well. The authors modified gold surfaces with hydrophobins, small fungal proteins. After the processing of the surface with hydrophobins, the hydrophilicity of the material improved, altering the WCA from 73.8° to 45.3°. Subsequently, hydrophilic proteins were immobilized on the gold electrode surface modified with hydrophobins preserving its bioactivity. The modified electrode was used to fabricate an amperometric choline biosensor showing a highly effective biocatalytic reaction, therefore exhibiting great potentials as a biosensor [25]. In a similar manner, Adeniyi Olugbenga et al. proved that the wettability modification of electrodes can facilitate the formation of biomolecules on the electrode surface and allow high-ordered immobilization of glucose oxidase on the graphene. Thus, the sensitivity of the material increased which is relevant to the fabrication of electrochemical biosensors [26].

6.1.5 Troubleshooting of Wettability Analysis Technique

The wettability in a material provides indirect information about the physical and chemical properties of a surface. For instance, the roughness and the presence of polar surface functional groups over a material surface can greatly impact the surface wettability. The measurement of contact angle of the droplet with the surface determines the wettability of a variety of specimens (e.g., metals, polymers, and carbon platforms). While the technique appears to be simple to operate, common errors may occur during the test mainly due to following wrong procedures. For instance, the accuracy of the data could be affected if the droplet is not dispensed from the middle of the needle, or the sample holder is inclined. Establishing a correct baseline allows the user to acquire an adequate fitted curve. Moreover, the camera might not be well-positioned, or the surface might be highly irregular that would expectedly result in inconsistent measurements. Furthermore, the values of the water contact angle will not be constant if the droplet is not uniform [17]. The light source is also vital in obtaining a clear image or video, hence, reliable data. If the lamp does not operate well, the user may check the power source or change the bulb. However, it is not recommended for the user to attempt at repairing the lamp [27]. Table 6.3 shows detailed troubleshooting for water contact angle measurement for wettability analysis as well as listing possible causes and solutions.

Table 6.3 Troubleshooting for WCA used for wettability analysis

Technique	Problem	Cause	Solution
WCA	WCA values are not similar for the same sample [17]	Droplet is non-consistent [17]	Use different zones of the sample or report about the uniformity of the sample if the problem persists [17]
		Vibrations occur during the experiment [17]	Remove the vibration source and repeat the analysis [17]
	The light lamp does not work [27]	The lamp socket is damaged [27]	Check the power source or change the bulb with the help of an expert [27]
	"Stick and slip" behavior [28]	NS	Use larger drop volumes [28]
	Results are inconsistent [17]	The curve-fitting method is inadequate [17]	Use a different fitting method such as polynomial fitting [17]
		Needle position is wrong [17]	Position the needle at the center of the droplet [17]
		The sample is placed in an incorrect position [17]	Move the sample platform to the middle of the recorded image [17]
	Viscous samples present dynamic effects [17]	Flow rate is extremely high [17]	Use a lower flow rate and increase it gradually along with each measurement [17]
	Baseline is not evident [17]	The camera is inclined upward [17]	Make sure the camera is in the right position [17]
	The sample is not in the correct position [17]	Sample platform is inclined [17]	Move the specimen platform from right to left until the distance among the platform and needle persists constantly [17]

NS not specific; *WCA* water contact angle

References

1. J. W. Drelich et al., Contact angles: history of over 200 years of open questions. Surface Innov. **8** (1–2), 3–27 (2020)
2. Y. Yuan, R. Lee, Surface science techniques, in *Springer Series in Surface Sciences Techniques*, vol. 51, no. 1 (2013). https://doi.org/10.1007/978-3-642-34243-1
3. T. Young, III. An essay on the cohesion of fluids, in *Philosophical Transactions of the Royal Society of London*, 95th ed. (1805), pp. 65–87
4. L. Rayleigh, On the tension of water surfaces, clean and contaminated, investigated by the method of ripples. Philos. Mag. Sci. 5 **30** (186), 386–400 (1890). https://doi.org/10.1080/14786449008620040
5. R.N. Wenzel, Resistance of solid surfaces to wetting by water. Ind. Eng. Chem. **28** (8), 988–994 (1936). https://doi.org/10.1021/ie50320a024

6. B.D. Cassie, S. Baxter, Wettability of porous surfaces. Trans. Faraday Soc. **40**, 546–551 (1944)
7. L. Gao, T.J. Mccarthy, How Wenzel and Cassie were wrong. Langmuir **23**(7), 3762–3765 (2007)
8. J.W. Drelich, Contact angles: From past mistakes to new developments through liquid-solid adhesion measurements. Adv. Coll. Interface. Sci. **267**, 1–14 (2019). https://doi.org/10.1016/j.cis.2019.02.002
9. R. Tadmor, P. Bahadur, A. Leh, H. N'guessan, R. Jaini, L. Dang, Measurement of lateral adhesion forces at the interface between a liquid drop and a substrate. Phys. Rev. Lett. **103** (26), 266101 (2009)
10. R.S. Hebbar, A.M. Isloor, A.F. Ismail, Contact angle measurement, in *Membrane Characterization* (2017), pp. 219–255
11. A. Lafuma, D. Quéré, Superhydrophobic states. Nat. Mater. **2**(7), 457–460 (2003)
12. I. Ahmad, C.W. Kan, A Review on development and applications of bio-inspired superhydrophobic textiles. Materials **9** (11) (2016). https://doi.org/10.3390/ma9110892
13. T.T. Chau, A review of techniques for measurement of contact angles and their applicability on mineral surfaces. Miner. Eng. **22**(3), 213–219 (2009)
14. L.R. Fisher, Measurement of small contact angles for sessile drops. J. Colloid Interface Sci. **72**(2), 200–205 (1979)
15. J.W. Song, L.W. Fan, Temperature dependence of the contact angle of water: A review of research progress, theoretical understanding, and implications for boiling heat transfer. Adv. Colloid Interface Sci. **288** (2021). https://doi.org/10.1016/j.cis.2020.102339
16. T. Zhao, L. Jiang, Contact angle measurement of natural materials. Colloids Surf. B **161**, 324–330 (2018). https://doi.org/10.1016/j.colsurfb.2017.10.056
17. T. Huhtamäki, X. Tian, J.T. Korhonen, R.H.A. Ras, Surface-wetting characterization using contact-angle measurements. Nat. Protoc. **13**(7), 1521–1538 (2018). https://doi.org/10.1038/s41596-018-0003-z
18. R.S. Hebbar, A.M. Isloor, A.F. Ismail, Contact angle measurements. Membrane Charact. 219–255 (2017). https://doi.org/10.1016/B978-0-444-63776-5.00012-7
19. A. Alghunaim, S. Kirdponpattara, B.M.Z. Newby, Techniques for determining contact angle and wettability of powders. Powder Technol. **287**, 201–215 (2016). https://doi.org/10.1016/j.powtec.2015.10.002
20. C. Aparicio, Y. Maazouz, D. Yang, Measuring wettability of biosurfaces at the microscale. Methods Mol. Biol. **811**, 163–177 (2012). https://doi.org/10.1007/978-1-61779-388-2_11
21. T. Xu, L.P. Xu, X. Zhang, S. Wang, Bioinspired superwettable micropatterns for biosensing. Chem. Soc. Rev. **48**(12), 3153–3165 (2019). https://doi.org/10.1039/c8cs00915e
22. S. Almohammed, S.O. Oladapo, K. Ryan, A.L. Kholkin, J.H. Rice, B.J. Rodriguez, Wettability gradient-induced alignment of peptide nanotubes as templates for biosensing applications. RSC Adv. **6**(48), 41809–41815 (2016). https://doi.org/10.1039/c6ra05732b
23. L. Jiang et al., Development of a fluorescent and colorimetric detection methods-based protein microarray for serodiagnosis of TORCH infections. Biosens. Bioelectron. **24**, 376–382 (2008). https://doi.org/10.1016/j.bios.2008.04.019
24. Y. Chen et al., Superwettable microchips with improved spot homogeneity toward sensitive biosensing. Biosensors and Bioelectronics **102** (July 2017), 418–424 (2018). https://doi.org/10.1016/j.bios.2017.11.036
25. Z. Zhao et al., Self-assembled film of hydrophobins on gold surfaces and its application to electrochemical biosensing. Colloids Surf. B **71**, 102–106 (2009). https://doi.org/10.1016/j.colsurfb.2009.01.011
26. A. Olugbenga et al., Acetylene-sourced CVD-synthesised catalytically active graphene for electrochemical biosensing. Biosens. Bioelectron. **89**, 496–504 (2017)
27. ChemInstruments, Instruction Manual for Contact Angle Meter Model (2005)
28. J.T. Korhonen, T. Huhtamäki, O. Ikkala, R.H.A. Ras, Reliable measurement of the receding contact angle. Langmuir **29**(12), 3858–3863 (2013). https://doi.org/10.1021/la400009m

Chapter 7
Characterization Techniques for Electrochemical Analysis

Euth Ortiz Ortega, Hamed Hosseinian, Ingrid Berenice Aguilar Meza, Andrea Rodríguez Vera, María José Rosales López, and Samira Hosseini

Abbreviations

AC	Alternating current
Ag^+	Silver ions
AMF	Autocrine motility factor
Au	Gold
AuNPs	Gold nanoparticles
BSA	Bovine serum albumin
C	Capacitance
CDMPV	Cyclic differential multipulse voltammetry
CE	Counter electrode
CEA	Carcinoembryonic antigen
CMPV	Cyclic multipulse voltammetry
CNTs	Carbon nanotubes
CSs	Microcarbon sphere
Cu-Au	Copper-gold
CV	Cyclic voltammetry
DA	Dopamine
DC	Direct current
DHF	Dengue hemorrhagic fever
DME	Dropping mercury electrode
DSS	Dengue shock syndrome
$E(t)$	Sinusoidal perturbations of potential
E_0	Amplitude of potential
EIS	Electrochemical impedance spectroscopy
ELLI	Electro-lateral flow immunosensor
E_p	Peak potential
E_{pa}	Peak anodic potential
E_{pc}	Peak cathodic potential

E. O. Ortega et al., *Material Characterization Techniques and Applications*,
Progress in Optical Science and Photonics 19,
https://doi.org/10.1007/978-981-16-9569-8_7

FAR	Folate receptor
FRA	Frequency response analyzer
FSCV	Fast-scan cyclic voltammetry
H_2SO_4	Sulfuric acid
$I(t)$	Sinusoidal current
I_0	Amplitude of current
i_{pa}	Anodic current
i_{pc}	Cathodic current
K_4 [Fe $(CN)_6$]	Potassium ferricyanide
LTS	Linear systems theory
LOD	Limit of detection
MUA	Mercaptoundecanoic acid
PEG	Polyethylene glycol
PGI	Phosphoglucose isomerase
RE	Reference electrode
R_{et}	Resistance to electron transfer
R_s	Resistance to the solution
SNR	Signal-to-noise ratio
SPGE	Screen-printed gold electrodes
SWNTs	Single-walled carbon nanotubes
UiO-66	Zirconium MOFs
V_1	Potential 1
V_2	Potential 2
WE	Working electrode
Z	Electrochemical impedance of electrode reaction
Z_w	Warburg impedance
ϕ	Phase shift
ω	Frequency
Ω	Ohms

7.1 Cyclic Voltammetry for Material Characterization

7.1.1 History of Cyclic Voltammetry

Cyclic voltammetry (CV) is an electroanalytical method which obtains qualitative information about electrochemical reactions by measuring the flow of current between electrodes. These are presented as current–voltage curves, also called voltammograms. In order to discuss this method, some concepts must first be introduced [1, 2]. Among the given concepts that were established for the development of electro-analysis were the fundamentals of electrolysis initiated by the discovery of Alessandro Volta's pile in 1800 (Fig. 7.1). A year later, in 1801, William Cruikshank

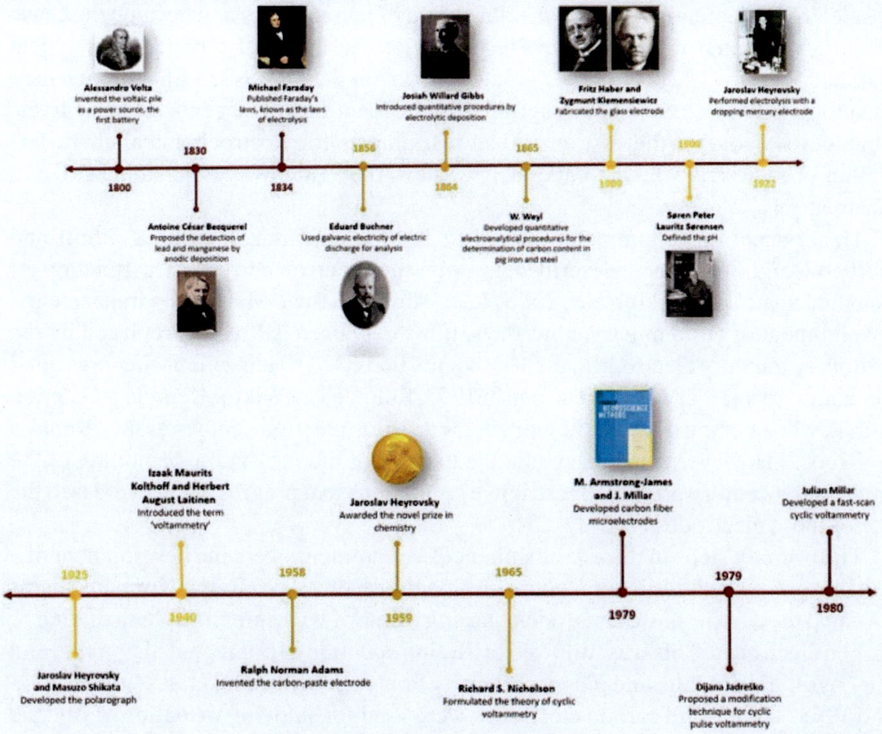

Fig. 7.1 History of CV from its beginning with the invention of the "voltaic pile" in 1800 until 1980, when Millar developed the fast-scan CV

suggested that current could be used as a parameter in the analysis of metals after having worked with electrolysis of aqueous solutions containing copper salts. In 1830, Antoine César Becquerel proposed the use of anodic deposition for the detection of lead and manganese. A few years later, in 1831, Michael Faraday published the laws of electrolysis. In the following decades, Eduard Buchner utilized galvanic electricity as a sensitive reagent and W. Weyl presented the mechanisms in order to identify carbon amounts in pig iron as steel after the iron electrolytic dissolution. However, quantitative electroanalytical determinations were only presented after 1864 when Josiah Willard Gibbs proposed copper precipitation by electrolysis [1]. Similarly, in 1865, Carl Luckow described copper determination by electrolytic depositions; thus, electrogravimetry for metal determination surged.

At the beginning of the twentieth century, the most outstanding development of electroanalytical techniques included the achievement of Fritz Haber and Zygmunt Klemensiewicz on glass electrodes, and the pH scale by Søren Peter Lauritz Sørensen [1]. In 1922, Jaroslav Heyrovsky performed electrolysis experiments with the dropping mercury electrode (DME). It was concluded that the results regarding the nature of the reduced species could be obtained from calculating the current while varying the electrode's potential. This coined the origin of voltammetry and became the cause of an international reputation for Heyrovsky as the father of the electroanalytical characterization [1, 2]. Three years later, Heyrovsky and Masuzo Shikata invented an automated device called polarographs for the recording of current–voltage curves. This tool was one of the first analytical instruments for electrochemical characterization of materials which, years later, was awarded Heyrovsky the Nobel Prize in Chemistry [1, 2].

The concept of voltammetry surged in 1940 when Izaak Maurits Kolthoff and Herbert August Laitinen described the measurement of current as a function of potential at a solid working electrode [1, 2]. Although the DME greatly impacted the development of voltammetry, nowadays, it is rarely used. DME was replaced by the stationary mercury electrode in the 1950s and 1960s [1, 2]. Heinz Gerischer presented the stationary mercury drop electrode in 1953, followed by Wiktor Kemula and Zenon Kublik who later introduced a hanging mercury drop electrode known as the "Kemula electrode." However, concerns about the toxicity of mercury at the beginning of the twenty-first century promoted restrictive regulations that progressively decreased the use of these electrodes [1, 3].

The primary steps in the establishment of voltammetry were the development of a suitable pre-concentration procedure and electrodes. In 1958, Ralph Norman Adams invented the carbon-paste electrodes, starting a time of widespread implementation of carbon electrodes. This was followed by the introduction of impregnated graphite and later pyrolytic graphite and glassy carbon by Francis J. Miller and H. E. Zittel in 1963 and 1965. These solid carbon electrodes were used for studying oxidation of organic molecules. Most recently, a vast versatility of strategies for modification of different electrodes has been investigated. The category of enzyme-modified electrodes is particularly relevant to the development of sensors. Nowadays, the modification of the electrode surface is possible with submonolayer nanoparticle coverage [1, 2, 4, 5].

Investigation of electrode surface's chemical modification around 1975 permitted better comprehension on how the surfaces could be adapted to develop the required applications. Arthur Thornton Hubbard, Theodore Kuwana, Royce W. Murray, among others, made initial efforts in covalently attaching different monolayers to electrodes. In the late 1970s, Murray, along with L. L. Miller, Allen J. Bard, and co-workers, modified electrodes by creating controlled polymer layers on the electrode surface [1, 2, 4]. In 1965, Richard S. Nicholson formulated the theory of cyclic voltammetry. In spite of its complex theory, it soon became one of the prevailing electrochemical techniques. It is often employed to obtain thermodynamic and kinetic information of electroactive compounds, either organic, organometallic, or inorganic species, especially when electrochemical methods are occasionally used [2, 6].

Computers greatly influenced electrochemical instrumentation as they served to generate potential programs. Voltammetric perturbation applied to the system could now be adjusted by researchers on modern software packages in electrochemical instrumentation. Therefore, in 2013, technique variations of the cyclic pulse voltammetry were proposed. Some of these included the cyclic multipulse voltammetry (CMPV) and cyclic differential multipulse voltammetry (CDMPV), resulting in a more complete characterization of the system and sound, and faster electrochemical calculations. The obtained signal enabled the qualitative analysis of the procedure from the visual examination of the voltammograms [2, 4]. Additionally, bulk kinetics could be used to identify authentic nano-electrocatalysis by simulating different experiments. These include voltammetry experiments on nanoparticle-modified electrodes in relation to surface coverage and voltammetry experiments on bulk material using a macroelectrode. Due to the diffusion profile being able to overlap and the complex interaction of the electrode surface geometry, the production of physically significant kinetic data from modified electrodes was a laborious drawback. Therefore, these simulation analyses could be used to conclude if the voltammetric response was experimentally adapted by the application of the nanomaterial [4]. The world is moving toward miniaturization; hence, the characterization domain is advancing toward quicker and more compact techniques, allowing voltammetry to proceed in the same manner [2]. In that path, selective reagents, electrochemical mediators, new materials, and modification strategies should be incorporated to improve the selectivity and the sensitivity of existing electrochemical sensors [3].

7.1.2 *Mechanism of Operation of Cyclic Voltammetry*

In order to conduct a CV experiment, some criteria must be met such as the three-electrode setup with a working electrode (WE), a reference electrode (RE), and a counter electrode (CE). To clarify this mechanism, there are some phrases that must be first defined. Reduction occurs in the electrode on the cathode, marking the cathode as the reducing electrode. A WE is an electrode where the reduction or oxidation electrochemical phenomena are taking place. CE functions as an electron source; hence, the current may flow from the external circuit through the cell. For instance, for each electron stopped at the WE by oxidation, a one-electron reduction procedure must occur at the CE. A RE is a stable electrode with a constant potential (despite the electrolyte solution it is dipped in), and it calculates the potential of the WE. An inorganic salt that is dissolved in an aqueous solution which is an ionic conductor is called an electrolyte solution. A potentiostat is an electronic amplifier designed to control the WE potential in a multiple-electrode electrochemical cell. The potentiostat usually contains a sensing component like a RE, a component for balancing the current flow like a CE, and various internal circuits that allow it to function in this capacity. By using a potentiostat and a three-electrode setup, it is possible to record a current–potential polarization curve for two purposes including: (1) having control over the voltage between the WE and the RE, and (2) calculating the flow of current between the WE and the CE.

For a bioanode, the above-mentioned is demonstrated in Fig. 7.2. This figure represents the voltage drop between the RE and the WE. This voltage drop is obtained via "R," the resistance generated by the electrolyte located between the WE and the RE. *R* is also referred to as the uncompensated resistance. As the current flows, according to Ohm's law, this resistance results in a decrease in potential in the solution. As the resistivity of the solution and the distance between the WE and the RE increase, the

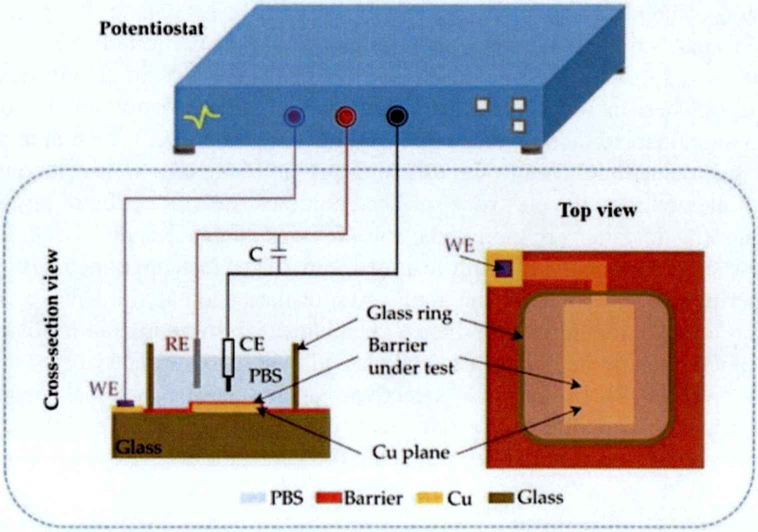

Fig. 7.2 A conventional setup with three electrodes of an electrochemical cell connected to a potentiostat

potential further decreases. Additionally, it is best to maintain the RE secluded from the ion migration between WE and CE to avoid incorrect potential measurement. In order to prevent a large decrease in potential between the WE and the RE, it is advised to position the RE and WE near each other. The effect of the uncompensated resistance, the resistance between the working electrodes, and the full equipotential surface through the tip of the reference electrode could be measured for different setups [7].

Additionally, CV can provide other relevant information such as the chemical rate constant of the redox potential and the electrochemical reaction rates. A very noticeable factor for chemical reactions to happen at a significant proportion is the scan rate which must be long enough. A current is measured between the WE and the CE, while a voltage will be measured between the RE and the WE. The voltammogram includes oxidative and reductive peaks, as is presented in Fig. 7.3 where a normal voltammogram is demonstrated to represent different peaks and components. The current rises from the capacitive or resting phase, while the applied potential hikes from Potential 1 (V_1) to Potential 2 (V_2). As a result, it approaches the electrochemical oxidative potential of the analyte. The two peak potentials of a wave are distinguished by referring to the peak occurring at the positive potential sweep, the peak anodic potential (E_{pa}). The WE potential where the maximum current in a voltammetric analysis is received is called the peak potential (E_p). In the reverse cycle, the current decreases, the applied potential surpasses the E_p point, and the applied potential moves toward the negative values. In this case, the measured current turns into highly negative until reaching the electrochemical reductive potential of the analyte. Moreover, when the applied potential exceeds the E_p value, a peak polar to the anodic peak, named peak

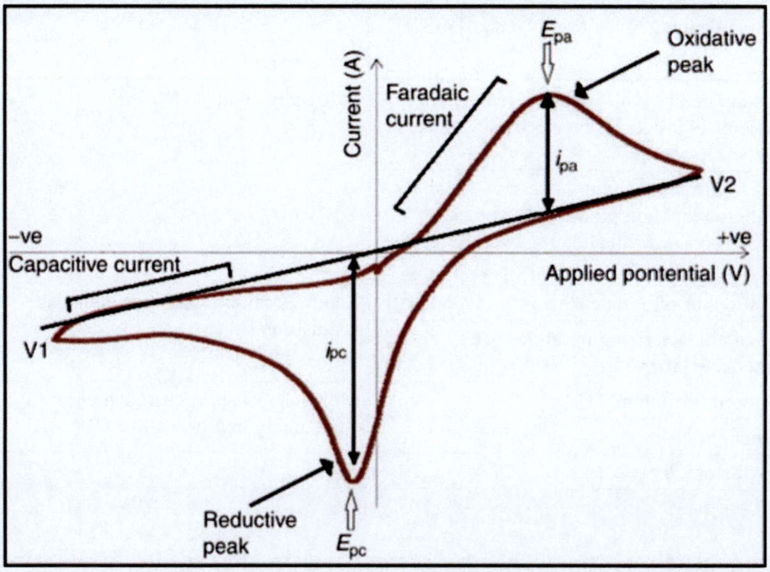

Fig. 7.3 Cyclic voltammogram presenting the peak cathodic potential (E_{pc}), peak anodic potential (E_{pa}), cathodic current (i_{pc}), and anodic current (i_{pa}). Reproduced (or adapted) with permission [8], © 2016, Elsevier

cathodic potential (E_{pc}), is formed and can be determined. Additionally, i_{pc} and i_{pa} are the differences between the cathodic and anodic currents from the resting current, respectively. While the process continues, the oxidative and reductive CV curves converge, indicating the reaction is nearing completion [8].

7.1.3 Advantages and Disadvantages of Cyclic Voltammetry

Cyclic voltammetry presents various advantages and a few main disadvantages. One of the advantages of CV is that it provides qualitative data about the features of electrochemical processes on electrode surfaces. Moreover, this technique yields significant data about the thermodynamics of oxidation–reduction and kinetics of adsorption processes, heterogeneous electron transfer, and coupled chemical reactions. This is possible since the researcher may conduct a reversible reaction in either direction [9, 10]. Different from other electrochemical methods, CV provides information through voltammograms. In addition, this method can characterize multiple species simultaneously under the condition that these species exhibit reduction and oxidation activity at different potentials. Furthermore, since pulse techniques were implemented, CV has achieved high sensitivity by deducting the influence of capacitive current [10]. Straightforward and largely accessible apparatus allow a simple operation condition [11].

Table 7.1 Advantages and disadvantages of CV

Advantages	Disadvantages
Provides qualitative data about the features of electrochemical processes on electrode surfaces [9, 10]	Indirect method [11]
Provides information about the thermodynamics of oxidation–reduction processes, kinetics of adsorption processes, and heterogeneous electron transfer and coupled chemical reactions [9, 10]	
Voltammograms offer extensive information [10]	Requires calibration for reliable data acquisition [11]
Capable of characterizing multiple species simultaneously [10]	
Simplicity of operation [11]	Molecules of same chemical class are challenging to differentiate [10]
High sensitivity [9]	

Cyclic voltammetry (CV)

Some constraints of this technique are that it is an indirect characterization method and careful calibration is required for reliable data acquisition [11]. Moreover, alternative methods of CV occasionally have difficulties in discerning molecules that belong to the same chemical class. For instance, fast-scan cyclic voltammetry (FSCV) cannot differentiate dopamine, norepinephrine, and epinephrine as they are catecholamines that produce comparable voltammetric responses [10]. A summarized list of these advantages and disadvantages is found in Table 7.1.

7.1.4 Applications of Cyclic Voltammetry

Electrochemical techniques have been employed in order to identify a vast range of biomolecules. In combination with nanotechnology, they can develop low-cost, ultrasensitive, reliable, simple preparation routes, reproducible, portable, and easy-to-use devices. In electrochemical sensors, a biological event can be transformed into a quantifiable electrochemical signal, such as changes in current, impedance, and/or potential. The possibility of rapid and real-time detection makes this technology a potential tool for biosensing. A biosensor coupled with any electrochemical technique is able to quantify from small molecules to viruses in the order of nanometers [12]. For instance, miRNA is a biomarker that has recently been used for timely cancer detection. Modified graphene oxides with gold nanoparticles allowed the development of a device which detects miRNAs within plasma samples with great reproducibility and performance [13]. Another example is the fabrication of a biosensor using enzymes as biorecognition sites [14, 15]. The presence of high levels of uric acid can cause cardiovascular disease and hypertension, while low levels of it may lead to Alzheimer's disease. Cyclodextrins immobilized on carbon platforms allowed the electrochemical detection of uric acid with a detection threshold as small as 0.06 μM [16]. Additionally, electropolymerized β-cyclodextrin on boron-doped reduced graphene oxide presented a limit of detection (LOD) of 0.6 nmol L^{-1} for flunitrazepam determination [17]. Another application of cyclodextrins is for the detection of spermine, a biomarker for cancer cells with a LOD of 0.001 mg L^{-1} which was achieved through benefiting from CV within an electrochemical setup [18].

The identification of protein biomarkers in the early stages of an illness could potentially pose a great challenge. Different cancer types including breast, ovarian, colon, and lung are associated with the serum protein called carcinoembryonic antigen (CEA). When the concentration of CEA is superior to 5 ng/ml, there is a high probability of developing a cancerous tumor. Timely detection of high CEA levels allows the individuals to receive appropriate attention, thus increasing the chances of preventing the disease. Conventional methods to identify CEA have drawbacks such as high costs, long response times, and complex procedures. Electrochemical techniques are attractive alternatives considering their portability, inexpensiveness, brief operation time, and ease of handling. Duy Thanh Tran et al. developed a sensing platform with high sensitivity and low DL for the electrochemical detection of CEA. This novel platform consisted of a combination of carbon nanotubes (CNTs) with microcarbon spheres (CSs) modified with copper–gold (Cu–Au) nanocrystals, and anti-CEA which was immobilized as a biorecognition site (Fig. 7.4a). The detection method was based on monitoring the oxidation–reduction reactions with potassium ferricyanide (K_4 [$Fe(CN)_6$]) as a redox couple. When CEA interacts with the

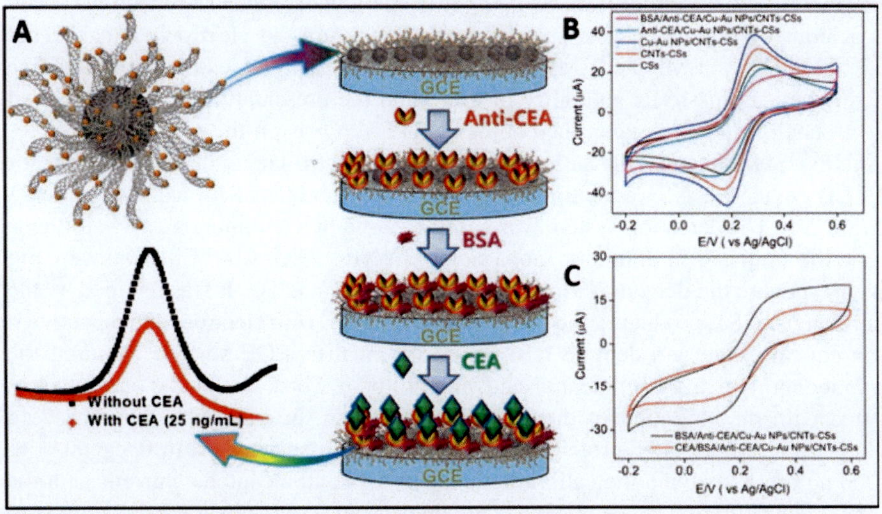

Fig. 7.4 **a** Fabrication of an electrochemical sensor for the identification of CEA. **b** CV curves of modified GCE in 0.1 M KCl with 5 mM $K_4[Fe (CN)_6]$ at a scan rate of 50 mV s^{-1} and **c** CV curves of BSA/anti-CEA/Cu-Au NPs/CNTs-CSs modified GCE with and without 25 ng ml^{-1} of CEA in 0.1 M KCl containing 5 mM K4[Fe $(CN)_6$] at a scan rate of 50 mV s^{-1}. Reproduced (or adapted) with permission [19], © 2018, Elsevier

sensing platform, the oxidation–reduction currents of the redox couple decreases. Figure 7.4b displays the electrochemical evaluation of each component present in the sensing platform and their effect on the performance of the redox reaction. The good electrical conductivity of the metal nanoparticles and the carbon structures favored the redox reaction, manifesting a hike in the peak currents of cathodic and anodic (blue and red lines). On the other hand, the presence of CEA and bovine serum albumin (BSA) on the platform surface decreased the electrochemical area, therefore affecting the performance of the redox couple (pink and green lines). The proposed sensing platform was successful in detecting CEA due to the decrease in peak currents associated with the redox couple in 25 ng ml^{-1} of CEA. The authors stated that CEA was adsorbed on the surface of the modified electrode forming an immunocomplex layer among CEA and anti-CEA (Fig. 7.4c) [19].

Dengue fever is a fatal disease that affects about 2.5 billion people which can further manifest into more acute phases such as dengue hemorrhagic fever (DHF) known as one of the most lethal symptoms. There are 4 different serotypes of dengue (DENV1, DENV2, DENV3, and DENV4) which could lead to dengue fever, and if not detected in time, may cause serious complications such as DHF or dengue shock syndrome (DSS). The flavivirus NS1 protein that circulates within the bloodstream from the starting point of the illness is utilized for the timely detection of dengue. Electrochemical sensors produce an electrical signal from a biological event, allowing the possibility of miniaturizing the system in a user-friendly, cost-effective, and portable fashion. Prima Dewi Sinawang et al. developed screen-printed gold electrode (SPGE) for the electrochemical detection of the NS1 protein where anti-dengue NS1 was immobilized, and polyethylene glycol (PEG)-stabilized was sued as a biorecognition site. The species used to measure the electrochemical signal was the radical TEMPO (TEMPO•) (Fig. 7.5a). The authors explained the choice of this radical due to its solubility in water and the presentation of a well-defined electrochemical signal when the conjugated TEMPO• is on the gold nanoparticles (AuNPs). Subsequently, the authors developed an electro-lateral flow immunosensor (ELLI) device for NS1 detection using the SPEG modified with anti-dengue NS1 (Fig. 7.5b). The detection system was based on a sandwich immunoassay as follows: First, the synthesized immunonanoparticles (AuNPs–PEG–Ab–TEMPO•) were put in contact with the dengue NS1 protein present in the analyte. It was retained by the anti-dengue NS1, producing the first immunocomplex with electrochemical activity. Subsequently, the anti-dengue NS1 anchored on the SPGE surface captured the immunocomplex to complete the sandwich protocol. Thus, the anodic current associated with the oxidation of the TEMPO• radical in the sandwich was measured using CV. Dengue NS1 protein concentrations were selected from 0 ng ml^{-1} to 1000 ng ml^{-1} to obtain the calibration curves. A rise in the anodic current with the greater concentration of dengue NS1 protein was observed, which led the authors to claim that the system was capable of detecting dengue NS1 protein at concentrations higher than 50 ng ml^{-1} (Fig. 7.5c) [20].

Functionalization of the electrode surface increases the performance of electrochemical sensing devices considering that the inclusion of functional groups improves the surface interaction with the analyte of interest. Keita Nishimura et al.

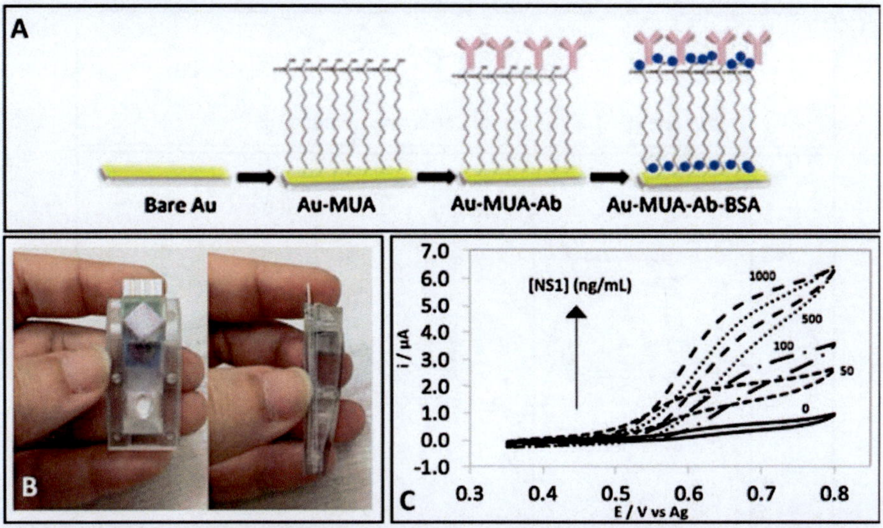

Fig. 7.5 **a** A schematic of the detailed procedure of screen-printed gold electrode (SPGE): gold (Au) surface, mercaptoundecanoic acid (MUA), and bovine serum albumin (BSA). **b** The electro-lateral flow immunosensor (ELLI) device. **c** Cyclic voltammograms of the detection of the NS1 protein over the lateral flow immunosensor by TEMPO• label on the formulation of electroactive immunonanoparticles. Reproduced (or adapted) with permission [20], © 2018, Elsevier

studied the effect of electrochemical functionalization with a 2 M sulfuric acid (H_2SO_4) solution on single-walled carbon nanotubes (SWNTs). The alterations in SWNTs morphology and electron transfer kinetics were evaluated, as well as the sensitivity changes in the detection of dopamine (DA). DA is a neurotransmitter that serves as a biomarker for Parkinson's disease and schizophrenia, in addition to being a stress marker. DA exhibits electrochemical activity by reversibly oxidizing to o-dopamine quinone (Fig. 7.6a). The authors stated that the current of the DA redox reaction increased at the functionalized electrode; hence, the sensitivity in DA detection also improved. This could be as a result of the functionalized surface impacting the electron transfer rate and the interaction between the electrode surface and the inner sphere redox species as with the DA (Fig. 7.6b) [21].

7.2 Electrochemical Impedance Spectroscopy for Material Characterization

7.2.1 History of Electrochemical Impedance Spectroscopy

Electrochemical impedance spectroscopy (EIS) is an electrochemical analytical method that can be utilized to observe biomolecular events at the surface of the

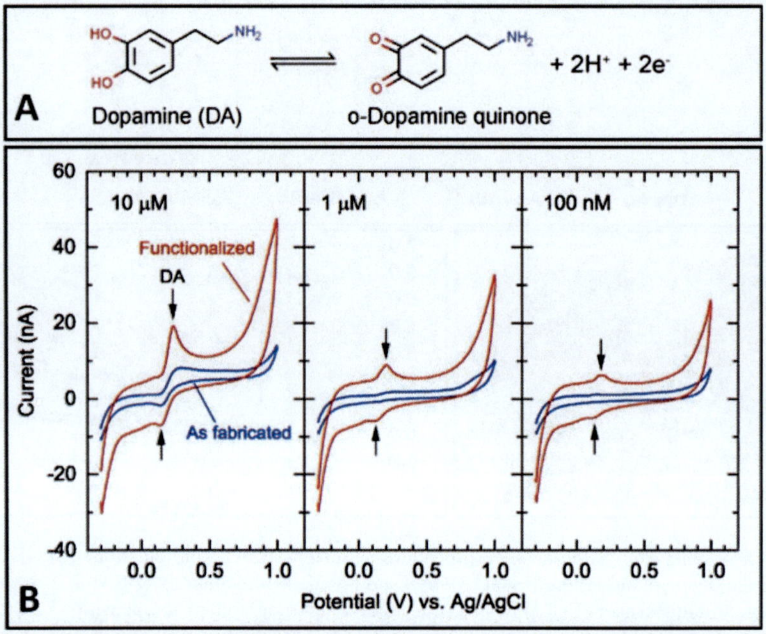

Fig. 7.6 **a** Redox reaction for dopamine. **b** Dopamine cyclic voltammograms obtained for as-fabricated and functionalized SWNTs with dopamine concentrations of (left) 10 μM, (middle) 1 μM, and (right) 100 nM. Reproduced (or adapted) with permission [21], © 2019, Elsevier

electrode through analyte interaction. EIS requires a large range of frequencies to measure the impedance properties of an electrochemical system. In comparison with other methods of biorecognition, electrochemical sensors are fast, portable, and cost effective [22, 23]. EIS was developed in 1894 by Oliver Heaviside, who coined the terms "impedance," "admittance," and "reactance" (Fig. 7.7). The origins of EIS date back to the late nineteenth century with the work of Heaviside, who introduced the linear systems theory (LST), formulated operational calculus, and established the operational impedance. In 1894, Walther Nernst measured the dielectric values of aqueous electrolytes and some organic fluids by means of the Wheatstone bridge. Later, in 1899, Emil Warburg extended the application of impedance to electrochemical systems by publishing an article on the impedance of diffusional transport of an electroactive species. In particular, the invention of the potentiostat in the mid-1900s and the introduction of frequency response analyzers in the 1970s were the major contributors to expansion of the field of EIS [22, 24, 25].

The implementation of EIS in electrochemical and corrosion mechanism research came to light when the potentiostat and the frequency response analyzer were developed as it enabled the examination of electrochemical interfaces at different frequencies and within the sub-millihertz range. The use of EIS was extended into biological systems by 1925 when Hugo Fricke obtained information of blood suspensions. Later, Kenneth Cole used EIS to collect data from urchin eggs and cell membranes

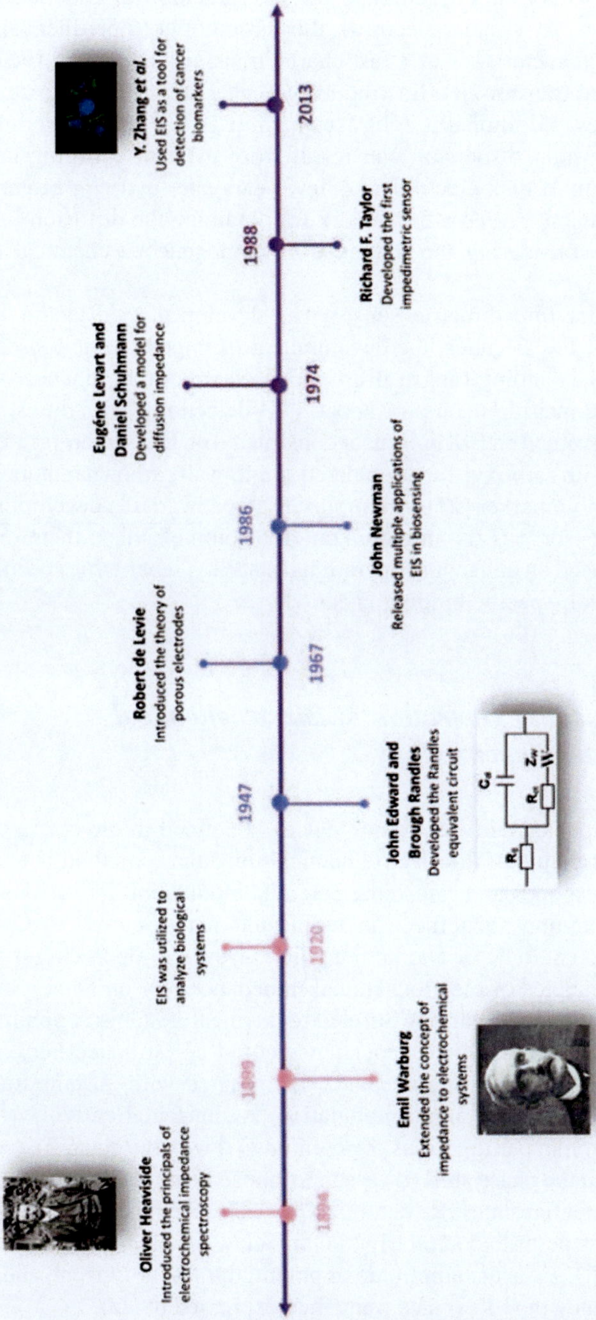

Fig. 7.7 History of EIS

(Fig. 7.7) [22, 24]. In 1947, John Edward Brough Randles developed Randles's equivalent circuit. This system facilitated the interpretation of EIS data obtained from solid interfaces. At a planar electrode, this circuit offers a pertinent simulation of the impedance characteristics of a fast charge transfer reaction. In 1967, Robert de Levie formulated transmission line models to calculate the impedance response of porous electrodes. Additionally, John Newman in 1970 demonstrated that high-frequency time-constant dispersion can result from the non-uniform current and potential distribution of disk electrodes. A few years later in 1974, Eugéne Levart and Daniel Schuhmann created a framework for obtaining the diffusion impedance of a rotating disk considering the influence of homogeneous chemical reactions [22, 24, 25].

In 1988, the first impedimetric sensor was developed by Richard F. Taylor et al. Over the next few decades, the development of impedimetric biosensors has progressed. In 2004, Dianping Tang et al. created an electrode that enhanced the sensitivity of impedance measurements for hepatitis B detection [26]. Almost a decade later, Zhang et al. invented an EIS biosensor constructed on five screen-printed carbon electrodes coated with carboxyl-functionalized single-wall carbon nanotubes in order to identify a cancer biomarker [27]. An important step toward the development of an impedimetric biosensor was the immobilization of biomolecules; therefore, future advancements focused on enhancing operational stability, sensitivity, specificity, and resistance toward non-specific binding [27].

7.2.2 Mechanism of Operation of Electrochemical Impedance Spectroscopy

The concept of impedance in electronic devices is defined in the context of resistance. The basic principle of the electrochemical impedance method is to calculate how an electrode responds to a sinusoidal potential modulation for various frequencies. Regularly, common phenomena including alternating current (AC) modulations are applied to cathodic or anodic potential, or even to open-circuit potential. The numerical principles of electrochemical impedance rely on Ohm's law where the dependency of potential current for electrochemical systems is nonlinear [28]. However, in some cases, the cited dependence could be estimated between 5 and 10 mV; thus, the calculation of impedance is executed with an amplitude of 5–10 mV under a sinusoidal potential modulation. A sinusoidal current, $I(t)$, can be induced from sinusoidal perturbations of potential $E(t)$ with the equal frequency (ω). This frequency with the phase shift (ϕ) is superimposed onto the steady-state current according to the potential state. The definition of electrochemical impedance of electrode reaction (Z) is described similarly to Ohm's law for physical electric circuits. According to Eq. 7.1, E_0 is the amplitude of potential, I_0 is the current, and $\omega = 2\pi f$ is the radial frequency in rad/s (f is a frequency expressed in Hz).

$$Z(\omega) = \frac{E(t)}{I(t)} = \frac{|E_0|\sin\,\sin(\omega t)}{|I_0|\sin\,\sin(\omega t - \phi)} = Z_0\frac{\sin\,\sin(\omega t)}{\sin\,\sin(\omega t - \phi)} \tag{7.1}$$

In a general sense, there are some components for the equivalent circuits of an electrochemical capacitor that maintain a capacitor connected to a resistance, an inductor, and an equivalent in series with a resistance. These are used to estimate the experimental impedance values with the distributed impedance elements, connected in series and/or in parallel. Impedance is the opposing force to electrical current in a circuit where its measurement unit is the same as resistance, ohms (Ω) [29]. Nevertheless, these differ considering resistance is defined as the opposition to the flow of electric current in the circuit, while impedance controls the flow of AC due to any components that are resistive, inductive, or capacitive. The key distinction between these two concepts is their behavior toward AC and direct current (DC). Under certain conditions, resistance can only be understood in an AC circuit: (a) The AC voltage is in phase with the current, (b) frequency independency is the main feature of the resistance, and (c) the resistance is applicable to all currents and voltages. As capacitive and/or inductive outcomes are noticeable with the majority of frequencies and the phase angle is not null in most cases, therefore resistance can be defined as the function of frequency.

Impedance demonstrates a quantitative concept of the opposition force to electrical current. A specified sinusoidal voltage is performed in the majority of EIS experiments by a potentiostat through a 3-electrode cell. This includes a solution of electrolyte harboring the researched molecule. The amplitude and load of the specified sinusoidal voltage are dependent on the type of molecular system of interest [30]. For instance, as the biological molecule structures are denatured, they are frequently influenced by smaller voltages than non-biological specimens. Another method for using EIS experiments within the biological applications is to functionalize the electrode surface. This would attract the analyte of interest to the surface of the electrode through an obstruction of a "molecularly wired" entrance process, thus affecting the conductivity of the system. After the composite and target analyte were immobilized on the electrode surface, impedance measurements can be conducted in a common redox probe (for instance, potassium ferricyanide (II), hexamine ruthenium (III) chloride, or a blank buffer solution) [31]. Consequently, when the load is conducted at a given frequency, a current will flow through the electrochemical cell which can be then obtained by the potentiostat. The registered flow current is converted by a software into an impedance measurement with an actual and predetermined element. The procedure would be duplicated for both components over a frequency range where different values are recorded [32].

As represented in Fig. 7.8a, in EIS experiments with potentiostat, the impulse is a small AC sine wave in the potential performed additionally to the DC potential. The behavior of the current to the electrochemical system includes an AC component translated into phase shifts from that of potential phase. The magnitude and phase of the AC mechanisms of the potential and current are considered as the phase. Additionally, kinetic information can be obtained from the magnitudes of the complex numbers E ~ and j ~. For example, their rate indicates the impedance of the circuit.

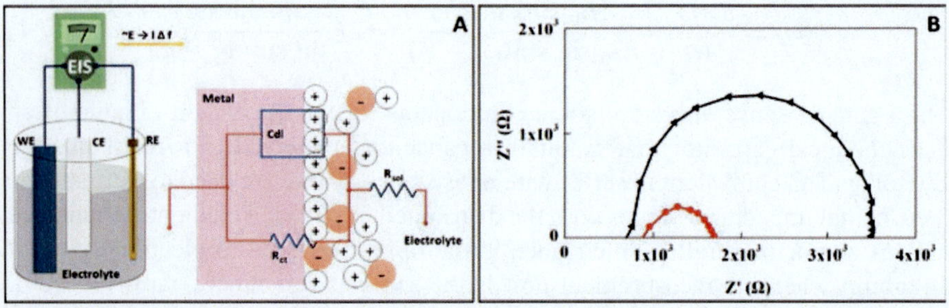

Fig. 7.8 **a** Illustration of the EIS test configuration. **b** Nyquist plot for a pure silicone-based material. Reproduced (or adapted) with permission [35], © 2014, American Physical Society

Experimentally, a function generator produces the small amplitude sine wave voltage (normally 5 mV rms), which is performed with a DC voltage in the electrochemical cell via a potentiostat. The reaction to this potential voltage is a current which overall contains both AC and DC components. The current is obtained through the current-to-voltage converter within the potentiostat or by calculating the voltage. Finally, a tool such as a frequency response analyzer (FRA) recognizes the amplitude and phase of the AC element of the current and the potential. Lastly, the amplitudes are converted into an impedance measure at the frequency of the sine wave which would be repeated at different frequencies in order to provide an impedance spectrum [33].

EIS spectra provide valuable information with respect to the electrochemical properties of the samples. The complex-impedance plane diagram, or Nyquist plot, maintains the polar plots and ensures the stability of the closed-loop control systems. In another word, the comprehensive frequency response of the open-loop transfer function is derived from Nyquist plots. It is common to plot $-Z_{Im}$ (also as $-Z''$ or $-Zj''$) on the y-axis, which results in fitting data into the first quadrant of a graph. Even though the Nyquist plot is useful for recognizing the characteristic features of the system, the frequency information is inherently lost [34]. It is important to identify the frequencies of vital measurements such as superior and inferior real axis intercepts, and the characteristic frequency of an arc, ω_c. This value displays a peak $-Z_{Im}$ for a system. A Nyquist plot example is shown in Fig. 7.8b.

7.2.3 Advantages and Disadvantages of Electrochemical Impedance Spectroscopy

There are two main factors which draw specific attention to EIS: First, EIS data can be utilized to identify physical properties, such as diffusion coefficients, chemical reaction rates, and microstructural characteristics of the electrochemical system. Second, the execution of an EIS analysis is rather straightforward. Moreover, EIS is a versatile technique that allows its application in a wide range of fields, including electrocatalysis and medicine [36]. Biomolecular events such as affinity interactions between peptides, receptors, nucleic acids, cells, and antibodies can occur at the surface of electrodes [23]. Such events, in turn, provide physical, chemical, and structural information of the surface at the micrometer scale [36]. Additionally, electric and dielectric properties of compounds under investigation can be obtained and analyzed using EIS [37]. Unlike other electrochemical techniques, EIS enables the measurement of the correlation between the current and applied voltage through the estimation of its transfer function within the frequency domain [36]. These measurements are possible for a wide range of frequencies, are characterized without requiring labels, and have high signal-to-noise ratio [23, 38, 39]. Rapid testing times, its sensitivity, ease of use, and non-destructive nature are among other known beneficial features of this technique (Table 7.2).

An important drawback of EIS is the ambiguity in the result interpretation. However, statistics and modeling can be implemented to overcome these issues [36]. Nonetheless, since the accuracy of the results depends on such fitting models, the ambiguity of the results can be aggravated [38]. An expensive instrumentation represents another limitation of EIS [39]. The application of EIS presents additional

Table 7.2 Advantages and disadvantages of electrochemical impedance spectroscopy

Advantages	Disadvantages
Wide range of applications, including the monitoring of biomolecular events at electrode surface [23, 36]	Susceptible to interpretation ambiguity [36]
Provides physical properties, chemical and microstructural information on surfaces [36]	
Provides information on electric and dielectric properties of samples [37]	Accuracy of results depends on fitting models [38]
Enables the estimation of the transfer function between current and applied voltage [36]	
Allows measurements at a wide range of frequencies [23, 38]	Expensive instrumentation [39]
Detection is label free [23]	
High SNR [23, 39]	Sensitive to surrounding environment thus requires Faraday cage [40]
Rapid test [38]	
Sensitivity [23]	Bulky experimental setups [40]
Ease of use [36]	
Non-destructive [37]	Requires theoretical simulation for data analysis [40]
Statistics and modeling can be used to overcome ambiguities in result interpretation [36]	
Signal-to-noise ratio (SNR)	

drawbacks, for instance, sensitivity to the surrounding environment, thus the need for a Faraday cage to mitigate noise. Moreover, the large experimental configurations and requiring theoretical simulation in order to analyze the data pose additional challenges [40].

7.2.4 Applications of Electrochemical Impedance Spectroscopy

The timely detection of diseases is a vital step for prevention and adequate treatments. EIS is an electrochemical technique that has been widely used in recent years as a detection tool considering it is easy to use, fast, and highly sensitive. The electrodes are modified/immobilized with biorecognition molecules that interact with the analytes of interest, hence altering the electrochemical response of the system. In this sense, the signal obtained could be attributed to (a) a non-faradaic process in which mainly the capacitance generated on the electrode/electrolyte interface is modified due to the interaction of the analyte with the surface of the modified electrode or (b) a faradaic process in which there is an electron transfer among a species present in the electrolyte and the modified electrode. The species that reacts electrochemically can be the analyte or a redox pair used indirectly by measuring the concentration of the analyte [23, 41, 42]. The advantages of EIS over other electrochemical techniques (e.g., CV) are that the biological species are not damaged during detection and the quantification signal is easy to obtain [43–45]. In addition, EIS allows the use of only two electrodes for detection, facilitating the fabrication of a laboratory on a chip (e.g., interdigitated electrodes) [46]. Some of the applications using non-faradic processes include the analysis of blood samples, detection of E. coli by aptamers, and detection of pre-osteoblast cells [47–49]. Moreover, biosensors based on faradaic processes for identification of the vascular endothelial growth factor, electrochemical detection of urea, cancer biomarkers, detection of NeutrAvidin, detection of active botulinum neurotoxin, and enzyme-based impedimetric biosensors are reported within the literature [50–56].

Lysozyme (concentration ~ 0.5–2.0 mg ml^{-1} in mammals) is a protein that functions in the human body for anti-histaminic, anti-viral, anti-inflammatory, and anti-tumor activities. Changes in the concentration of this protein can refer to occurrence or progression of different illnesses including leukemia and meningitis indicated with an increase of lysozyme in urine or serum. Wei Chen et al. developed an electrode capable of identifying lysozyme in a simple and quick manner. The detection system consisted of a gold electrode modified with the anti-lysozyme aptamer (biorecognition site) and electrodeposited AuNPs to enhance electrical conductivity (Fig. 7.9a). EIS principals were used to monitor the presence of the lysozyme protein. The biomolecular receptor was immobilized on the Au surface and acts as a barrier for electron transfer and an electrical insulator. Consequently, if there was a redox couple in the analyte, the interaction between the protein and anti-protein will

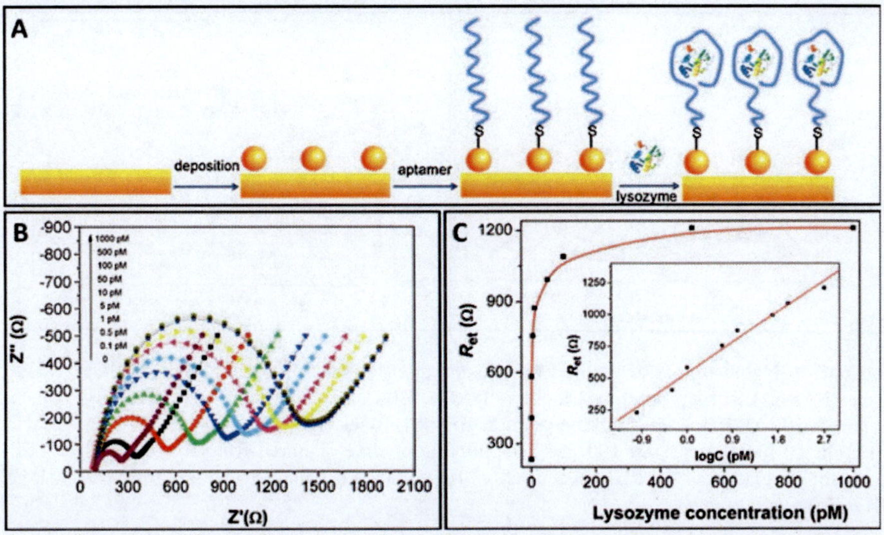

Fig. 7.9 a Illustration of an aptasensor with the procedure. **b** Nyquist plots of the aptamer/AuNPs/gold electrode following the interaction with a range of lysozyme concentrations between 0.1 and 500 pM. **c** The R_{et} dependence of the aptamer/AuNPs/gold electrode on the concentration of lysozyme. The inner graph presents a linear R_{et} with the logarithm of lysozyme concentration over the range of 0.1–500 pM. Reproduced (or adapted) with permission [43], © 2011, Elsevier

prevent the electrical transfer along the redox couple and the electrode. The Nyquist plots obtained at different concentrations of lysozine indicated how the resistance to electron transfer (R_{et}) increased with increasing protein concentration from 0.1 to 1000 pM (Fig. 7.9b). The calculated limit of detection (LOD) for this lysozyme aptasensor was 0.01 pM, and a linear trend of R_{et} with the logarithm of the protein concentration at the range of 0.1–500 pM resulted in an $R_{et} = 539.676 + 269.426 \log C$ (Fig. 7.9c) [43].

The development of efficient tools for the detection of tumor markers allows timely treatment of patients. An extracellular protein such as autocrine motility factor (AMF) stimulates the active migration of tumor cells and can be found in urine and blood serum. An important parameter in the motility of tumor cells is the interaction of AMF with its receptor AMFR/gp78 which has been associated with different types of cancer including prostate, breast, and bone among others. Devillers et al. presented an electrochemical biosensor based on F6P enzyme immobilization on gold (biorecognition phase) for the identification of a model protein named phosphoglucose isomerase (PGI) from rabbit muscle. EIS served as an electrochemical technique to indirectly detect the cancer biomarker AMF through the electrochemical activity of the redox couple. The authors used multiple concentrations from 10 fM to 100 nM of the model protein and observed in the Nyquist plots that increasing the enzyme concentration increased the resistance to electron transfer among the redox couple and the electrode surface (Fig. 7.10a). This behavior was attributed to the

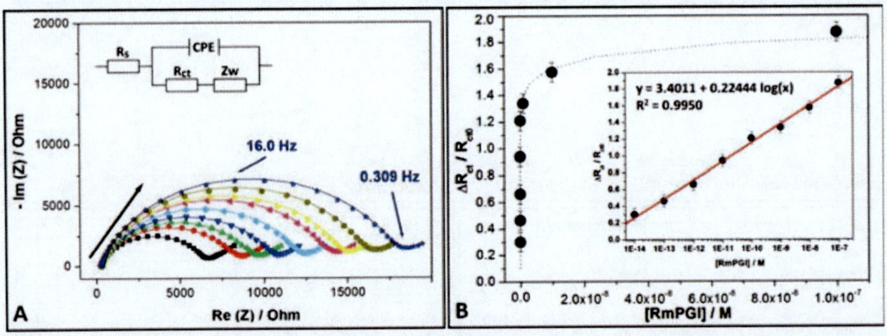

Fig. 7.10 **a** Nyquist plots of the PGI biosensor response (Black square: biosensor alone, Black circle: 10 fM, Black up-pointing triangle: 100 fM, Black down-pointing triangle: 1 pM, Black diamond: 10 pM, Black left-pointing pointer: 100 pM, Black right-pointing pointer: 1 nM, Black hexagon: 10 nM, Black star: 100 nM). **b** Calibration curve acquired from EIS data of the F6P biosensor in relation to the PGI concentration. Reproduced (or adapted) with permission [44], © 2017, Elsevier

fact that the interaction on the electrode surface among the F6P enzyme and the PGI protein prevents the redox couple from diffusing toward the electrode and reacting electrochemically. From the data obtained, the authors calculated a LOD of 6.6 fM over a range of PGI concentrations from 10 fM to 1 nM (Fig. 7.10b) [44].

Effective screening and capture of cancer cells aid patients to receive timely treatment and increase their chances of curing the disease. It is known that the concentration of folate receptor (FAR) correlates to the appearance of cancerous tumors; for example, ovarian tumors could present 34.31 pmol FAR/mg protein. Liping Du et al. developed an electrochemical biosensor for FAR-positive cells (HeLa cells) where the recognition site was folic acid immobilized on zirconium MOFs (UiO-66) using a gold electrode (Fig. 7.11a). The capture of HeLa cells on the surface of UiO-66 influences the system impedance since the characteristics of the electrode/electrolyte interface (e.g., resistance to electron transfer) are altered. When the concentration of the captured HeLa cells increased, the diameter of the semicircle in the Nyquist plot also increased assuming the HeLa cells adhered to the surface of the modified electrode blocked the diffusion of the redox probe to the surface (Fig. 7.11b). The calculated LOD was 90 cells ml^{-1} which was obtained from the changes in electron transfer resistance using Randles's equivalent circuit where Z_w is the Warburg impedance (diffusion redox probe), C is the capacitance (electrode/electrolyte interface), and R_s is the resistance to the solution (Fig. 7.11c) [45].

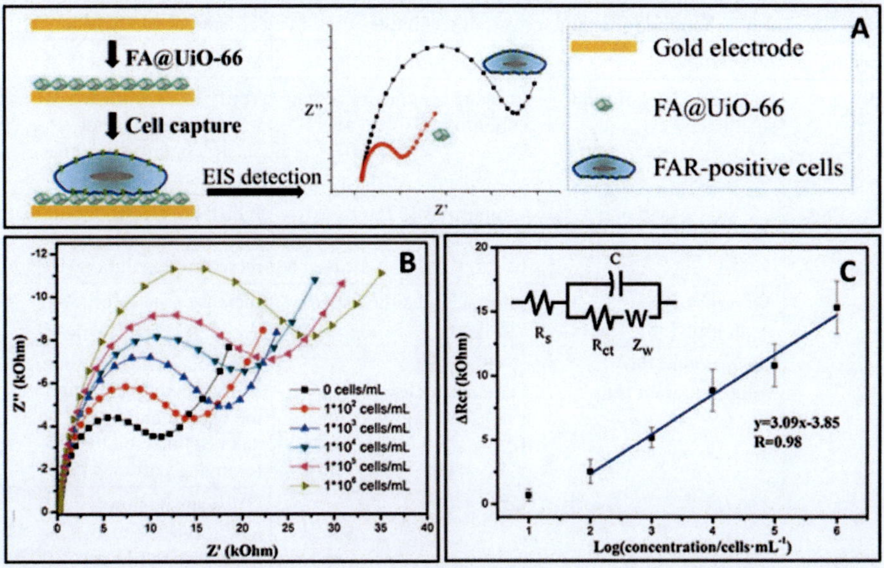

Fig. 7.11 **a** Elaboration of the FA@UiO-66-modified electrode to detect FAR-positive cells. **b** Nyquist plots of the gold electrode-functionalized FA@UiO-66 with various concentrations of HeLa cancer cells. **c** EIS calibration curves of FA@UiO-66-modified sensing electrodes; inside is the electrical equivalent diagram. Reproduced (or adapted) with permission [45], © 2019, Elsevier

7.3 Troubleshooting of Electrochemical Analysis Techniques

Voltammetry experiments measure the resulting current of an applied potential sweep on an electrochemical system. CV differs only in the fact that a linear potential is inversely increased once the linear potential sweep reaches the set value and subsequently returns to the initial potential [57, 58]. To perform a correct cyclic voltammetry experiment, the solvent and the analyte must be unreactive and must not be redox active in the selected scan window. Analytes can also be adsorbed to the electrode surface; therefore, before analyzing its reactivity, its homogeneity in solution should be assessed (Table 7.3) [57].

If the scan rate, the speed where the applied potential is scanned, is inadequately fast, it can lead to a voltammogram with no peak separations. This is due to the formation of a layer of the reduced analyte at the surface of the electrode, known as a diffusion layer [58]. The WE, which might be altered according to the experiment, is made from a redox-inert material. The desired electrochemical event occurs on its surface; thus, it is necessary to keep an extremely clean and well-defined area. Additionally, to ensure the kinetics of the reaction, it is important to have a greater surface area on the CE than the surface area of the WE (Table 7.3) [58].

The reference electrode must be maintained in good condition with an internal reference compound of a known potential. An issue that may present when using an

Table 7.3 Troubleshooting for characterization techniques used in electrochemical analysis

Technique	Problem	Cause	Solution
CV	Unseemly background scan [57]	Electrochemically active solution [57]	Verify that the scan window range is not electrochemically active for the solution [57]
		Impurities [57]	Polish the electrodes and glassware in order to recreate the solution [57]
	No peak-to-peak separation [58]	Surface-adsorbed species [58]	Increase scan rates [58]
	Distortion of the voltammogram [58]	WE surface contamination [58]	Electrode's surface should be polished (according to the type of electrode), or it can be subjected to removing particles [58]
			CV scans in simple electrolyte are also necessary to remove species [58]
	Increased peak-to-peak separation in the voltammogram in an electrochemically reversible experiment [58]	Ohmic drop [58]	Reduce the size of the WE to decrease the current or use low scan rates [58]
			Increase the electrolyte concentration [58]
			Place the working and reference electrode closer [58]
EIS	Unclear interpretation of the resulting spectra [25]	Non-uniform current flow [25]	Use an electrode where the current and potential distribution is constant [25]
	Signal-to-noise ratio, stochastic errors [25]	Selection of experimental parameters [25]	Increase integration time at each frequency as many cycles as required [25]
			Increase the modulation amplitude [25]
	Error in the impedance response [25]	Transient [25]	Apply a one- or two-cycle delay between the change of frequency and impedance measurement [25]
			Disregard the first obtained frequency when applying regression techniques [25]

CV Cyclic voltammetry; *EIS* electrochemical impedance spectroscopy; *WE* working electrode

AgCl/Ag electrode is the silver ion (Ag^+) leakage into the rest of the analyte solution which could lead to interference with the electrochemical measurements [58]. An inert atmosphere should be prepared at the assembled cell. Due to the electrochemical response of analytes, these can be altered by the presence of oxygen; thus, the electrolyte solution should be treated with an inert gas before the electrochemical experiment (Table 7.3) [57, 58].

Optimization of the parameter for EIS settings maximizes the values of the obtained spectrum. However, the parameter settings may be inapplicable to all impedance devices [25]. In impedance measurements, each frequency involves the transition of time-domain signals to a complex value. Therefore, if the integration time at each frequency is increased, the chances of errors can be minimized as well. The number of cycles required for every frequency, in turn, corresponds to the number of the calculation [25]. For systems that have a broad range of potential, the non-arbitrary miscalculations may be greatly mitigated by means of a wide adjustment amplitude. Another error in impedance measurements is a transient that could be observed in the change from one frequency to another; therefore in the difference between the frequency and impedance, a delay of one or two cycles is recommended. Additionally, a start-up transient might corrupt the first frequency of measurement; thus, it is suggested to ignore the first measurement (Table 7.3) [25]. Table 7.3 offers a number of possible errors when dealing with electrochemical characterizations of a sample, the possible causes, and the recommended solutions to address such problems.

References

1. K. Lubert, History of Electroanalytical Methods, pp. 1937–1946 (2010). https://doi.org/10.1002/elan.201000087
2. A.J. Bard, C.G. Zoski, Voltammetry restrospective. Anal. Chem. 346–352 (2000)
3. J.M. Díaz-cruz, N. Serrano, C. Pérez-ràfols, C. Ariño, M. Esteban, Electroanalysis from the Past to the Twenty-First Century : Challenges and Perspectives, pp. 2653–2661 (2020)
4. C. Batchelor-mcauley, E. Kätelhçn, E.O. Barnes, R.G. Compton, E. Laborda, A. Molina, Recent Advances in Voltammetry, pp. 224–260 (2015). https://doi.org/10.1002/open.201500042
5. M. Armstrng-james, J. Millar, Carbon fibre microelectrodes. J Neurosci. Methods **1**, 279–287 (1979)
6. D. Guziejewski, Electrode mechanisms with coupled chemical reaction—Amplitude effect in square-wave voltammetry. J. Electroanal. Chem. **870**, 114186 (2020). https://doi.org/10.1016/j.jelechem.2020.114186
7. F. Harnisch, S. Freguia, A basic tutorial on cyclic voltammetry for the investigation of electroactive microbial biofilms. Chem. Asian J. **7**(3), 466–475 (2012). https://doi.org/10.1002/asia.201100740
8. O. Guy, K. Walker,-Silicon carbide biotechnology, and undefined 2016, in Graphene Functionalization for Biosensor Applications. Elsevier
9. J. Hoyos-Arbeláez, M. Vázquez, J. Contreras-Calderón, Electrochemical methods as a tool for determining the antioxidant capacity of food and beverages: A review. Food Chem. **221**, 1371–1381 (2017). https://doi.org/10.1016/j.foodchem.2016.11.017
10. J.G. Roberts, L.A. Sombers, Fast-scan cyclic voltammetry: Chemical sensing in the brain and beyond. Anal. Chem. **90**(1), 490–504 (2018). https://doi.org/10.1021/acs.analchem.7b04732

11. J. Sworakowski, How accurate are energies of HOMO and LUMO levels in small-molecule organic semiconductors determined from cyclic voltammetry or optical spectrosco. Synth. Met. **235**, 125–130 (2018). https://doi.org/10.1016/j.synthmet.2017.11.013
12. J.R. Anusha, B. Chul, K. Yu, C.J. Raj, Electrochemical biosensing of mosquito-borne viral disease, dengue : A review. Biosens. Bioelectron. **142**(June), p. 111511 (2019). https://doi.org/10.1016/j.bios.2019.111511
13. M. Azimzadeh, M. Rahaie, N. Nasirizadeh, K. Ashtari, H. Naderi-manesh, An electrochemical nanobiosensor for plasma miRNA-155, based on graphene oxide and gold nanorod, for early detection of breast Cancer. Biosens. Bioelectron. (2016). https://doi.org/10.1016/j.bios.2015.09.020
14. C. Pan, H. Wei, Z. Han, F. Wu, L. Mao, Enzymatic electrochemical biosensors for in-situ neurochemical measurement. Curr. Opin. Electrochem. (2019). https://doi.org/10.1016/j.coelec.2019.12.008
15. J. Yoon et al., Flexible electrochemical glucose biosensor based on GOx/gold/MoS 2/gold nano fi lm on the polymer electrode. Biosens. Bioelectron. **140**(May), p. 111343 (2019). https://doi.org/10.1016/j.bios.2019.111343
16. K.P. Aryal, H.K. Jeong, Functionalization of β-cyclodextrin into ambient plasma modified carbon nanotube-thermally reduced graphite oxide for electrochemical sensing of uric acid. Mater. Chem. Phys. **238**(July), 121899 (2019). https://doi.org/10.1016/j.matchemphys.2019.121899
17. M. Hossein, Z. Norouzi, M. Mahdi, Using a nanocomposite consist of Boron-doped reduced graphene oxide and electropolymerized β-cyclodextrin for Flunitrazepam electrochemical sensor. Microchem. J. **156**(May), 104994 (2020). https://doi.org/10.1016/j.microc.2020.104994
18. S. Shrivastava, P. Kumar, N. Verma, Spermine biomarker of cancerous cells voltammetrically detected on a poly (β-cyclodextrin)—electropolymerized carbon film dispersed with Cu—CNFs. Sensors Actuators B. Chem. **313**(January), 128055 (2020). https://doi.org/10.1016/j.snb.2020.128055
19. D. Thanh, V.H. Hoa, L. Huu, N. Hoon, J. Hee, Cu-Au nanocrystals functionalized carbon nanotube arrays vertically grown on carbon spheres for highly sensitive detecting cancer biomarker. Biosens. Bioelectron. **119**(July), 134–140 (2018). https://doi.org/10.1016/j.bios.2018.08.022
20. P. Dewi, L. Fajs, K. Elouarzaki, J. Nugraha, R.S. Marks, Chemical TEMPO-based immunolateral flow quantitative detection of dengue NS1 protein. Sensors Actuators B. Chem. **259**, 354–363 (2018). https://doi.org/10.1016/j.snb.2017.12.043
21. K. Nishimura, T. Ushiyama, N. Xuan, M. Inaba, S. Kishimoto, Y. Ohno, Enhancement of the electron transfer rate in carbon nanotube flexible electrochemical sensors by surface functionalization. Electrochim. Acta **295**, 157–163 (2019). https://doi.org/10.1016/j.electacta.2018.10.147
22. J.V Rushworth, A. Ahmed, P.A. Millner, D.J. Pike, N.A. Hirst, J.A. Goode, Impedimetric Biosensors for Medical Applications: Current Progress and Challenges. Momentum Press, New York [New York] (2013)
23. L.G. Zamfir, M. Puiu, C. Bala, Advances in electrochemical impedance spectroscopy detection of endocrine disruptors. Sensors (Switzerland) **20**(22), 1–21 (2020). https://doi.org/10.3390/s20226443
24. D.D. MacDonald, Reflections on the history of electrochemical impedance spectroscopy. Electrochim. Acta **51**(8–9), 1376–1388 (2006). https://doi.org/10.1016/j.electacta.2005.02.107
25. M. Orazem, B. Tribollet, Impedance Spectroscopy (2008)
26. D. Tang, R. Yuan, Y. Chai, J. Dai, X. Zhong, Y. Liu, A novel immunosensor based on immobilization of hepatitis B surface antibody on platinum electrode modified colloidal gold and polyvinyl butyral as matrices via electrochemical impedance spectroscopy. Bioelectrochemistry **65**(1), 15–22 (2004). https://doi.org/10.1016/j.bioelechem.2004.05.004
27. A.J. Tkac et al., Electrochemical impedance spectroscopy-based biosensors: Mechanistic principles, analytical examples for assay of protein cancer biomarkers and challenges towards commercialization. Chem. Electro Chem. https://doi.org/10.1002/celc.201800848

28. "The Study of Thin Films by Electrochemical Impedance…—Google Scholar."
29. "C. Alexander and M. Sadiku, Fundamentals of Electric…—Google Scholar."
30. B. Hirschorn, I. Ibrahim, M.E. Orazem, H. Takenouti, B. Tribollet, Effect of large perturbation amplitudes on the impedance response of an electrochemical system. ECS Trans. **13**(13), 81–100 (2019). https://doi.org/10.1149/1.3004031
31. E.P. Randviir, D.A.C. Brownson, D.K. Kampouris, Electrochemistry of Q-graphene carbon dioxide utilisation view project lithium-ion battery view project. pubs.rsc.org. https://doi.org/10.1039/c2nr31823g
32. E.P. Randviir, C.E. Banks, E.P. Randviir, Electrochemical Impedance Spectroscopy: An Overview of Bioanalytical Applications. pubs.rsc.org. https://doi.org/10.1039/c3ay26476a
33. D.A. Harrington, Electrochemical Impedance Spectroscopy. Norwegian University of Science and Technology (NTNU) (2004)
34. M. Orazem, B.T.-N. Jersey, Electrochemical impedance spectroscopy. iopscience.iop.org (2008)
35. Y. Liu et al., Percolation mechanism through trapping/de-trapping process at defect states for resistive switching devices with structure of Ag/SixC1−x/p-Si Percolation mechanism through trapping/de-trapping process at defect states for resistive switching devices with. Artic. J. Appl. Phys. **116**(6), 64505 (2014). https://doi.org/10.1063/1.4893016
36. F. Ciucci, Modeling electrochemical impedance spectroscopy. Curr. Opin. Electrochem. **13**, 132–139 (2019). https://doi.org/10.1016/j.coelec.2018.12.003
37. J. Qi, G.E. Thompson, Comparative studies of thin film growth on aluminium by AFM, TEM and GDOES characterization. Appl. Surf. Sci. **377**, 109–120 (2016). https://doi.org/10.1016/j.apsusc.2016.03.115
38. C. Pastor-Fernández, K. Uddin, G.H. Chouchelamane, W.D. Widanage, J. Marco, A comparison between electrochemical impedance spectroscopy and incremental capacity-differential voltage as Li-ion diagnostic techniques to identify and quantify the effects of degradation modes within battery management systems. J. Power Sources **360**, 301–318 (2017). https://doi.org/10.1016/j.jpowsour.2017.03.042
39. D. Harrington, Chapter 1. Introduction, in Electrochemical Impedance Spectroscopy (2004), pp. 1–12
40. G. Luka et al., Microfluidics Integrated Biosensors: A Leading Technology towards Lab-on-a-Chip and Sensing Applications. mdpi.com (2015). https://doi.org/10.3390/s151229783
41. S. Brosel-oliu, N. Abramova, N. Uria, A. Bratov, Analytica Chimica Acta Impedimetric transducers based on interdigitated electrode arrays for bacterial detection e A review. Anal. Chim. Acta (xxxx) (2019). https://doi.org/10.1016/j.aca.2019.09.026
42. A. Chakraborty, D.N. Tibarewala, A. Barui, 5. Impedance-Based Biosensors. Elsevier Ltd (2019)
43. Z. Chen, L. Li, H. Zhao, L. Guo, X. Mu, Electrochemical impedance spectroscopy detection of lysozyme based on electrodeposited gold nanoparticles. Talanta **83**(5), 1501–1506 (2011). https://doi.org/10.1016/j.talanta.2010.11.042
44. M. Devillers, L. Ahmad, H. Korri-Youssoufi, L. Salmon, Carbohydrate-based electrochemical biosensor for detection of a cancer biomarker in human plasma. Biosens. Bioelectron. **96**(January), 178–185 (2017). https://doi.org/10.1016/j.bios.2017.04.031
45. L. Du, W. Chen, J. Wang, W. Cai, S. Kong, C. Wu, Folic acid-functionalized zirconium metal-organic frameworks based electrochemical impedance biosensor for the cancer cell detection. Sensors Actuators, B Chem. **301**(June), 127073 (2019). https://doi.org/10.1016/j.snb.2019.127073
46. A.Z. Murad, Interdigitated Electrodes as Impedance and Capacitance Biosensors: A Review, vol. 020276 (2017). https://doi.org/10.1063/1.5002470
47. S. Analysis, Interdigitated Sensor Optimization for Blood Sample Analysis (2020)
48. G.N. Abdelrasoul et al., Analytica Chimica Acta DNA aptamer-based non-faradaic impedance biosensor for detecting. Anal. Chim. Acta **1107**, 135–144 (2020). https://doi.org/10.1016/j.aca.2020.02.004

49. A.G. Montaño-figueroa, S.E. Wheelis, B.M. Hedden, N.H. Alshareef, Detection of apoptotic and live pre-osteoblast cell line using impedance- based biosensors with variable electrode design. Biosens. Bioelectron. **128**(December 2018), 37–44 (2019). https://doi.org/10.1016/j.bios.2018.11.057
50. M. Kim, R. Iezzi Jr., B.S. Shim, D.C. Martin, Impedimetric Biosensors for Detecting Vascular Endothelial Growth Factor (VEGF) Based on (PEDOT)/Gold Nanoparticle (Au NP) Composites, vol. 7, no. April, pp. 1–11 (2019). https://doi.org/10.3389/fchem.2019.00234
51. F.L. Migliorini, R.C. Sanfelice, L.A. Mercante, R.S. Andre, L.H.C. Mattoso, D.S. Correa, Urea impedimetric biosensing using electrospun nanofibers modified with zinc oxide nanoparticles. Appl. Surf. Sci. **443**, 18–23 (2018). https://doi.org/10.1016/j.apsusc.2018.02.168
52. N.I. Khan, A.G. Maddaus, E. Song, A Low-Cost Inkjet-Printed Aptamer-Based Electrochemical Biosensor for the Selective Detection of Lysozyme (2018). https://doi.org/10.3390/bios8010007
53. L. Han, P. Liu, A.P. Valery, A. Liu, A label-free electrochemical impedance cytosensor based on specific peptide-fused phage selected from landscape phage library. Nat. Publ. Gr. (October 2015), 1–10 (2016). https://doi.org/10.1038/srep22199
54. B. Hong, A. Sun, L. Pang, A.G. Venkatesh, D. Hall, Y. Fainman, Integration of Faradaic Electrochemical Impedance Spectroscopy into a Scalable Surface Plasmon Biosensor for in Tandem Detection, vol. 23, no. 23, pp. 813–820 (2015). https://doi.org/10.1364/OE.23.030237
55. J. Halliwell, A.C. Savage, N. Buckley, C. Gwenin, Electrochemical impedance spectroscopy biosensor for detection of active botulinum neurotoxin. Sens. Bio-Sensing Res. **2**, 12–15 (2014). https://doi.org/10.1016/j.sbsr.2014.08.002
56. J. Guan, Y. Miao, Q. Zhang, Impedimetric Biosensors, vol. 97, no. 4, pp. 219–226 (2004)
57. JoVE Science Education Database, Analytical chemistry. Cyclic Voltammetry (CV). JoVE, Cambridge, MA (2021)
58. K.J. Rountree, B.D. Mccarthy, E.S. Rountree, T.T. Eisenhart, J.L. Dempsey, A Practical Beginner's Guide to Cyclic Voltammetry (2017). https://doi.org/10.1021/acs.jchemed.7b00361

Chapter 8
Characterization Techniques for Chromatography Analysis

Ingrid Berenice Aguilar Meza, Euth Ortiz Ortega, Hamed Hosseinian,
Andrea Rodríguez Vera, María José Rosales López, and Samira Hosseini

Abbreviations

μPACs	Micropillar array columns
μ-TAS	Micrototal analysis system
ACS	Activated charcoal strip
AEX	Anion-exchange chromatography
ASE	Accelerated solvent extraction
BEH	Bridged ethyl hybrid
CEX	Cation-exchange chromatography
CIC	Chelation ion chromatography
CM	Conditioned medium
CPCs	Cardiomyocyte progenitor cells
DAD	Diode array detection
ECD	Electron capture detector
ErA	Erwinia chrysanthemi l-asparaginase
EVs	Extracellular vesicles
Fab	Fragment antigen-binding
Fc	Fragment crystallizable
FID	Flame ionization detector
FS columns	Fused silica columns
GC	Gas chromatography
GC×GC	Two-dimensional gas chromatography
GC×GC–TOFMS	Time-of-flight mass spectrometry
GPC	Gel permeation chromatography
HISD	High-intensity short duration
HPLC	High-performance liquid chromatography
HSDM	Headspace single-drop microextraction
HT-GPC	High temperature-gel permeation chromatography
IC	Ion chromatography

E. O. Ortega et al., *Material Characterization Techniques and Applications*,
Progress in Optical Science and Photonics 19,
https://doi.org/10.1007/978-981-16-9569-8_8

IC/MS	Ion chromatography-mass spectrometry
IEC	Ion-exclusion chromatography
IEX	Ion-exchange chromatography
IPC	Ion pair chromatography
LC	Liquid chromatography
LILD	Low-intensity long duration
LLE	Liquid–liquid extraction
mAbs	Monoclonal antibodies
MEMS	Microelectromechanical systems
MW	Molecular weight
MWD	Molecular weight distribution
M_n	Number average molecular weight
M_v	Viscosity average molecular weight
M_w	Average molecular weight
M_z	Average molecular weight
M_{z+1}	Higher average molecular weight
NS	Not specific
OAs	Organic acids
PLOT	Porous layer open tubular
PS	Polystyrene
PTM	Post-translational modification
PTV	Programmed temperature vaporizer
TB	Tuberculosis
QC	Quality control
RPLC	Reversed-phase liquid chromatography
SCOT	Support coated open tubular
SEC	Size-exclusion chromatography
SEC-EVs	Extracellular vesicles isolated by size-exclusion chromatography
SFC	Supercritical fluid chromatography
SFE	Supercritical fluid extraction
TD	Sorbent thermal desorption
TLC	Thin-layer chromatography
UC	Ultracentrifugation
UC-EVs	Extracellular vesicles isolated by ultracentrifugation
WCOT	Wall-coated open tubular
ZIC	Zwitterionic ion chromatography

8.1 Gas Chromatography for Material Characterization

8.1.1 History of Gas Chromatography

In 1941, Martin and Synge conducted a set of experiments regarding amino acid separation and published an article titled "A new form of chromatogram employing two liquid phases" in which they commented that the mobile phase could be gas instead of liquid. They consequently discovered liquid partition chromatography; however, it was not until 1952 that Martin and Synge made a breakthrough in the field. In 1941, Hesse used absorption chromatography to separate molecules with a molecular mass below 200 Da. This study allowed him to observe that small molecules were poorly adsorbed on the surface of the adsorbent due to the presence of solvents; hence, he suggested that gases could make better adsorbent candidates than liquids. Hesse succeeded in separating cyclohexane and benzene using gas chromatography (GC) principles with a distillation setup. In his subsequent publications, however, he named his method adsorption distillation, not chromatography. Cremer's research group used equipment based on a silica-filled column to separate a mixture of gases, thus developing the first gas chromatograph at the University of Innsbruck in 1944. However, their work was not published until 30 years later when it first appeared in "Chromatographia" [1–3] (Fig. 8.1).

During the 1940s and 1950s, various leading researchers such as Cremer, Hesse, and Phillips worked on the development of gas adsorption chromatography. In 1952, Martin and James invented gas–liquid chromatography through an experiment that involved separating volatile fatty acids using nitrogen gas (mobile phase) and silicone oil/stearic acid supported on diatomaceous earth (stationary phase). They received the Nobel Prize in Chemistry for the invention of gas–liquid chromatography. To date, it is still the most extensively used analytical technique in modern chemistry. In 1955, American and British companies built the first commercial GC with a thermal conductivity cell as a detector coupled with a column for the separation and detection of analytes. During the 1950s, the oil industry widely used GC to analyze the components present in petroleum followed by fields such as biochemistry and food industry where the technique was greatly explored. Around those years, the British Petroleum Institute organized "the International Symposium on Chromatography" to discuss the further advancements of this technique. By 1958, three discoveries were presented at this event that had significantly improved GC's performance: separation by capillary columns, the development of the flame ionization detector (FID), and the electron capture detector (ECD) which used the argon ionization detector as a precursor. The katharometer, a detector constructed by Ray, was the most used in GC until the invention of the FID, characterized by its low cost and high sensitivity compared with the katharometer. The FID provided low detection limits and response times, while it detected almost all organic compounds. By using ECD, analytes containing aromatic compounds or halogens can be detected which is particularly useful in environmental analysis. The operation of the capillary column was demonstrated by Golay in 1958. This column overcame the problems presented by packed columns,

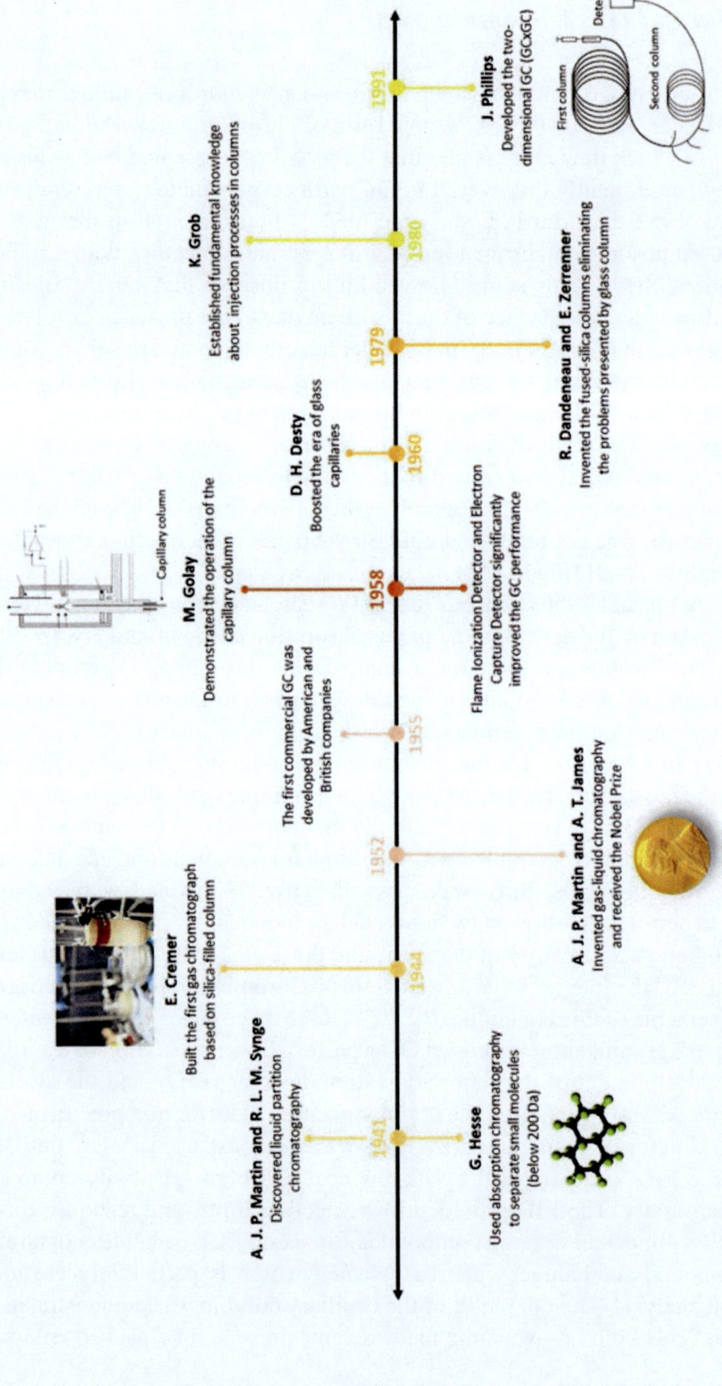

Fig. 8.1 History of GC

for instance, pressure drop. With the improvements of the 1950s, GC was used not only for hydrocarbon analysis but also for the analysis of fatty acids in medicine, environmental analysis, pharmacology, chemical industry, flavor and fragrance analysis, and toxicology, among others [1–3].

In 1960, Desty revolutionized the era of glass capillaries by building equipment capable of manufacturing spiral glass capillaries. However, the coiled glass capillaries were not well accepted at first since they broke easily and had low gas rates, and the deposition of the stationary phase on the inner walls of the column was deficient. It was not until 1979 when Dandeneau and Zerrenner developed the fused silica columns (FS columns) to avoid the troubles presented in glass columns including the interaction between glass and highly polar samples. These columns were treated with a protective layer of polyimide on the outer wall, and the low amounts of metal oxides made them inert, while the low wall thickness served their flexibility. The studies of fatty acid esters and fatty acids were very popular in the decade of 1960. Moreover, the reaction kinetics was studied by Purnell using GC in 1962. Around 1963, Grobs developed the idea of avoiding sample loss during vaporization by directly injecting the analyte into the inlet of the column. Griffiths proposed a method for samples with several volatile components, and the temperature in the column could gradually increase during the analysis. The developed methodology was officially published in 1966 [1–3].

Samples that are highly diluted required a stable solvent and sophisticated injection techniques. In the 1980s, Grobs generated valuable insights on injection processes and column heating during the feeding of liquid samples into the system (e.g., programmed temperature vaporizer (PTV)). During that decade, the in situ cross-linking of polymer systems was used to immobilize the stationary phase in films with a thickness of about 5 μm. To overcome the problems of automation, cost, speed of analysis, accuracy, and reliability of GC, it was necessary to rely on computer technology to handle the task of analyzing and storing the data obtained, as well as electronic control of the temperature and flow rate of the gases. In 1991, the maximum separation of components was carried out by Phillips using a system called GCxGC, where the separation of essential oil, fuels, and air pollutants was effectively performed. The GCxGC was considered the most innovative technological development in GC after the discovery of FS columns. With this technological progress, it was possible to separate complex samples. In 1993, solid-phase microextraction was commercially introduced based upon a concept developed by Belardi and Pawliszyn in 1989. This discovery further facilitated sample analysis without the need for solvent [1–3].

8.1.2 *Mechanism of Operation of Gas Chromatography*

Chromatography analysis involves a group of separation techniques that allow the identification and quantification of chemical substances from a mixture. The sample, or the analyte, runs through a matrix where the separation takes place. Chromatography separation occurs when a sample is presented in a flowing mobile phase with a cross section with a stationary phase [4].

The GC is a physicochemical method characterized by a separation of a homogeneous mixture with the help of a gas flow by distribution among two phases (stationary and mobile). The stationary phase will remove molecules that are inclined to stay at a lower velocity; in contrast, the mobile phase will favor molecules that have the tendency to move forward. Therefore, separation occurs from frequent sorption and desorption processes as the samples move along the stationary phase. The key mechanism of the separation is based upon the intermolecular interaction of the solutes migrating through a column located on the way of the stationary phase [5, 6]. The stationary phase usually consists of a porous, granular powder that has sorbents with a high surface area in the form of a homogeneous layer filled within a column. The mobile phase of chromatography is always filled with a gas such as hydrogen (H_2), helium (He), nitrogen (N_2), or argon (Ar) that are commonly used. The composition of the mobile phase does not significantly contribute to the separation procedure as the gases are relatively inert. The mobile phase travels through the layers at a constant average linear velocity prompted by capillary forces [5, 7].

8.1.2.1 Instrumentation

A GC instrument is exemplified in Fig. 8.2, which comprises four main modules: The gas supply which provides the necessary gas is the first module; followed by the injection device; typically, a gas sampling valve is used for gases. The column where the separation takes place is the third module, while the fourth module is the detector which is commonly more than one (mass-dependent or concentration-dependent). A final vital module, which might be considered external to the GC system, is the chromatography data analysis system. All modules are controlled by different temperature-control systems that transfer the information to one or more microprocessing systems [8].

The gas is the medium that carries the compounds from the injector along the column and leads them to the detector (Fig. 8.2). The flow of the gas is controlled by

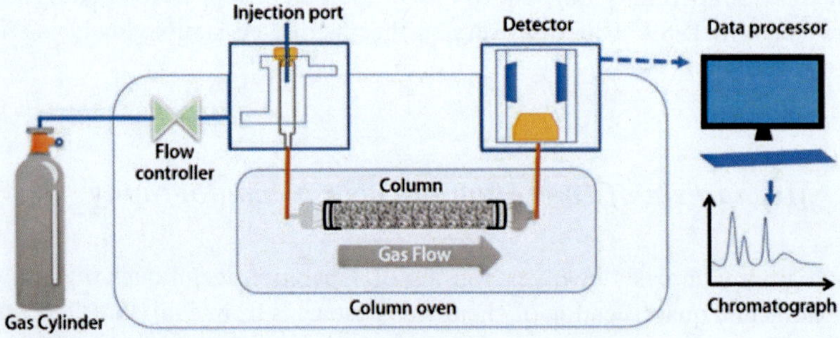

Fig. 8.2 Schematic of a GC system

pressure regulators and gas metering valves. The solutes move at the same average speed as the gas. Since diverse compounds have varied affinities to the stationary phase, compounds separate according to the window of time they take to travel through the stationary phase. The injection port is attached to the GC column, which provides the samples with temperatures sufficient to ensure vaporization of all components. The detector is directly connected to the opposite end of the column. The detector records the chromatogram as a response to the eluted arriving samples [7, 9].

Injection

Injection of the sample is critical in a GC analysis. It can contribute to the band, or peak, broadening. Inadequate injection techniques may decrease column resolution, hence impacting the quantitative results. Different methods can be used to inject the samples into the GC column including split injection, split-less injection, direct injection, and on-column injection. The type of injector design will depend on the column within the system [7, 9]. The sample is introduced into the carrier gas and is mixed with it. Subsequently, the sample is inserted into the injection port and held at high temperature, thus vaporizing the sample. Consequently, the carrier gas flows and sweeps the vaporized sample into the column [8, 9].

Separation Columns and Column Oven

Columns, without a doubt, are the most important part of the chromatography systems where the separation actually takes place. There are two categories of columns, packed and open tubular, used for GC differentiated by the way the stationary phase is held within the column. The choice of the column depends on the target compounds, their molecular weight, the number of components, and the instrument configuration [8–10].

Packed columns contain the stationary phase in powder form. The powder could be incorporated via coating in a liquid stationary phase to fill the space or a powder that is directly in the column. Packed columns are used to inject larger volumes without congesting the column which is beneficial to tracing contaminants in gases of high purity. Typically packed columns are 2–5 m in length with an internal diameter of 2 mm [9, 11].

In open tubular columns or capillary columns, the stationary phase is either coated or attached to the inside surfaces of the column. The internal diameter of the column is from 0.1 to 0.53 mm and is commonly made from fused silica and deactivated metals. Open tubular columns offer higher efficiency and greater separation capabilities than packed columns. Depending on the position and function of the stationary phase, capillary columns are classified as: support coated open tubular (SCOT) columns, porous layer open tubular (PLOT), and wall-coated open tubular (WCOT) columns [9, 11].

The column is mounted within an oven that has a precise and stable temperature regulator. The heat rate can be carefully controlled during the process to obtain efficient results. At the primary phase of analysis, the early eluting components are separated by adjusting the temperature low and are systematically increased to further separate the components that are rather more retained. The temperatures are predetermined and set on the data system and within microprocessor-controlled, as the analysis moves forward [8].

Detectors

The detector in the GC system monitors the separation and responds to the components and/or fractions of the components, while they elute from the column. The volume of the mobile phase must not exceed the detector's volume as the analyte peak will be broadened, the concentration will be reduced, and it will became more difficult to have an effective detection. Since the detector chamber allows compounds to expand into a larger volume, this can also act as a mixing chamber. Therefore, an effective detector should benefit from reasonable sensitivity, a broad linear dynamic range, and a small cell volume to avoid the distortion of the GC peak [7, 9, 10].

Multiple detector types for GC systems exist with distinctive operating parameters and performances. Table 8.1 provides a comparison of the characteristics of GC detectors. In most cases, the detector output is analog and is further electronically translated to a digital signal to be processed by a computer [8, 9].

8.1.3 Advantages and Disadvantages of Gas Chromatography

As GC was introduced, it swiftly found popularity as it was a timely invention at the time when analytical control was a major need in the petrochemical industries. GC constitutes the technique for separating volatile materials. It provides efficient complete analysis and high-resolution results within minutes and even seconds [13, 14]. Additionally, it is a sensitive and non-destructive method for sample analysis. This non-destructive nature not only prevents damage to samples but also allows online coupling of additional characterization techniques. The highly accurate quantitative analysis that GC offers, which is among the 1–5% of relative standard deviations, permits its use in a wide range of applications. Moreover, only a small quantity of sample (at the μL scale) is needed for this characterization. As listed in Table 8.2, other favorable aspects of this technique include a reliable column temperature control, simple instrumentation handling, and a relatively low cost of implementation [13, 15].

One of the limitations of GS is that its application is constrained to volatile samples. Furthermore, spectroscopy systems, such as mass spectroscopy, are required for peak identity confirmation. In addition, large samples (above 1000 Daltons) are complicated to analyze with GC, while it is not adequate for thermally unstable

Table 8.1 Characteristics of different detectors

Detector		Principle	Selectivity	Range
Universal	Thermal conductivity detector (TCD) [9]	Measures alteration of thermal conductivity among the carrier gas and the elution carrier gas [9, 12]	All compounds except for the carrier gas [9, 12]	10 ppm [9, 12]
	Flame ionization detector (FID) [9]	Measures ions created when the compound is burned in H_2/O_2 flame at 2000 °C [9]	Organic compounds [9]	0.1 ppm [12]
	Pulsed discharge helium ionization detector (PDHID) [12]	Measures the energy used for the ionization of analytes eluting from the column [12]	All compounds except for helium and neon [12]	0.07 ppm [12]
Selective high sensitivity	Electron capture detector (ECD) [9]	Measures changes in electron current caused by a reaction of an organic compound with electrons [9]	Organic halogen compounds; electronegative compounds [12]	0–01 ppb [12]
	Flame photometric detector (FPD) [12]	Uses a hydrogen-rich flame. Compounds entering the flame decompose, and the detector measures the light emitted by the formed radicals [12]	Sulfur compounds; organic phosphorus compounds; organic tin compounds [12]	10 ppb [12]
	Sulfur chemiluminescence detector (SCD) [12]	Measure the chemiluminescence as a result of the reaction of the combusted eluate with ozone (O_3) [12]	Sulfur-containing compounds [12]	1 ppb [12]

ppm parts per million; *ppb* parts per billion

samples (the upper-temperature limit for GC is ~380 °C). For that reason, samples must have a substantial vapor pressure (60 torrs or greater) within that temperature range. It is noteworthy, however, that the solutes regularly do not surpass the boiling points of 500 °C [13]. Another important drawback is that the conventional instrumentation is bulky. This results in high power consumption and low heating

Table 8.2 Advantages and disadvantages of GC

Advantages	Disadvantages
Separates volatile materials [13, 14]	Constrained to volatile samples [13]
Rapid analysis (within minutes or seconds) [13]	Spectroscopy is required for peak identity confirmation [13]
Efficient and high resolution [13]	Analysis of samples above 1000 Daltons is complicated [13]
High accuracy of quantitative analysis (RSDs of 1–5%) [13]	Not adequate for thermally unstable samples [13]
Non-destructive, sensitive, simple instrumentation handling, low cost [13]	Conventional instrumentation is bulky, not suitable for in situ analysis [14, 15]
Allows online coupling and small sample size [13]	Has low heating rates and requires high power consumption [14, 15]

RSD relative standard deviation; *GC* gas chromatography

rates and precludes in situ analysis. However, alternative GC systems, such as miniature laboratory-on-a-chip GC systems, resolve those challenges presented by the voluminous instrumentation of the conventional method [14, 15].

8.1.4 Applications of Gas Chromatography

As an ever-growing technique, GC has proven to be a useful tool for analyzing complex samples. Multiple reports explore the diverse range of applications of GC including pharmaceutical and cosmetics industries as well as environmental analysis. Commonly, samples that can be analyzed by GC are volatile. In the human body, blood, saliva, breath, and other types of body products that contain organic volatiles can be subjected to GC analysis. The knowledge of the mixture's components and the concentration of the elements within the mixture is of great importance in the field of medicine. This section offers a number of applications for GC with the main focus on biomolecular identification [16–19]. The study of metabolomics, obtaining fingerprints from pathogenic microorganisms, and profiling tissues and fluids to identify the volatile patterns within them, in turn, permits timely and adequate diagnosis and treatment [20]. GC can also be used for biomarker identification in pneumonia, tuberculosis, cystic fibrosis, asthma, and lung cancer. Hill et al. proposed a two-dimensional (2D) gas chromatography (GC×GC) separation technique for the diagnosis of tuberculosis (TB). This method was reported to be rapid, straightforward, and non-invasive for instantaneous diagnosis and concentration identification of species attributed to tuberculosis (Fig. 8.3) [21]. This technique successfully detected a number of biomarkers within breath that allowed differentiating TB from other respiratory infections (Fig. 8.3).

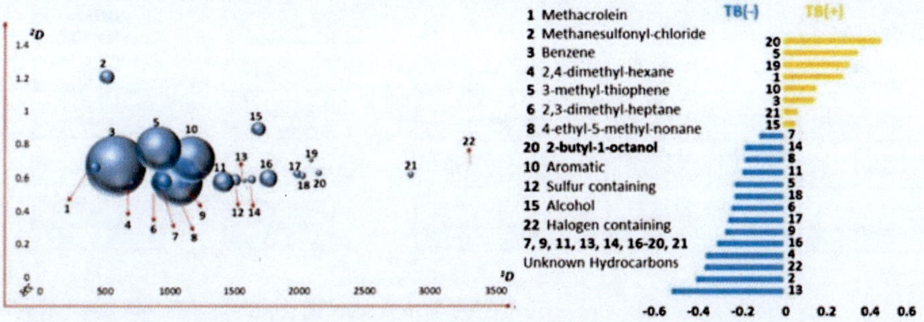

Fig. 8.3 Bubble illustration of discriminatory elements attributing to positive TB (TB+) and negative TB (TB−). Reproduced (or adapted) from [19], ©2018, with permission from Elsevier

A number of reports focus on urine [22], serum, and tissues [23] to explore the impact of diet, exercise, as well as metabolic procedures on an individual's health. In that regard, Cox et al. used GC together with time-of-flight mass spectrometry (GC× GC–TOFMS) to demonstrate a chemical profile of kidney tissues as well as serum samples. This coupled approach facilitated the biorecognition of approximately twice as many metabolites within a serum than what other techniques would be capable of achieving. Furthermore, 35 arterial lesion biomarkers have shown a strong correlation with diet-induced atherosclerosis that not only allowed expanding knowledge on the progression of this condition but also shed light on the sex specificity and heritable characteristics of this problem [23]. Moreover, Groessl et al. utilized a GC×GC–TOFMS to analyze steroid metabolites in urine which plays a key role in multiple physiological activities and hormonal disorders. Although varied in type, steroids commonly have similar chemical structures and are available in small quantities, hence the need for high-resolution and ultrasensitive analytical methods for identification [22]. The study reported GC×GC–TOFMS to be capable of identifying and quantifying 40 steroids within human urine in addition to 10 progesterone metabolites found in newborn urine. Furthermore, the study showed that the method can be also used for untargeted analysis which is highly beneficial for biorecognition of unknown metabolites.

Forensic discipline leverages scientific methods and tools to address legal matters and arguments. Forensic chemistry is the knowledge of analyzing a vast number of forensic samples aimed at extracting and interpreting the data with enough science-based evidence and rigor to be presented in civil and judicial settings. GC is a capable analytical technique that can have multiple uses in forensic research. Sample collection and preparation are of great importance in GC analysis, and incorrect methodology may impact the integrity of GC analysis. When handling traces and/or ultratraces of a given analyte within biological, environmental, fire debris, and explosive residues, it is important to collect samples of sufficient volume and valid nature. For that reason, in forensic chemistry, a number of protocols are developed for sample collection, storage, and analysis that should be strictly followed prior to GC [24]. Figure. 8.4 demonstrates some of the main application areas

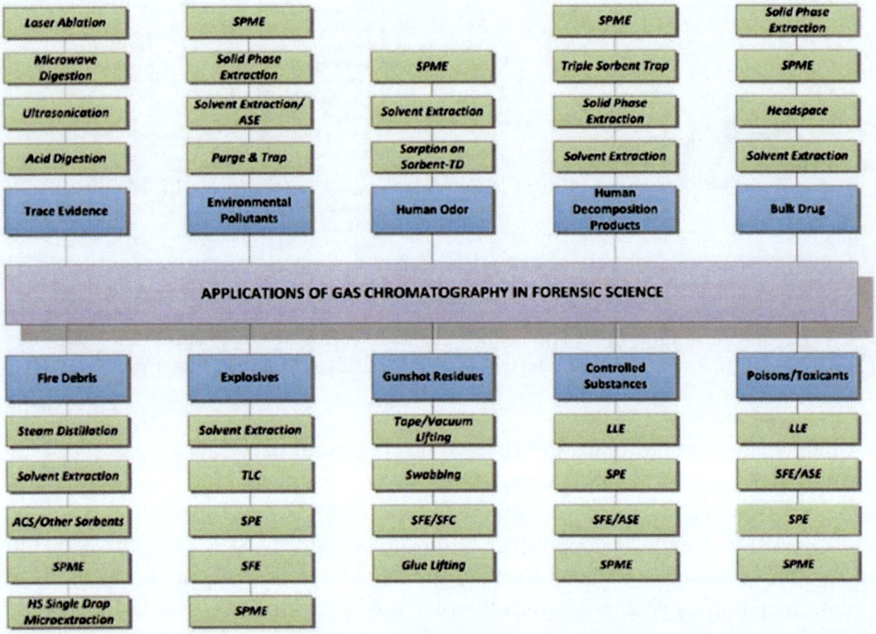

Fig. 8.4 Main applications of GC in the forensic domain. Reproduced (or adapted) from [25], ©2012, with permission from Elsevier

of GC in forensic sciences. Some of the most common uses of GC within this domain can be listed as follows: drug screening and analysis, profiling human odor, tracing evidence, artificial environmental contamination, toxicology, ignitable and explosives examination.

8.2 High-Performance Liquid Chromatography for Material Characterization

8.2.1 History of High-Performance Liquid Chromatography

Liquid chromatography (LC) is a class of chromatography aimed at analyzing individual components of a mixture that are separated in solution. In specific, high-performance LC (HPLC) is a useful separation technique widely used for complex polymer characterization. Figure 8.5 presents the evolution of science until the invention of HPLC and provides an overview of the significant contributions that have led to the development of this technique [5, 26, 27].

In the early 1900s, Tswett invented chromatography to separate plant pigments. Years later, in 1941, Martin and Synge shared their advancements in an article on

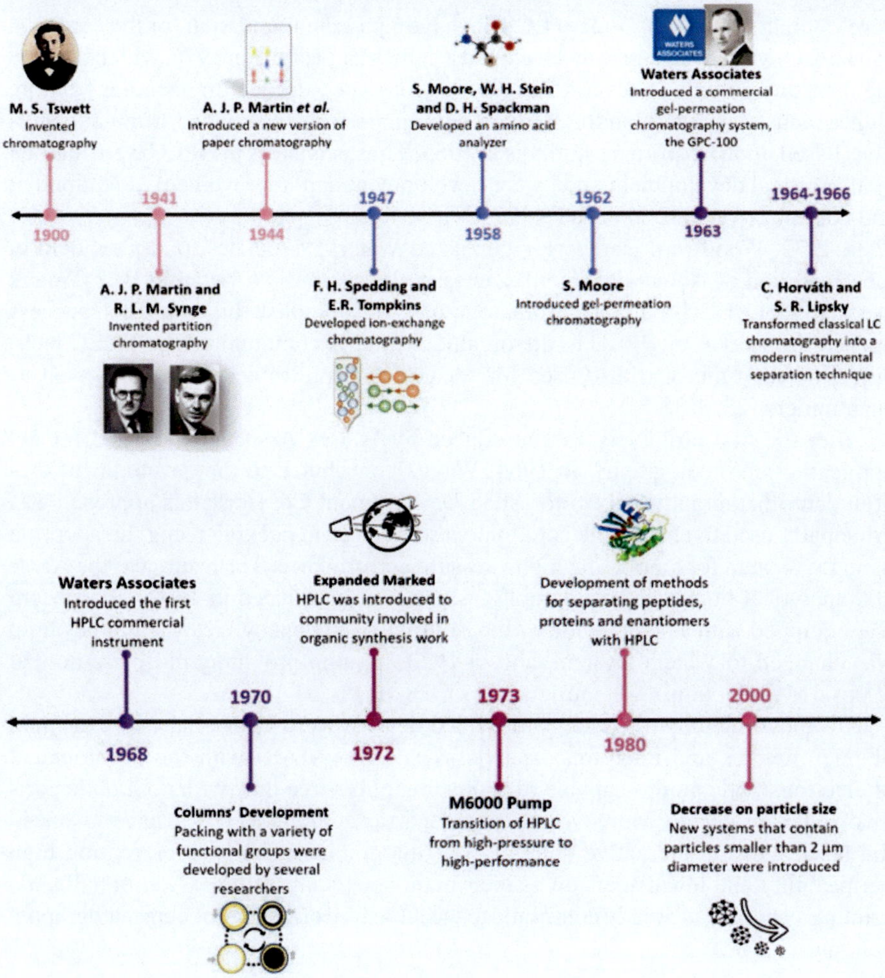

Fig. 8.5 Timeline of relevant events for the development of HPLC

liquid partition chromatography where they described that a mixture of small particles and high pressure lead to an efficient separation. Another historically important event in the development of HPLC was the progress in ion-exchange chromatography in 1947 along with the amino acid analyzer in 1958. The advances of gel permeation chromatography at the beginning of the 1960s were an achievement for early liquid chromatographs. However, the development of gas chromatography played a major role in the foundations of LC [27–29].

HPLC was proposed and primarily developed in the 1960s. Prior to this development, scientists applied the knowledge of GC to improve liquid column chromatography. Giddings in 1963 was the first to establish the theoretical basis for the changes required for LC to be performed successfully. In 1964, Horváth and Lipsky began developing a system for modern LC which later laid the foundation for the commercial HPLC. Moreover, Horváth developed a pellicular packing in 1976, which became the mainstay column in several laboratories for over a decade. In a similar fashion, Huber focused his work on the study of column performance under different conditions based upon his primary findings in 1965. Thus, advances in HPLC were marked by theoretical development and by the development and improvement of equipment and columns over several decades [27, 28].

In 1972, Woodward Hamberger contacted Waters Associates to request help in the separation of isomers of intermediates in the synthesis of vitamin B12. Waters used a model ALC-100 liquid chromatography to accomplish this task. Their success opened the market of HPLC to the organic synthesis community. The HPLC technique, by the time, was also used for separating peptides and proteins as well as enantiomers [28, 30].

After the first prototype was introduced by Waters Associates in 1963 for gel permeation chromatography, in 1964, Waters contributed to the development of a liquid chromatography system for Shell Development Co. Horvath's approach was principally used to change the column to a solid column packing using the adoption of an LC system for fabricating high-pressure performance. The result was the ALC-100 analytical liquid chromatography, which was introduced in 1968. The system was equipped with a pump made by the Milton Roy Company. Years later, the pump was changed to Waters' system, and in 1973, a pump providing pulseless flow at 600psi (M6000 pump) was introduced [27, 28, 31].

Advances in pumping systems for HPLC devices were further driven by the field of microfluidics and micrototal analysis systems (μ-TAS) with the development of electroosmotic pumps capable of producing pulse-free flow with no mobile parts involved in the design. Moreover, micropillar array columns (μPACs) have advanced the fabrication of alternative columns via lithography techniques presenting high permeability and low dispersion as their main beneficial features. Additionally, 3D printing systems can greatly contribute to the increased efficiency of chromatographic columns [32, 33].

8.2.2 Mechanism of Operation of High-Performance Liquid Chromatography

8.2.2.1 Liquid Chromatographic Methods

As previously mentioned, LC is one of the main chromatography methods in which the movable phase is a liquid; thus, a solution mixture is separated into its discrete constituents [5, 26]. Common procedures in HPLC comprise a liquid phase that is mobile and travels through a stationary phase. Different HPLC techniques can be categorized based upon their retention, separation, or operation mechanism. The major modes of classification are adsorption, partition, ion-exchange, size-exclusion, and affinity chromatography presented in Table 8.3 [4, 5, 26].

Table 8.3 Separation methods used in HPLC

Separation method	Principle	Stationary phase	Mobile phase	Retention mechanism
Adsorption [26, 34]	Based on the molecular exchange between the liquid mobile phase and the neutral, solid stationary phase [4, 26, 34]	Polar with hydroxyl groups. Porous surface covered with an active bonded phase [4, 35]	Nonpolar solvent [4]	Polarity; retention of molecules on the solid stationary phase [4]
Partition [26, 34]	The solute molecules are distributed among two non-miscible liquid phases [4, 26, 34]	A nonpolar liquid is immobilized on a solid support. Usually with alkyl, phenyl, cyano-, or amino groups [4, 35]	Polar solvent [4]	Polarity; hydrogen bond formation [4]
Size exclusion [26, 34]	Separation according to the hydrodynamic volume of substances occurs by using a molecular filter [26, 34]	Porous structures [4, 34]	Organic solvent [34]	Gel filtration and gel permeation [4, 34]
Affinity [26]	Lock and key mechanism using the specific binding of protein [4, 26]	Biological complement immobilized on the solid support [36]	A target biological compound specific to the complementary segment [36]	Reversible attachment of a protein to a ligand covalently bounded to a solid surface [36]
Ion exchange [26, 34]	The exchange process among an ion species in the solution and another ion species within the stationary phase takes place through ionic interactions [34]	Ionic or counterionic groups [34]	Ionic or anionic species [34]	Cation-exchange or anion-exchange chromatography [34]

8.2.2.2 Instruments

The HPLC has been the premier method for the characterization of polymers: a technique that pumps a solution mixture mechanically through a column and guides it through the stationary phase. Common instruments involved in this analytical technique are injectors, pumps, columns, and detectors. Figure 8.6 shows a schematic representation of a basic HPLC system [5, 26].

Mobile Phase

In HPLC systems, the right selection of a mobile phase is crucial to its effective operation. Moreover, the inherent characteristics of the analytes, the stationary phase, and the detection method to determine the type of solvents that are adequate for the system. The selected solvent is aimed to offer separation of the component mixtures within the minimum required time, and the delivery of analytes to the detection point without affecting the measurement. An ideal mobile phase is a solvent that elutes analytes within shorter retention time periods [4].

Stationary Phase

The thermodynamic aspects of the separation process can be controlled by the chemical class of the stationary phase. Thus, by adjusting the physical and chemical characteristics of the stationary phase, the performance of the HPLC can be manipulated. The packing material of the column would directly impact the chemical and physical interaction with the analyte of interest, while such interaction can be further influenced by changing the solvent strength. When the interaction between the stationary phase and analyte increases, the retention of solutes also increases [5, 26].

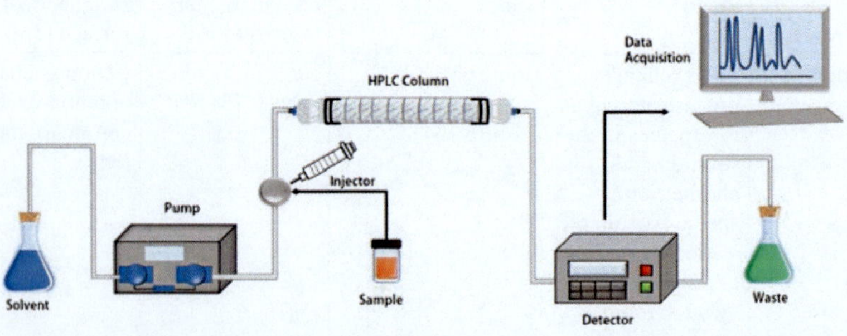

Fig. 8.6 Schematic representation of an HPLC system

8.2.2.3 Mechanism of Retention

In HPLC, solutes are injected in the mobile phase which travels through the column with the help of a pump to inject the solutes at the head of the chromatographic column. The pressure exerted by the pump moves the mobile phase along the column, and the eluded molecules that leave the column are then detected by a certain type of detector, according to the physicochemical properties of the solution. Finally, the received signal will be processed to produce the chromatogram [34].

The displayed peaks in the chromatograph correspond to diverse components that exist within the mixture. The height and peak specifications of the chromatogram are direct functions of multiple elements including the concentration of the compound within the mixture, the quantity of the injected samples, in addition to the detection sensitivity of the system. Therefore, the system's efficiency can be adjusted by frequent optimization of these factors to obtain better separation outcomes and minimize band broadening. In HPLC, the variables of selectivity and capacity of the column are principally designed and developed by the manufacturers, while the efficiency and resolution can be optimized by the chromatography technicians. Table 8.4 provides general insights on different formulations used to control the efficiency and resolution along with other essential parameters involved in HPLC [26, 34].

8.2.3 Advantages and Disadvantages of High-Performance Liquid Chromatography

While a wide variety of fields use HPLC as a powerful technique for the separation and quantification of mixture components, this chromatography technique, like others, has both advantageous and disadvantageous features. HPLC is known to be a versatile technique as it can be applied in a wide extent of applications [38, 39]. Moreover, it is suitable to analyze various samples, whereby between 60 and 80% of the existing compounds can be successfully separated, detected, and identified [38]. Another advantageous characteristic of this technique is the precision of the analysis (of <0.5% relative standard deviations) and rich quantitative outcomes [38, 39]. HPLC features a simple operation process using a unique automated system. The data system embedded within the instrument permits analysis and report of the outcomes without user's attendance and autosamplers [38]. In addition, HPLC provides high-resolution results and constitutes a sensitive characterization technique capable of performing detection within nanogram, picogram, and femtogram levels while facilitating rapid analysis and a significant sample recovery rate [38, 40].

Despite its advantages, HPLC requires an expensive implementation and large amounts of solvent. The high sensitivity of columns to pH changes in mobile phases represents another limitation of the method [41]. Since some analytes do not absorb UV or are challenging to ionize, the detection is not universal for HPLC [38]. Another disadvantage is the large number of variables that need to be adjusted prior

Table 8.4 Equations used in HPLC to control its variables

Name	Equation	Specifics	Terms
Capacity factor (retention factor) [26, 37]	$k_R = \frac{t_R - t_0}{t_0} = \frac{V_R - V_0}{V_0}$	Determines retention time of the solute by measuring the strength of the interaction between the sample and the packing substance [26]	k_R = capacity factor t_R = retention time t_0 = holdup time V_R = retention volume V_0 = void volume [26, 37]
Selectivity [26, 37]	$\propto = \frac{t_2 - t_0}{t_1 - t_0} = \frac{k_2}{k_1}$	Describes the effectiveness of a system to separate two compounds by measuring the variance in retention periods of two given peaks [26, 37]	$\propto$ = selectivity of the column t_2 = retention time of component 2 t_1 = retention time of component 1 k_2 = capacity factor of component 2 k_1 = capacity factor of component 1 [26, 37]
Resolution [26, 37]	$R = \frac{1}{4} \frac{\propto - 1}{\propto} \left(N^{\frac{1}{2}}\right) \frac{k}{1+k}$	Characterizes peak separation; $R>1$ is considered a good peak separation [26, 37]	N = efficiency of the column $\propto$ = selectivity k_R = capacity factor [26, 37]
Efficiency [26, 37]	$N = \left(\frac{t_R}{\sigma}\right)^2 = \frac{t_R^2}{\sigma^2}$	Explains peak broadening in correlation with retention. The extent of peak broadening is determined by σ [26, 37]	N = efficiency t_R = retention time σ = width of the Gaussian curve [26, 37]

to each run which, in turn, requires training [38, 39]. Additionally, the analysis of complex mixtures is rather more demanding. However, it remains relevant to highlight that existing columns and instrumental developments have significantly reduced the constraints of HPLC [38]. Table 8.5 summarizes the advantages and disadvantages of this powerful analytical method.

8.2.4 Applications of High-Performance Liquid Chromatography

It was initially believed that HPLC would act as a complementary technique to GC. Nowadays, however, this powerful analytical method has nearly substituted GC in various disciplines. The properties of the samples are the main determinants for

Table 8.5 Advantages and disadvantages of HPLC

Advantages	Disadvantages
Versatility, high sensitivity, and rapid analysis [38–40]	Expensiveness [41]
Suitable for diverse samples (60% to 80% of all existing compounds can be analyzed) [38]	Requires large amounts of solvent [41]
Precise (<0.5% RSD) and rich quantitative outcomes [38, 39]	Columns sensitive to extreme pH in the mobile phases [41]
Simple operation [38]	Detection is limited to ionizable components [38]
High resolution [38, 40]	A large number of variables must be adjusted before each run [38, 39]
Significant sample recovery [38]	Requires trained technicians [38]
Automated operation [38]	Analysis of complex mixtures is rather demanding [38]

HPLC high-performance liquid chromatography; *RSD* relative standard deviation

the incorporation of the liquid mobile phase which, in turn, allows a great deal of flexibility in sample analysis. The data that could be collected via HPLC involves resolution, compound identification and quantification, and chemical separation, and, in specific cases, sample purification. This analytical technique has various applications in diverse fields such as pharmaceutics, environmental monitoring, forensic research, and clinical diagnosis, to name a few [42].

In recent years, HPLC has attracted considerable attention in the isolation and purification of peptides and proteins. Proteins such as insulin can be detected, and the concentration can be accurately measured via biological assays. Although specific and effective, biological assays fall short in detecting contaminants or decomposition components in a mixture. HPLC can be a great candidate for such assessments. For instance, the straightforward HPLC technique equipped with diode arrays was found to be highly effective in insulin detection (50–500 μg/mL). The study suggests the method to be accurate, sensitive, and stable over time which marks it as an excellent candidate for pharmaceutical preparations. Figure 8.7 represents the triplicated chromatograms for insulin and the blank sample [43].

The field of forensic studies is perhaps the most benefited area of research from HPLC analyses. From detection to the identification of biological traces and drug and poison recognition, a wide range of forensic reports were published in the literature highlighting identification of acute or lethal toxins, workplace drug check, change of driver's performance under drug's influence, and doping test for sports.

Heroin is a diacetyl derivative of morphine that is considered to be an unlawful drug obtained from the opium poppy. HPLC is a useful method in the detection of heroin's metabolites in addition to identifying other synthetic drugs including methadone and tramadol, and semisynthetic drugs including buprenorphine opioids. In 2007, HPLC coupled with diode array detection (DAD) was proposed by Fernandez et al. to detect heroin, methadone, cocaine, and other metabolites within

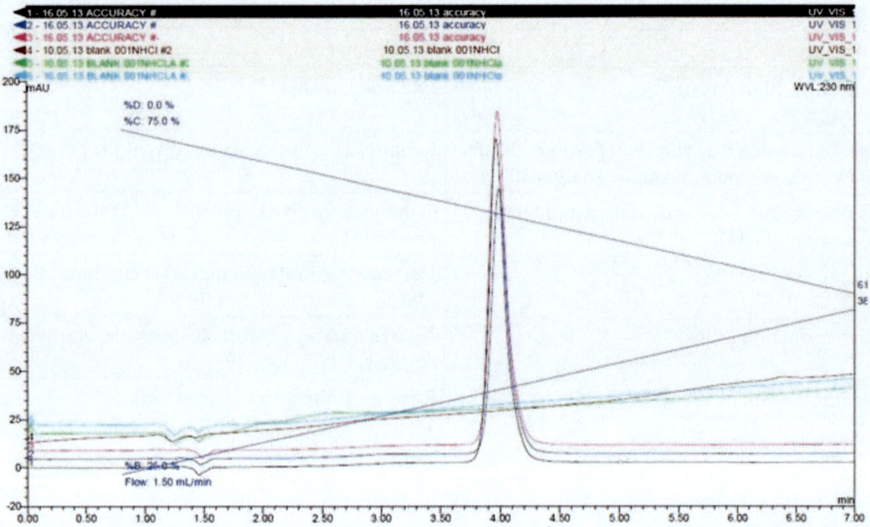

Fig. 8.7 Sample chromatograms for insulin detection as opposed to blank samples. Reproduced (or adapted) from [43], ©2018, with permission from the Journal of Advances in Biotechnology

the plasma. Unlike GC–MS, HPLC–DAD involved no earlier analyte derivatization and could be used for the analysis of non-volatile components as well. In addition to detecting the above-mentioned analytes, this technique has also proven to be effective for monitoring the commitment of drug addicts to the methadone detoxification programs. Figure 8.8 demonstrated the chromatogram (λ = 285 nm) of drug-free plasma spiked with 10 mg/mL of each analyte [44].

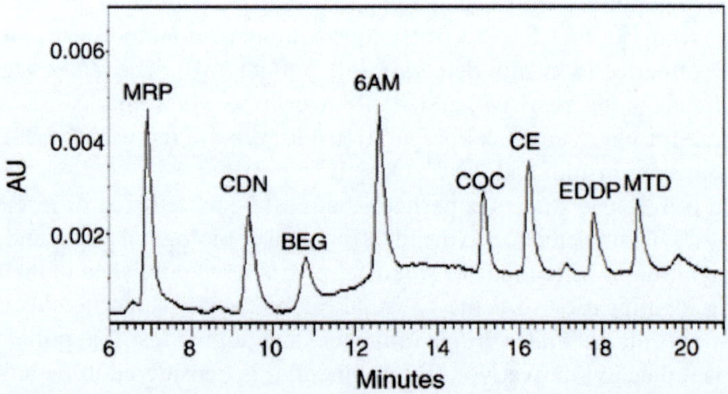

Fig. 8.8 Chromatogram from spiked plasma (conc.: 10 mg/mL; l = 285 nm). Reproduced (or adapted) from [44], ©2006, with permission from Elsevier

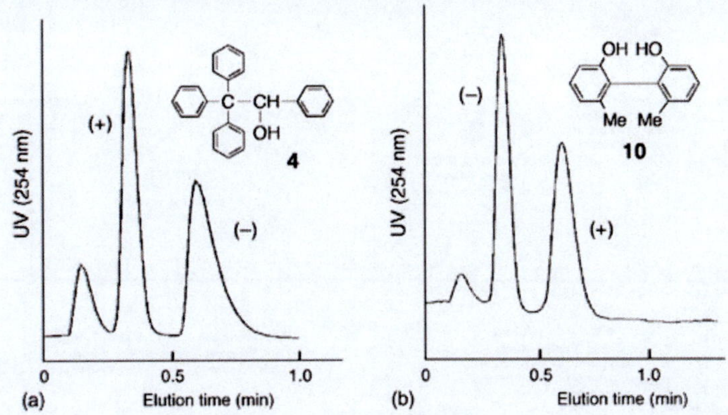

Fig. 8.9 The enantioseparation of **a** 1,2,2,2-tetraphenylethanol and **b** 2,2′-dihydroxy-6,6′-dimethylbiphenyl via a customized capillary column. Reproduced (or adapted) from [46], ©2006, with permission from Elsevier

Historically, the majority of bioanalytical chiral drug analyses were established based upon the use of HPLC with UV or fluorescence detection as well as mass spectrometry (MS) analysis. In specific, for examining complex matrices, the sensitive and selective nature of MS is highly favorable and acts as a great complementary element to HPLC analysis in offering a complete set of information. This coupled methodology saves operation time and expenses as it avoids column switching and thorough sample cleanup processes. In that perspective, a timely examination of enantiomers for both chiral purity testing and enantioselective checking is essential [45]. In 2006, Chankvetadze et al. used a customized capillary column that consisted of monolithic silica modified with amylose tris(3,5-dimethylphenylcarbamate) and conducted enantioseparations in approximately 1 min. This successful coupled analytical method examined two chiral compounds (1,2,2,2-tetraphenylethanol and 2,2′-dihydroxy-6,6′-dimethylbiphenyl) in a highly effective manner (Fig. 8.9) [46].

8.3 Ion Chromatography for Material Characterization

8.3.1 History of Ion Chromatography

Ion chromatography (IC) was primarily presented in 1975, a technique that was laid upon decades of foundational advancement. Rare soils were first separated by displacement ion-exchange chromatography around the mid-1940s followed by the work of Kraus and Nelson who later demonstrated the metal ion separation by anion-exchange chromatography with a focus on chloride, fluoride, nitrate, or sulfate complexes [47, 48].

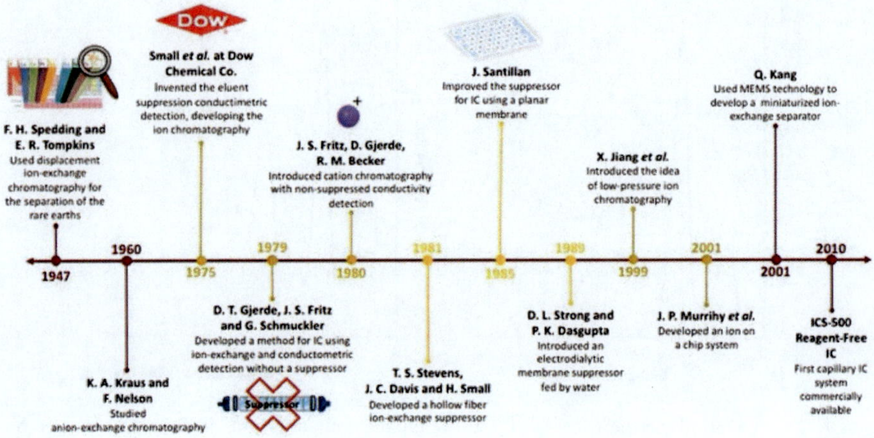

Fig. 8.10 Timeline of the development of IC

As part of Dow Chemical Co., Small, Stevens, and Bauman created a milestone in ion chromatography development by conceiving a system that leveraged suppressed conductivity detection for separation (Fig. 8.10). This system needed a large suppressor column that required frequent regeneration and broadening of the peak [47, 49].

Four years later, in 1979, Gjerde et al. published a technique for anion chromatography without the need for suppressed conductivity detection. The system used an eluent that had significantly lower conductance; therefore, the separated sample ions were easily identified by solely using a conductivity detector. The following year, a comparable technique offered cation chromatography [47, 49].

In 1981, Stevens et al. introduced a membrane suppressor which was used within the chromatography system and the concept of chemically regenerated eluent suppressor was born from their advancement. Later, membrane devices were linked to electrochemically regenerated devices that used only water and an electrical current to keep the system functioning. In this way, previous shortcomings of the suppressor were addressed and the evolution of the suppressor took place [49–51].

In search of miniaturizing the IC systems, Jiang et al. proposed a low-pressure ion chromatography (LPIC). Furthermore, separation using a microchip-based system was examined in 2001 when the first microchip for ion separation was announced. In the same year, Kang et al. used microelectromechanical systems (MEMS) to develop a miniaturized ion-exchange separator, and in 2010, the first capillary IC system was commercially produced and introduced to the market [51, 52].

The trends of IC developments continued with new designs of ion exchangers to increase their capacity, the reduction of particle size to increase the column's efficiency, and the coupling of sensitive and selective spectroscopic and spectrometric techniques. Moreover, the incorporation of nanoparticles to the columns for dynamic coating was widely explored thereafter [52].

8.3.2 Mechanism of Operation of Ion Chromatography

The term IC is given to a collection of techniques used for ion mixture separation. IC utilizes liquid chromatography for the separation process; therefore, the mobile phase is a liquid of small charged solutes. IC principally operates based on electrostatic interactions that occur between oppositely charged sites/groups within the stationary phase. Separations by IC involve the use of ion-exchange resins. The separation mechanism utilizes the reversible ion interchange between a functionalized insoluble resin (the ion-exchange component or stationary phase) and an ionizable material in the solution. In the separation process, the ion exchanger is brought to equilibrium with eluent ions by pumping the eluent through the column [53–56].

Figure 8.11 presents different techniques used in IC. The most common methodologies for IC separations are conducted by ion-exchange chromatography (IEX). The mechanisms used by the IEX technique can be categorized into two groups. The first is non-suppressed ion chromatography, in which an ion-exchange column is directly joined to the detector. The second group is the suppressed ion chromatography which consists of an ion-exchange chromatography, an eluent suppressor, and

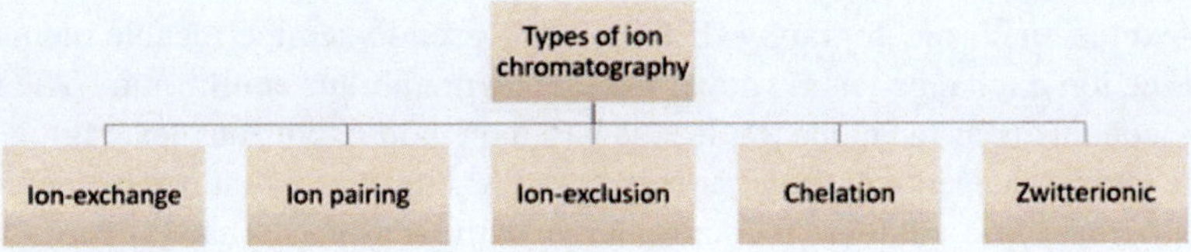

Fig. 8.11 Chromatographic methods using IC

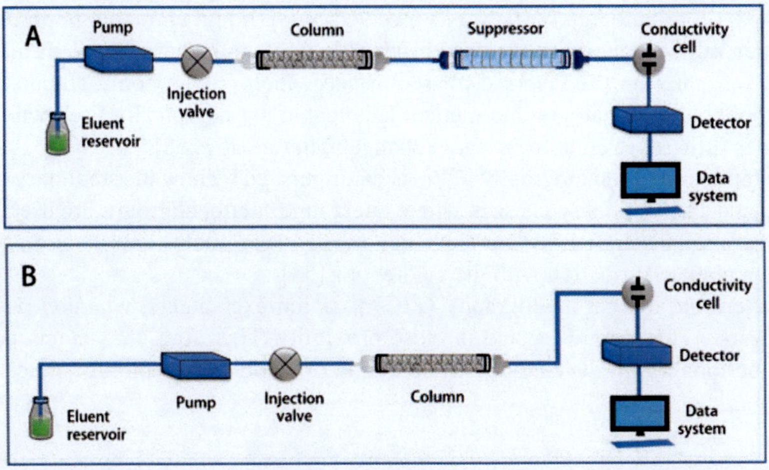

Fig. 8.12 Schematic representation of an IC: **a** suppressed and **b** non-suppressed

a conductometric detector. Figure 8.12 shows a schematic representation of the major components of IEX [55–57].

The mechanism of separation used in IEX benefits from changes in ion affinity for the fluid ion-exchange phase that allows ion separation based on ionic interaction with the stationary phase and eluent. The stationary phase, also known as ion exchanger, has ionic sites mounted on a polymer backbone equipped with a mobile counterion charged oppositely. When the solution that surrounds the ion-exchange is free of charge, the counterions stay where they are within the ion-exchange material. If the counterions and solution have the same ion type, it leads the ions into a dynamic equilibrium, hence swiftly traveling from the aqueous phase to the ion-exchange phase [56, 57].

However, if ions are of different types, the ion of the highest affinity would accumulate within the ion-exchange material while those with the lowest affinity would expectedly concentrate in the surrounding aqueous phase. Noteworthy, as the name suggests, the counterion is swapped as a result of the ion-exchange course. Therefore, ions that move faster through the column will reach the detector first, for which the detector provides separate bands. Ion-exchange resin can act as a cation-exchange resin if negatively charged or as an anionic exchange when positively charged [56, 57].

Ion-exclusion chromatography (IEC) forms a pseudo-semipermeable membrane around the ion-exchange resin, known as Donnan membrane equilibrium. The separation mechanism, as the name implies, is based on "ion-exclusion" leveraging electrostatic repulsion. Neutral species are separated when they are retained by the resins' pores, whereas ionic solutes of the same charge as the stationary phase undergo repulsion from the resin surface, preventing the permeation into the stationary phase. The retention control is manipulated by the pH of the mobile phase. If the pH decreases, the retention increases [53, 57].

Ion pair chromatography (IPC) is a reversed-phase liquid chromatographic technique that utilizes ion pair reagents to produce in situ ion-exchange sites via adsorption of ion pairs on the reversed-phase surface which is commonly composed of a hydrophobic chromatographic media. This technique is typically used when the analyte is difficult to elute from ion-exchange materials [57, 58].

Chelation ion chromatography (CIC) is distinguished by how the stationary phase interreacts. The stationary phase is where chelating functional groups are used, and, apart from ion separation by ion-exchange means, it promotes coordination bonds based on chelate formation with the solute ions [53].

Zwitterionic ion chromatography (ZIC) uses ion exchangers, whereby positive and negative charges are located in close proximity. Therefore, this technique has dual functionality for the concurrent separation of anionic and cationic solutes [53].

8.3.2.1 Instrumentation

Columns

Ion Exchangers

The most significant features of the ion exchangers are their porous matrix and how the functional groups (charged exchange groups) are generated within them. These features control the mass transfer kinetics within the chromatographic column. Ion exchangers can be categorized into the following groups: (i) ion exchangers with grafted functional groups; (ii) ion exchangers with grafted ionogenic polymer layers; (iii) polymer-coated ion exchangers; (iv) and agglomerated ion exchangers or dynamically modified ion exchangers. The structure and preparation method in each type explains the category and its functions [53].

Materials used in IEX can be divided into four groups: (i) strong base anion-exchange components with quaternary ion sites; (ii) strong acid cation-exchange materials with sulfonate ion-exchange sites; (iii) weak acid cation-exchange components with carboxylate or phosphonate ion-exchange sites; and (iv) weak base anion-exchange materials with primary, secondary, or tertiary amine ion-exchange sites. Nowadays, the majority of exchange materials incorporate carboxylic acid ion-exchange sites [57].

Suppression Columns

The function of the suppressor system is to chemically decrease the background conductivity of the electrolytes within the eluent while increasing the conductivity of the sample ions. The suppressor is mounted immediately after the column [57]. The basic suppressor functions are based on the dialysis reactions that occur in ion-exchange membranes. Modern suppressors have a flat sheet as a membrane. The eluent contacts one side of the membrane; meanwhile, the regenerant solution streams in the opposite direction at the other side of the membrane. The sandwich configuration equipped with gaskets helps defining the anticipated flow paths. The eluent flows in the central layer with ion-exchange membrane sheets above and below its current. The regenerant, however, streams in a countercurrent track over the configuration's outer layers. Mesh screens are composed of polymeric ion-exchange materials that are introduced to the eluent cavity and those of the flowing regenerant solution [55]. Post-column reactions that result in a decrease in the background conductance of the eluent are also known as suppression reactions [55].

Eluent

The eluent solution in IC is principally water; however, this can diverge from dilute electrolytes to complex multicomponent buffer solutions based on the operating IC technique. Eluents can be classified using the list below [53, 57].

- Anion-exchange eluents for suppressed IC.
- Anion-exchange eluents in non-suppressed IC.
- Cation-exchange eluents in suppressed IC.
- Cation-exchange eluents in non-suppressed IC.
- Eluents in ultraviolet–visible detection IC of transition metals.
- Ion pair eluents in IC.
- Ion-exclusion eluents in IC.

Eluent Generator

Eluent generators produce the IC eluent as an ion source by means of a deionized water flow generated by the IC pump. It is located before the sample injector. The concentration of produced ions is a function of the ions' current to a charged membrane and the flow rate of the receiver solution. Since the two parameters can be precisely controlled, the eluent can be produced with accuracy as well [53, 57].

Detection

Detection methods are diverse for IC, and some of the most common techniques are listed in Table 8.6. The majority of IC applications operate based on conductivity detectors since ions are conductive in a solution which can, in turn, cover a wide range of components. Ions can be detected with direct or alternating current modes. However, direct current results in oxidation or reduction of analyte ions during conductivity assessment. Thus, alternating current is used in conductometric detections since it can overcome the anomalous measurements of direct current by quickly changing the direction of the electric field, canceling errors correlated to oxidation and reduction [53, 56, 57].

8.3.3 Advantages and Disadvantages of Ion Chromatography

The principle of the IC analytical method is that the substances are separated and preferentially distributed among a stationary and a mobile phase. With the advancements of the technology over the years, the IC has become much more sensitive, while the technique still holds a number of benefits and shortcomings (summarized in Table 8.7). One of the beneficial features of IC is that it allows the determination and separation of a vast variety of ionic components within samples including cations, anions, large/small ions, and organic/inorganic ions [59].

Furthermore, the high degree of selectivity which is confirmed by the appropriate choice of separation and detection systems simplifies the identification of unknown sample components. In addition, high sensitivity detection has been achieved by the implementation of efficient stationary phases and microprocessor technology. This facilitated the detection of ionic constituents at a significantly low concentration

Table 8.6 Detection methods used for IC

Detection	Subtype detection	Principle	Remarks
Conductometric [53]	Non-suppressed conductivity [53]	The conductivity of the elute and the transient bands of elute ions are assessed [53]	Most commonly employed technique [53, 57]
	Suppressed conductivity [53]	The suppressor reduces the background conductivity [53]	
Electrochemical [53]	Charge detector [53]	Measures the charge to which the signal is relative [53]	Changes with the residence time of analyte ions within the change between the anion and cation exchanges [53]
	Amperometry detection [53]	Through applying a potential solute undergo oxidation and reduction when passing over the working electrode [53]	Used for electroactive solutes [53]
Spectroscopic [53]	Photometric detection [53]	The measurements occur by having absorbing eluent ions in the system [53]	Operates only with solutes that have a substantial absorbance profile [53]
	Post-column reaction detection [53]	A chemical reaction between the analyte and a chromophore takes place, and the species are converted after passing the column [53, 56]	The system reacts with most metals and quickly forms kinetics [56]

Table 8.7 Advantages and disadvantages of IC

Advantages	Disadvantages
Determination of various ionic components with simultaneous detection [59–61]	Frequent incompatibility of samples with IC system [59]
High selectivity and sensitivity [59–61]	Requires cautious sample preparation [59, 60]
Low cost, rapid analysis (≈5–15 min), and large dynamic working range [59–61]	High acidity or alkalinity may require further workload and augmented analysis time [59]
Safety and environmental friendliness [61]	Loss of analytes might be observed in small sample concentration [59]

IC Ion chromatography

range (microgram/Liter) without the need for pre-concentration. The ability to simultaneously separate and detect multiple sample components is also an advantageous feature offered by IC [59–61].

Perhaps, one of the main strengths of IC is the rapid analysis of the samples which takes 5–15 min for a complete test [59–61]. This, in turn, allows a greater quantity of samples to be analyzed in each session, hence positively impacting the manufacturing costs of products. The recent introduction of highly efficient separator columns has further contributed to this rapid analysis [60]. In addition, the high stability of separator columns is an outstanding feature of this analytical method. The resin materials (e.g., styrene-based polymers) that are filled within the column enable high pH stability. This, in turn, permits strong acids and bases to be analyzed as eluents expanding the application areas of IC [60, 61]. Other remarkable characteristics of IC include its great dynamic working range, the low consumption of sample solution and reagents, the availability of low-cost systems, and its safe and environmentally friendly nature that commonly uses water-soluble compounds (e.g., $Na_2CO_3/NaHCO_3$, HCl, HNO_3) [59–61].

The incompatibility of samples with the IC system is one of the most frequently reported limitations of IC. Additionally, in specific cases, sample preparation may require careful and detailed processing. For instance, significantly low analyte concentrations and high acidity or alkalinity occasionally limit the overall analysis of ionic components, while highly accentuated differences in ion concentration limit simultaneous ion quantification [59, 60]. This involves further workload and augmented analysis time, and may contribute to error in sample analysis including conversion, loss of analytes, and contamination [59].

8.3.4 Applications of Ion Chromatography

As a well-established characterization method, IC is a great candidate for analyzing various sample types in their amphoteric form. Benefiting from the stationary and mobile phases, IC presents a vast range of applications in different research and industry lines. The non-suppressed IC is recognized as a generally less sensitive system than suppressed IC which marks it as a more suitable choice for analyzing samples that contain medium-to-high solute concentrations (1–100 mg/L). For lower solute concentrations, advanced suppressed IC can be a better choice as it accurately quantifies inorganic anions in aqueous media (10 μg/L −10 mg/L). Both IC systems (suppressed and non-suppressed), however, have been reportedly utilized characterization of organic and inorganic cations in addition to alkali and alkaline metal ions as well as organic amines [62].

Among various metabolites, organic acids (OAs) play a crucial role as hosts of diverse regulatory pathways. Having a polar nature makes OAs challenging to characterize by many of the existing characterization techniques. In one of the reports of the literature, C18 reversed-phase chromatography was used for quantifying ionic or polar metabolites [63], where derivatization was used as a method for enhancement of the retention of such specimens. In the study of metabolomics, IC has also been shown to effectively retain multiple polar metabolites [64, 65]. Petucci et al. in 2016 proposed that when IC is coupled with MS, it can act as a complementary separation method to HPLC for retention and quantification of polar OAs. The key advantage of IC in this study is the separation of highly polar OAs that could not be otherwise retained by reversed-phase chromatography [66]. Generally, ion chromatography-mass spectrometry (IC/MS) demonstrates a wide range of applications for quantifying various polar compounds. In another study, the IC/MS was applied to measure the OA alterations within quadriceps muscles for the purpose of comparing sedentary mice with fatigued mice. The fatigued mice were subject to treadmill exercise following the low-intensity long duration (LILD) or high-intensity short duration (HISD) scheme. As illustrated in Fig. 8.13, this combined method clearly indicates baseline for narrow Gaussian OA peaks with almost no background noise, and the following were separated from the analyzed samples: monocarboxylic acids such as lactic acid, dicarboxylic acids such as succinic acid, and tricarboxylic acids such as citric acid. The result of this study showed that 11 out of 28 OAs presented lower concentrations within the mouse quadriceps muscle homogenate.

The pharmaceutical industry is closely connected to various characterization techniques for analyses of the newly developed drugs. Advances in therapeutic and biosimilar proteins demand solid chromatography methods to thoroughly investigate multiple involving factors prior to release. These include identity, heterogeneity,

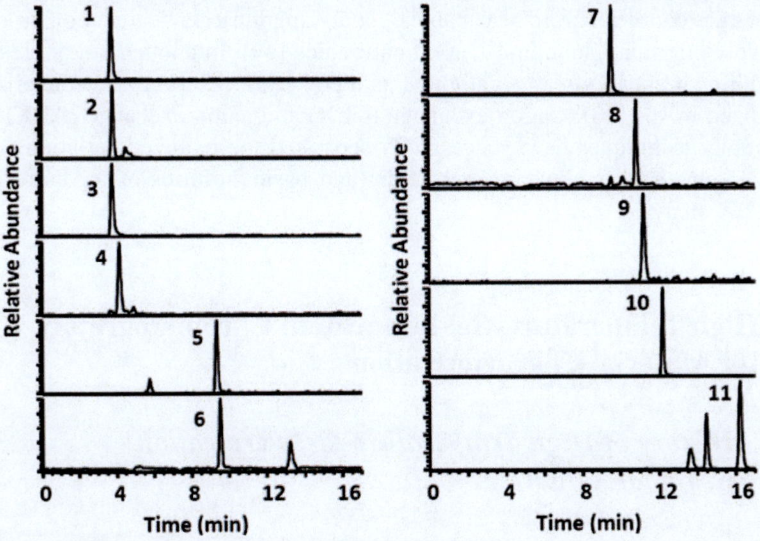

Fig. 8.13 Ion chromatography coupled with mass spectroscopy for analysis of metabolites in mouse quadriceps muscle. The consecutive elution peaks correspond to (1) lactate, (2) pantothenate, (3) 3-hydroxyisobutyrate, (4) 2-hydroxybutyrate, (5) hippurate, 96) succinate, 97) malate, 98) fumarate, (9) pyruvate, (10) citrate, (11) 2-ketoglutarate. Reproduced (or adapted) from [66], ©2016, with permission from ACS publication

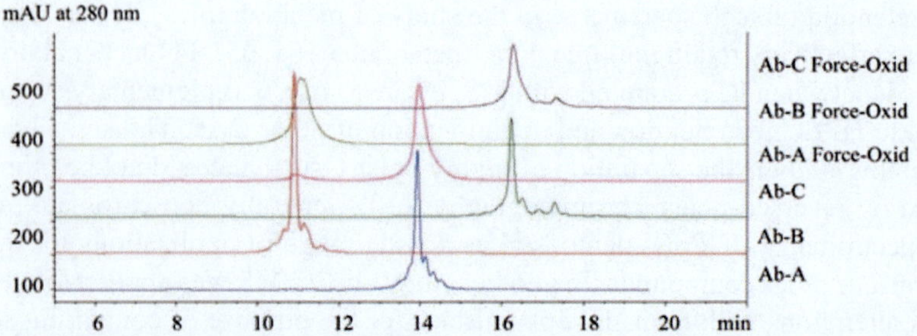

Fig. 8.14 Cation-exchange HPLC performed on XOMA 3AB antibodies showing multiple isoforms of mAbs including Ab-A, Ab-B, and Ab-C. Reproduced (or adapted) from [70], ©2010, with permission from Elsevier

impurity, and activity of the compounds that have to be carefully assessed. A diverse range of analytical approaches, including IEX and reversed-phase liquid chromatography (RPLC), can serve this purpose. For instance, IEX can be a great candidate for separating protein charge variants and isoforms. This technique can support processes such as (i) intact protein analysis, (ii) partially digested large fragments of protein analysis, and (iii) tryptic digest analysis and peptide mapping [67].

A monoclonal antibody (mAb) is an identical copy of an antibody that can bind to a complementary antigen in a very specific manner. In 1997, Moorhouse et al. were among the first to suggest IEX can be used for mAb detection [68]. Reports of the literature confirm IEX to be capable of analyzing intricate degradation processes that involve immunoglobulin-1 (IgG1) antibodies [69]. In another study presented by Teshima et al., IEX was highlighted as a powerful method for characterization of oxidized mAbs [70]. Cation-exchange (CEX) and anion-exchange (AEX) chromatography techniques have successfully separated the oxidized variations of the mAbs. Figure 8.14 exhibits several acidic and basic isoforms of mAbs resulting from CEX analysis.

8.4 High Temperature-Gel Permeation Chromatography for Material Characterization

8.4.1 History of High Temperature-Gel Permeation Chromatography

In the early 1930s, the porosity property of zeolites was named by McBain as a "molecular sieve" which laid the foundation of the high temperature-gel permeation chromatography (HT-GPC). Around 1944, Synge and Tiselius were the first to observe the separation of small molecules according to their molecular size using zeolites with relatively small pores. In 1959, Porath and Flodin used cross-linked

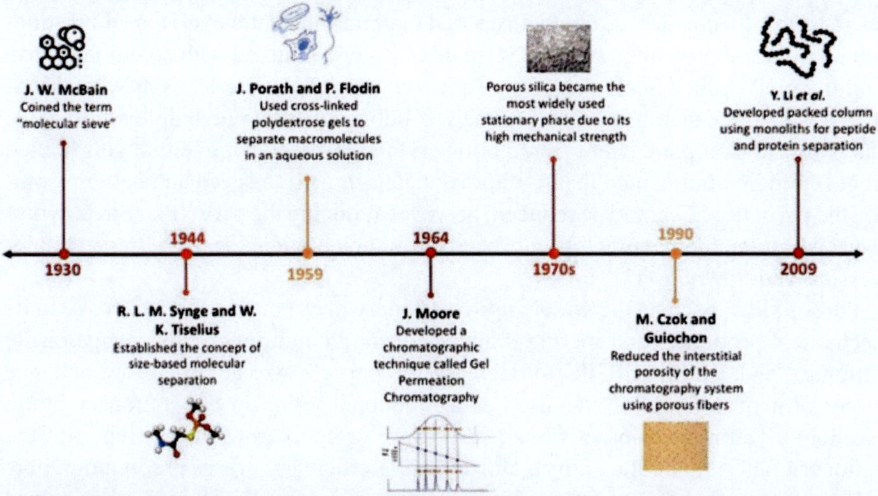

Fig. 8.15 History of HT-GPC

polydextrose gels to separate macromolecules in an aqueous solution and demonstrated that it was possible to separate molecules based on their sizes (Fig. 8.15). When these gels became commercially available, scientists began to employ them to separate biomolecules using GPC at low pressures. Moore first developed the GPC technique in 1964, and he worked for the Dow Chemical Company and used synthetic polymers on cross-linked polystyrene (PS) as a separation medium in a mobile organic phase. Over time, this technique was called by different names such as size-exclusion chromatography, steric-exclusion chromatography, liquid-exclusion chromatography, gel filtration chromatography, and restricted-diffusion chromatography. This chromatographic technique soon became popular for the separation of species such as polymers, aggregates, micelles, and single molecules according to their size dissolved in solution. In addition, it could obtain the molar mass of polymers and analyze aggregation phenomena. The technique was also used for the analysis of phenomena where a change in the hydrodynamic volume was recorded including polymerization, degradation, hydrolysis, and refolding of proteins (Fig. 8.15) [71, 72].

In 1955, Lindqvist and Storgards were successful in separating peptides from amino acids using packed columns containing starch. This study was the first to demonstrate that biomolecules can be separated using size-exclusion processes. Around 1956, Lathe and Ruthven used the "molecular sieve" effect to separate a range of compounds comprising peptides and proteins through a packed column containing corn starch. Two years later, in 1958, Clark carried out a series of studies in which he demonstrated that alcohols that form sugar can be removed by a potent cation-exchange resin. Nonetheless, starch limited the rate at which the molecules separated due to its low mechanical strength. Proposed materials that later overcame this limitation by minimal protein interaction were dextrans cross-linked with epichlorohydrin (commercially called Sephadex). Subsequently, several polymeric resins were developed to substitute starch, e.g., polyvinylpyrrolidone and polyvinylethylcarbitol gels. The Bio-Rad Company later offered polyacrylamide-based commercial gels (called biogel). Flodin commented that to improve data acquisition speed and better resolution the particle size should be reduced; however, reducing the particle size in polymer resins is limited by manufacturing processes as such particles are compressed under pressure and flow [72, 73].

Porous silica became the widely used stationary phase during the 1970s due to its mechanical properties and inertness under different conditions. This change of the stationary phase allowed GPC to decrease the particle size and therefore enhance its performance. The diol was used as a functional group in the stationary phase to reduce hydrophobic interactions and prevent proteins from interacting with the stationary phase (silanol activity). However, the efficiency of this chromatographic technique remarkably increased with the development of the bridged ethyl hybrid (BEH) particles, since this advancement involved reduced particle size (1.7 μm), high mechanical strength, as well as drastically reduced ionic interactions. The columns were packed with porous particles that facilitated the sample separation. In 1990, Czok and Guiochon used porous fibers reducing the interstitial porosity by 15% to 18%, even though this advancement did not necessarily translate to the efficiency of the system, as fibers have possibly prevented the radial dispersion within the packed column. In 2009, Li et al. developed monoliths and used them for the separation of biopolymers which permitted the separation of proteins and peptides of above 670,000 Da [73].

8.4.2 Mechanism of Operation of High Temperature-Gel Permeation Chromatography

Gel permeation chromatography (GPC) is commonly used under the term of size-exclusion chromatography (SEC), since the procedure involves separating the components based upon the size of the fragments. GPC is also a type of high-performance LC that can be used for separating polymer chains according to their sizes in addition to its application in determining the molecular weight (MW) and the molecular weight distribution (MWD) of the polymers. The term of high temperature-gel permeation chromatography (HT-GPC) is used when referring to processes that involve high temperature (up to 250 °C) for characterization [72, 74]. Since the system operates at high temperatures, the mobile phase's viscosity is reduced, the analytes diffuse in a more effective manner, and therefore, the system exhibits a higher resolution compared with a simple SEC system. As previously mentioned, HT-GPC is utilized when there is a need for high temperatures. Compared with a

GPC system, the HT-GPC analysis can be carried out for samples that are resistant to degradation and do not initiate reactions during the sample preparation step [74, 75].

8.4.2.1 Chromatography System

A GPC or SEC system is typically composed of a solvent transfer pump, a sample injector, a set of columns, one or more detectors, and software to control the different segments of the tool and analyze and display the results. A regular SEC system can be modified and converted to an HT-GPC system by adding a column oven and a restrictor tubing to control the backpressure [74, 76].

Columns

The columns are, indeed, the heart of chromatography equipment since it is in their interior where separation actually takes place. On GPC systems, columns are packed with a porous cross-linked gel polymer packing where the passing molecules interact with the polymer pores dependent on their size. Several types and sizes of pores for columns are available. For HT-GPC systems, the columns can stand high temperatures, and in order to acquire acceptable results, sets of several columns (three or four) are commonly installed [5, 10].

8.4.2.2 Separation Process

To start a SEC characterization analysis, first, the polymer sample has to be dissolved in a solvent, and then the molecules curl themselves to form a coil with a size dependable on the molecular weight. Higher molecular weight polymers coil up to form larger spheres. The coiled-up molecules are captured by the mobile phase and are introduced to the GPC column (stationary phase). The column is packed with semisolid polymer particles that cross-link to form a gel that retains the molecules according to their size. Once the mobile phase carries the polymer molecules through the column, the larger polymer coils would pass straight while the smaller ones would be temporarily retained by the pores (Fig. 8.16) [10, 76, 77].

Lastly, the components are separated based on their sizes into discrete chromatographic bands and a detector responds to their signals by producing corresponding peaks for different bands. The peak size is proportionate to the concentration of the molecules present within the mixture [77].

8.4.2.3 Calibration

It is necessary to calibrate the HT-GPC system to have adequate precision and linearity which are determined by the upper and lower limits of the calibration curves.

Fig. 8.16 Mechanism of separation in a GPC column

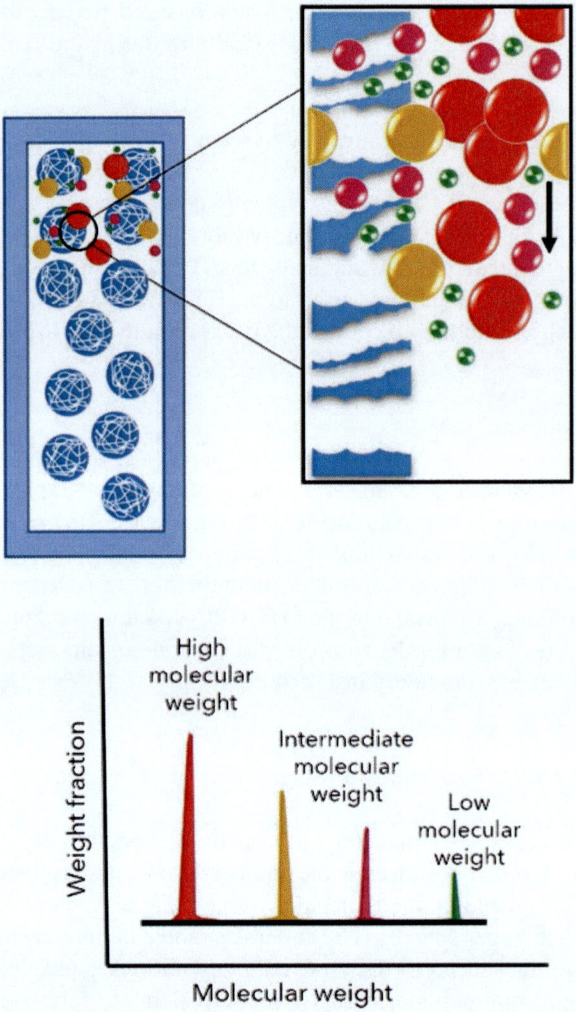

The purpose of calibration is to define the correlation between MW and retention volume in a selective permeation range of the column. For this purpose, standard polymer solutions of known MW and narrow MWD are used to obtain a calibration curve. The solution is injected into the GPC system, and the calibration curve is subsequently plotted on the y-axis using the logarithmic value of MW and on the x-axis the elution volume [5, 10, 77]. In HT-GPC systems, parameters such as retention volume, MW calibration errors, and solvent should be considered as a function of the separation temperature. However, the analysis time is reduced in HT-GPC which, in turn, results in increased separation efficiency [10].

8.4.3 Advantages and Disadvantages of High Temperature-Gel Permeation Chromatography

The characterization of molecular and macromolecular species is mainly done using the HT-GPC technique since the molecular weight and mass sample distribution can be reliably determined through this technique. HT-GPC is a fast and high-resolution technique for analyzing aggregation phenomena as well as the identification of polymeric compounds. By using higher temperatures, the diffusion increases and the viscosity decreases, thus increasing the speed of data acquisition. Samples commonly analyzed with this chromatographic technique are micelles, single molecules, aggregates, and polymers as they are separated according to their size in solution. HT-GPC uses the elution times to produce data in one of the following forms: average molecular weight (M_z), weight average molecular weight (M_w), viscosity average molecular weight (M_v), higher average molecular weight (M_{z+1}), and number average molecular weight (M_n) [74, 78–81].

When the characteristics of the sample do not rely only on the molecular weight, for example, the particle size, using the HT-GPC technique has certain limitations. High molecular weights, insoluble polymers, or samples of complex composition also pose challenges when analyzing the specimen using HT-GPC. The resolution of the obtained peaks directly depend on the molecular weight of the sample. Additionally, pre-treatment is necessary to eliminate unwanted compounds that complicate the interpretation of the results. The operating conditions and experimental setup may vary in different laboratories (e.g., the mobile phase and the column). Common advantages and drawbacks encountered when analyzing samples by HT-GPC are listed in Table 8.8 [71, 74, 81, 82].

8.4.4 Applications of High Temperature-Gel Permeation Chromatography

Extracellular vesicles (EVs) are lipid bilayer-delimited elements that contain several biological components such as proteins and RNA projecting the cell content from which they are secreted. EVs impact physiological and pathological processes by facilitating intercellular communication [83]. For that reason, they attract considerable attention in therapeutic areas such as cell-free methods for drug delivery and regenerative medicine. Multiple methods for isolating EVs are known including differential ultracentrifugation (UC) that separates EVs from other extracellular components based on high-speed sedimentation. Another technique for EVs isolation was recently proposed by Nordin et al. which relies on ultrafiltration and SEC principles [84]. When UC-EVs and SEC-EVs were compared, the EVs were found to be rather more intact using the latter technique since no centrifugation at high speeds was used. Mol et al. in 2017 studied whether the isolation method would negatively

Table 8.8 Advantages and disadvantages regarding HT-GPC

Advantages	Disadvantages
Theoretical plate height tends to be low [78]	Calibration is not easy [82]
The exclusion range and resolution of the technique are high [74, 78, 81, 82]	Determining the volume is complicated [82]
Fast analysis [74, 78, 80–82]	Signals by detectors with different sensitivities and widths are difficult to correlate [81, 82]
The technique has high process control and can be scaled to microsystems [78]	For large molecules, the packaging does not offer reasonable mechanical properties [82]
Determines the mass distribution and molecular weight of the sample [79] [81]	If the molecular weight is high, the resolution decreases [81, 82]
The sample does not interact with the stationary phase [80]	
Interactions between macromolecules can be analyzed [80]	Samples must be soluble [82]
Requires small sample size [74]	At high pressures, the detectors do not have good stability [74]

impact EVs' functionality. In this study, the EVs derived from cardiomyocyte progenitor cells (CPCs) were isolated using UC and SEC. A schematic illustration of the UC and SEC isolation protocols is shown in Fig. 8.17. CPC-based EVs are known for their proangiogenic properties and hence are widely studied for cardiac-related therapies. The results of this study favored the application of SEC over UC hinting that the cells show a more healthy profile and make better candidates for therapeutics testing [85].

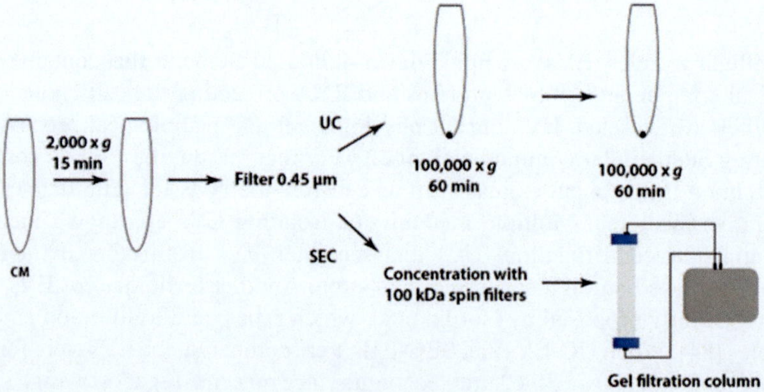

Fig. 8.17 Schematic summary of EVs isolation protocols using UC and SEC. Reproduced (or adapted) from [85], ©2017, with permission from Elsevier

Biotechnological approaches play a significant role in the development of various pharmaceutical products including monoclonal antibodies, vaccines, enzymes, and cytokines with high specificity and potentials that can be crucial in the treatment and control of lethal diseases such as inflammation, infections, cardiovascular diseases, cancer, and immune disorders. To ensure the efficacy and safety of biopharmaceuticals, they have to be thoroughly examined at different development stages ranging from primary steps of the formation, such as amino acid sequencing and post-translational modification (PTM), to secondary and tertiary formation of proteins for the right assembly of multiunit proteins [86]. The evaluation process of protein-based microheterogeneity products can be complex and challenging for the large size and multifaceted structure of the involving proteins. In the case of mAbs, for instance, high MW aggregates of heavy chains may act counterproductive to the mission of the protein and be considered as an impurity or unwanted components that may give rise to immunogenicity [87]. It has been revealed that the characterization of higher MW aggregates is not reproducible using conventional SEC columns. To overcome this limitation, hydrophobic coated-silica SEC columns were suggested as a substitute to achieve reproducible and high-resolution results. Figure 8.18 demonstrates an acceptable consistency between replicated analyses of a biopharmaceutical enzyme, Erwinia chrysanthemi l-asparaginase (ErA) octamer, and monomer. In this study, a ProSEC 300S column with hydrophobic coated-silica was used [88].

8.5 Troubleshooting of Chromatography Analysis Techniques

To identify different gas compounds within a sample, the GC is a useful chromatographic technique that provides the user with excellent accuracy and precision. In general, a mixture of gases is injected into a column and the data obtained is plotted in correlation with the time or volume of the carrier gas. This technique is highly beneficial in pharmaceutical and environmental analyses for the identification of environmental pollutants, residual solvents, and drug products. However, the GC could present several shortcomings during its operation, mainly with respect to operating conditions and equipment maintenance. The troubleshooting of GC is presented in Table 8.9 [89–91].

Within analytical laboratories, and clinical and industrial setups, the HPLC is considered to be a vital chromatography technique for quantitative and qualitative analyses of target samples as it provides high resolution and short time for analysis. This technique identifies molecules present in complex chemical and biological systems and separates molecules with similar physical and chemical properties. The species present in the sample are separated according to the affinity they have to the molecules present within the column. Nevertheless, there are several drawbacks when handling HPLC that mostly depend on equipment components (pumps, seals, detectors, data system, etc.) and/or sample preparation for the mobile phase. One of the recommendations when dealing with HPLC is to clean the column thoroughly

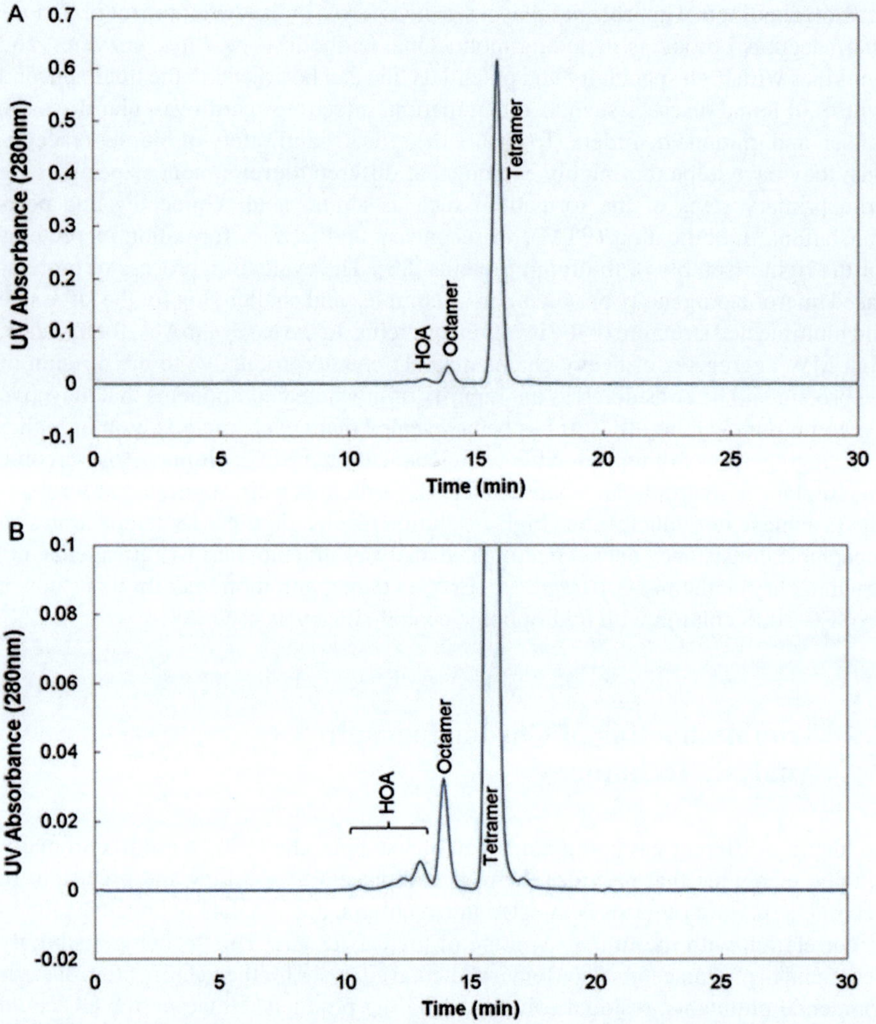

Fig. 8.18 SEC chromatograms of ErA obtained from a coated-silica column in (**a**) full-scale and (**b**) zoomed-in. Reproduced (or adapted) from [88], ©2017, with permission from Elsevier

and frequently. Table 8.9 addresses some of the main challenges users face when working with HPLC, as well as their possible causes and solutions [92].

A commonly employed analytical method for the identification of ions present in various samples is IC. One key element that defines the quality of the results obtained by IC is the pretreatment of the samples. Some components present in sample matrices must be removed prior to IC analysis as they prevent effective separation resulting in poor quantification. Depending on the structure, some organic or inorganic ions may be retained in the column decreasing its capacity, hence affecting the retention time.

Table 8.9 A summarized troubleshooting guide for GC, HPLC, IC, and HT-GPC

Technique	Problem	Cause	Solution
GC	The shape of the earlier-eluting peaks is broad and tailing [89]	The chosen ratio [89]	Split the ratio depending on the column diameter. The lowest recommended ratios are 1:20–1:25 for 0.20 mm; 1:15–1:20 for 0.25 mm; 1:10–1:12 for 0.32 mm; and 1:3–1:5 for 0.53 mm [89]
	A hump peak appears previous to the analyte peak [89]	Degradation of some compound in the column [89]	Decrease both the temperature and the retention time [89]
	Extra peaks appear at the chromatogram [89]	Some species might be condensed in the column [89]	Heat the oven at the highest temperature for 15–30 min and subsequently at 40–50 °C for 60 min [89]
	The peak capacity is deficient [90]	NS	Reduce the column diameter, select a bigger column length, or a mixture of both [90]
	The peaks do not appear [91]	The column is cracked [91]	If the damage is at the edges of the column, it could be carefully cut and reinstalled. On the contrary, if the crack is in the center, it must be replaced [91]
		The temperature in the column is very low [91]	Check the temperature controls [91]
		The carrier gas is not running [91]	Verify if the column has suffered any damage or if there is a sufficient carrier gas flow rate [91]
	Injectors are leaking [91]	The connections of the gas chromatograph are in a bad condition or are not properly installed [91]	Verify all the gas lines in the equipment [91]
	The retention time is too long [91]	Capillaries, tubes, or column are blocked [91]	Verify the flow at inlet and outlet [91]
			Clean the gas lines [91]
		The gas flow rate is very slow [91]	Increase the gas flow rate [91]
	The chromatogram is noised [91]	The column is dirty [91]	Clean the column with a special solvent [91]

(continued)

Table 8.9 (continued)

Technique	Problem	Cause	Solution
		The combustion gases or flow rates are not appropriate [91]	Verify and reset the values of the gas conditions [91]
	The baseline is drifted [91]	Septum bleed is occurring [91]	Use lower injector temperature to analyze the specimen [91]
HPLC	The absent peaks [92]	The detector lamp is switched off [92]	Make sure that the detector lamp is switched on [92]
		There is no mobile phase flow [92]	Verify there is enough liquid in the sampler vials, and air bubbles are not present in the sample [92]
	The HPLC column does not present liquid flow [92]	The pump is not working, or the liquid line is blocked [92]	Turn on the pump and check that the mobile phase has no air [92]
	Detectors are leaking [92]	The cell sealer is damaged, or the cell window is broken [92]	Check and replace components that are in poor condition [92]
	The pressure is low or close to 0 [92]	There are leaks in the system [92]	Verify all the connections in the system, and if necessarily open and then reconnect everything. Also, inject the solvent at twice the previous flow rate [92]
	Inappropriate detector response [92]	Response time is very low, or the noise is very high [92]	Check the response time is correct. Generally, the response time is 1/4 of the peak width at half-altitude of the narrowest peak [92]
		The data points are few [92]	Choose at least 20–30 points regarding reproducible peak integration [92]
	Poor resolution [92]	The mobile phase is polluted. Also, incorrect columns or temperatures are used [92]	Verify and replace the mobile phase. Change the column or the temperature of operation [92]
IC	Retention time is progressively decreasing [93]	Organic or inorganic ions are absorbed to the column [93]	The ions can be removed from the sample using resins (e.g., polyvinylpyrrolidone) [93]
	Low peak efficiency [93]	There is a higher concentration of matrix ions than the ion-exchange column [93]	The conditions under which the chromatography is performed must be changed [93]

(continued)

Table 8.9 (continued)

Technique	Problem	Cause	Solution
			Change the column [93]
	The peak area or peak response shows a low reproducibility [93]	Pulsed or direct current may cause fouling of the detector electrode [93]	Prior to analysis, remove compounds such as fats, proteins, and surfactants from the sample [93]
	Poor peak resolution [94]	Similar retention times [94]	Change the chromatographic conditions [94]
			Pretreatment of the matrix sample may help [94]
	Peak heights reduced [95]	Sample injection problem [95]	Check that the valves are correctly functioning [95]
	Baseline shows deviations [95]	Detector's problem [95]	Pre-treat the sample to avoid impurities [95]
	Low linearity [95]	Electrolytic suppression [95]	Membranes should be regenerated or cleaned [95]
	Air bubbles [95]	Pump's problem [95]	Each time freshwater or eluent is used, the pumps must be purged [95]
HT-GPC	Pressure rapidly increases in the system [96]	The column is damaged, or blockage occurs in the system [96]	Replace or clean the column [96]
	Noise in the chromatogram [96]	Something is trapped in the refractometer cell [96]	Clean the refractometer cell [96]
	The pressure increases [97]	System or column failures [97]	Clean or replace the filters [97]
			Samples should be filtered before use [97]
			Ensure that none of the sample components precipitate during injection [97]
	Peaks do not appear [97]	The injection system does not work properly [97]	Make sure that the injection system is not leaking [97]
	Unexpected peak shapes [97]	The molar mass of the sample or system conditions (tubing length, type of the column) [97]	Use a combination of columns [97]
	Peak splitting [98]	The injection valves are partially clogged [98]	Replace the rotor sealer [98]
	Ghost peaks [98]	The total permeation limit can produce some components that may elute [98]	Change the solvent [98]

(continued)

Table 8.9 (continued)

Technique	Problem	Cause	Solution
	Molecular weight variations [98]	Column's aging [98]	Verify the pump flow rate [98]

NS Not specific; *GC* gas chromatography; *HPLC* high-performance liquid chromatography; *HT-GPC* high temperature-gel permeation chromatography; *IC* ion chromatography

When the sample has an ion concentration that exceeds that of the ion-exchange column, either due to a low sample pH or weak acidity of the column, it causes a low peak acquisition. In Table 8.9, the troubleshooting of the IC technique is presented [93–95].

Calculating the molecular weight of some polymer compounds is complicated when they have low solubility in commonly used organic solvents. HT-GPC is an interesting tool for the characterization (e.g., molar mass) of macromolecules such as proteins and polymers as it is a non-destructive technique. Detectors, injection systems, separation columns, and pumping systems are the main components found in HT-GPC equipment. Despite the advancement of technology in recent years, there are some complications that can occur during the operation of the equipment that affects the quality of the results. Failures in the injection system are among the most frequently observed; as a consequence, the peaks may present strange behaviors [96–98]. The most common problems and their possible causes are summarized in Table 8.9, and a possible solution is presented for each error.

References

1. I.G. Kolomnikov, A.M. Efremov, T.I. Tikhomirova, N.M. Sorokina, Y.A. Zolotov, Early stages in the history of gas chromatography. J. Chromatogr. A **1537**, 109–117 (2018). https://doi.org/10.1016/j.chroma.2018.01.006
2. K.D. Bartle, P. Myers, History of gas chromatography. TrAC Trends Anal. Chem. **21**(9–10), 547–557 (2002). https://doi.org/10.1016/S0165-9936(02)00806-3
3. K. Dettmer-Wilde, W. Engewald, (eds.), *Practical Gas Chromatography, a comprehensive reference*. (Springer, 2014)
4. S.C. Moldoveanu, V. David, Basic Information about HPLC, in *Essentials in Modern HPLC Separations* (Elsevier, 2013), pp. 1–51
5. N.S. Muralisrinivasan, Chromatographic techniques, in *Basic of Polymers. Volume II, instrumental methods of testing* (2019), pp. 77–105
6. L.M. Blumberg, Theory of gas chromatography, in *Gas Chromatography*, ed. by C.F.B.T.-G.C. Poole (Elsevier, Amsterdam, 2012), pp. 19–78
7. P. de Coning, J. Swinley, Overview and theory, in *A practical Guide to Gas Analysis by Gas Chromatography* (2019), pp. 8–32
8. P. de Coning, J. Swinley, The gas chromatograph, in *A Practical Guide to Gas Analysis by Gas Chromatography* (2019), pp. 77–98
9. F.W. Karasek, R.E. Clement, Gas chromatography, in *Basic Gas Chromatography—Mass Spectrometry Principles and Techniques*, ed. by F.W. Karasek, R.E.B.T.-B.G.C., M. S. Clement (Elsevier, Amsterdam, 1988), pp. 5–40

10. M.G. Neira-Velázquez, M.T. Rodríguez-Hernández, E. Hernández-Hernández, A. R.Y. Ruiz-Martínez, Polymer molecular weight measurement, in *Handbook of Polymer Synthesis, Characterization, and Processing* (2013), pp. 355–366
11. P. de Coning and J. Swinley, "GC Colums for gas analysis," in *Practical guide to gas analysis by gas chromatography*, 2019, pp. 207–214
12. P. de Coning, J. Swinley, GC detectors, in *Practical guide to gas analysis by gas chromatography* (2019), pp. 248–289
13. H.M. McNair, J.M. Miller, Introduction. Basic Gas Chromatogr. 1–13 (2009). https://doi.org/10.1002/9780470480106.ch1
14. F. Haghighi, Z. Talebpour, A. Sanati-Nezhad, Through the years with on-a-chip gas chromatography: a review. Lab Chip **15**(12), 2559–2575 (2015). https://doi.org/10.1039/c5lc00283d
15. A. Wang, H.D. Tolley, M.L. Lee, Gas chromatography using resistive heating technology. J. Chromatogr. A **1261**, 46–57 (2012). https://doi.org/10.1016/j.chroma.2012.05.021
16. M.E. Machado, Comprehensive two-dimensional gas chromatography for the analysis of nitrogen-containing compounds in fossil fuels: a review. Talanta **198**, 263–276 (2019)
17. J. Aspromonte, K. Wolfs, E. Adams, Current application and potential use of GC × GC in the pharmaceutical and biomedical field. J. Pharm. Biomed. Anal. **176**, 112817 (2019).https://doi.org/10.1016/j.jpba.2019.112817
18. A.A. Elbashir, H.Y. Aboul-Enein, Multidimensional gas chromatography for chiral analysis. Crit. Rev. Anal. Chem. **48**(5), 416–427 (2018). Taylor and Francis Ltd. https://doi.org/10.1080/10408347.2018.1444465
19. M. Beccaria et al., Preliminary investigation of human exhaled breath for tuberculosis diagnosis by multidimensional gas chromatography—time of flight mass spectrometry and machine learning. J. Chromatogr. B Anal. Technol. Biomed. Life Sci. **1074–1075**, 46–50 (2018). https://doi.org/10.1016/j.jchromb.2018.01.004
20. F.A. Franchina et al., Towards the use of breath for detecting mycobacterial infection: a case study in a murine model. J. Breath Res. **12**(2) (2018). https://doi.org/10.1088/1752-7163/aaa016
21. M. Nasir, H.D. Bean, A. Smolinska, C.A. Rees, E.T. Zemanick, J.E. Hill, Volatile molecules from bronchoalveolar lavage fluid can 'rule-in' Pseudomonas aeruginosa and 'rule-out' Staphylococcus aureus infections in cystic fibrosis patients. Sci. Rep. **8**(1), 826 (2018). https://doi.org/10.1038/s41598-017-18491-8
22. A. Bileck, S.N. Verouti, G. Escher, B. Vogt, M. Groessl, A comprehensive urinary steroid analysis strategy using two-dimensional gas chromatography-time of flight mass spectrometry. Analyst **143**(18), 4484–4494 (2018). https://doi.org/10.1039/c7an01990d
23. B.B. Misra, R.P. Upadhayay, L.A. Cox, M. Olivier, Optimized GC–MS metabolomics for the analysis of kidney tissue metabolites. Metabolomics **14**(6) (2018). https://doi.org/10.1007/s11306-018-1373-5
24. A. Kabir, K.G. Furton, Applications of gas chromatography in forensic science, in *Gas Chromatography* (Elsevier, 2021), pp. 745–791
25. A. Kabir, K.G. Furton, Applications of gas chromatography in forensic science. Gas Chromatogr. 563–604 (2012). https://doi.org/10.1016/B978-0-12-385540-4.00025-0
26. A. Weston, P.R. Brown, High-performance liquid chromatography, in *High performance liquid chromatography & capillary electrophoresis: principles and practices* (1997)
27. L.R. Snyder, J.W. Dolan, Milestone in the development of liquid chromatography, in *Liquid chromatography: Fundamentals and Instrumentation*, ed. by P.R. Haddad, C. Poole (2013), pp. 2–14
28. L.S. Ettre, J.V. Hinshaw, Modern liquid chromatography, in *Chapters In The Evolution Of Chromatography* (Imperial College Press, London, 2008), pp. 371–403
29. A.H. Gordon, A.J.P. Martin, R.L.M. Synge, A study of the partial acid hydrolysis of some proteins, with special reference to the mode of linkage of the basic amino-acids. Biochem. J. **35**(12), 1369–1387 (1941). https://doi.org/10.1042/bj0351369

30. Y.I. Yashin, A.Y. Yashin, Contribution of russian experts to the development of high-performance liquid chromatography. J. Anal. Chem. **75**(10), 1252–1263 (2020). https://doi.org/10.1134/S1061934820100159
31. R.E. Majors, Historical Developments in HPLC and UHPLC column technology: the past 25 Years. LC-GC North Am. **33**(11), 818–840 (2015), [Online]. Available: http://0-search.ebscohost.com.biblioteca-ils.tec.mx/login.aspx?direct=true&db=asn&AN=110949128&lang=es&site=eds-live&scope=site
32. F. Rahimi, S. Chatzimichail, A. Saifuddin, A.J. Surman, S.D. Taylor, A.S. Reyhani, A review of portable high—performance liquid chromatography : the future of the field ? Chromatographia 1165–1195 (2020)
33. A. Zotou, An overview of recent advances in HPLC instrumentation. Cent. Eur. J. Chem. **10**(3), 554–569 (2012). https://doi.org/10.2478/s11532-011-0161-0
34. S.C. Moldoveanu, V. David, Equilibrium Types in HPLC, in *Essentials in Modern HPLC Separations* (Elsevier, 2013), pp. 85–114
35. S.C. Moldoveanu, V. David, Stationary phases and their performance, in *Essentials in Modern HPLC Separations* (Elsevier, Waltham, MA, 2013), pp. 191–362
36. S.C. Moldoveanu, V. David, Retention mechanisms in different HPLC types, in *Essentials in Modern HPLC Separations* (Elsevier, Waltham, MA, 2013), pp. 145–190
37. S.C. Moldoveanu, V. David, Parameters that characterize HPLC analysis, in *Essentials in Modern HPLC Separations* (Elsevier, Waltham, MA, 2013), pp. 53–83
38. M.W. Dong, 1.1.4 Advantages and limitations, in *Modern HPLC for Practicing Scientists* (Wiley, 2006), pp. 1–14
39. P.K. Sahu, N.R. Ramisetti, T. Cecchi, S. Swain, C.S. Patro, J. Panda, An overview of experimental designs in HPLC method development and validation. J. Pharm. Biomed. Anal. **147**, 590–611 (2018). https://doi.org/10.1016/j.jpba.2017.05.006
40. A.M. Sabir, M. Moloy, P.S. Bhasin, A review: HPLC method development and validation. Int. Res. J. Pharm. **4**(4), 39–46 (2016). https://doi.org/10.7897/2230-8407.04407
41. C.M. Loescher, D.W. Morton, S. Razic, S. Agatonovic-Kustrin, High performance thin layer chromatography (HPTLC) and high performance liquid chromatography (HPLC) for the qualitative and quantitative analysis of calendula officinalis-advantages and limitations. J. Pharm. Biomed. Anal. **98**, 52–59 (2014). https://doi.org/10.1016/j.jpba.2014.04.023
42. M. Dong, *HPLC and UHPLC for Practicing Scientists* (2019)
43. A. Iyire et al., Development, optimisation, validation and inter-laboratory verification of a reversed phase HPLC method for quantification of human recombinant insulin. J. Adv. Biotechnol. **7**(1), 984–998 (2018). https://doi.org/10.24297/jbt.v7i1.7192
44. P. Fernández, L. Morales, C. Vázquez, A.M. Bermejo, M.J. Tabernero, HPLC–DAD determination of opioids, cocaine and their metabolites in plasma. Forensic Sci. Int. **161**(1), 31–35 (2006). https://doi.org/10.1016/j.forsciint.2005.10.016
45. D. Corradini, *Handbook of HPLC* (2016)
46. B. Chankvetadze, C. Yamamoto, M. Kamigaito, N. Tanaka, K. Nakanishi, Y. Okamoto, High-performance liquid chromatographic enantioseparations on capillary columns containing monolithic silica modified with amylose tris(3,5-dimethylphenylcarbamate). J. Chromatogr. A **1110**(1–2), 46–52 (2006). https://doi.org/10.1016/j.chroma.2006.01.076
47. J.S. Fritz, Early milestones in the development of ion-exchange chromatography: a personal account. J. Chromatogr. A **1039**(1), 3–12 (2004). https://doi.org/10.1016/j.chroma.2003.12.068
48. F. Nelson, R.M. Rush, K.A. Kraus, Anion-exchange studies. XXVII. adsorbability of a number of elements in HCl-HF solutions 1,2. J. Am. Chem. Soc. **82**(2), 339–348 (1960). https://doi.org/10.1021/ja01487a021
49. H. Small, Ion chromatography: an account of its conception and early development. J. Chem. Educ. **81**(9), 1277–1284 (2004)
50. T.S. Stevens, The membrane suppressor: a historical perspective. J. Chromatogr. A **956**(1), 43–46 (2002). https://doi.org/10.1016/S0021-9673(02)00142-5
51. S. Wouters, P.R. Haddad, S. Eeltink, System design and emerging hardware technology for ion chromatography. Chromatographia **80**, 689–704 (2017)

52. P.R. Haddad, P.N. Nesterenko, W. Buchberger, Recent developments and emerging directions in ion chromatography. J. Chromatogr. A 456–473 (2007)
53. P.N. Nesterenko, B. Paull, Ion Chromatography, in *Liquid chromatography: Fundamentals and Instrumentation*, 2nd ed., S. Fanali, P. R. Haddad, C.F. Poole, M.-L. Riekkola (Elsevier, 2017), pp. 205–244
54. F.J. Dechow, Ion exchange, in *Separation and Purification Techniques in Biotechnology* (William Andrew Publishing/Noyes, 1989), pp. 163–194
55. P.R. Haddad, Liquid chromatography: mechanism: ion chromatography, in *Handbook of methods and instrumentation in separation science*, vol. 1, ed. by I.D. Wilson, C.F. Poole (Elsevier, 2009), pp. 505–514
56. J.P. Lemmon, Ion Chromatography, in *Encyclopedia of Materials: Science and Technology* 2nd edn. K.H.J. Buschow, R.W. Cahn, M.C. Flemings, B. Ilschner, E.J. Kramer, S. Mahajan, P.B.T.-E., M.S.T. Veyssière, (Oxford: Elsevier, 2001), pp. 4280–4283
57. C.A. Pohl, T.F. Scientific, Ion Chromatography. **10** (2019). https://doi.org/10.31399/asm.hb.v10.a0006630
58. J. Stahlberg, Liquid chromatography: mechanisms: Ion-pair chromatography, in *Handbook of methods and instrumentation in separation science*, vol. 1, (Elsevier, 2009), pp. 524–532
59. W. Frenzel, I. Markeviciute, Membrane-based sample preparation for ion chromatography—techniques, instrumental configurations and applications. J. Chromatogr. A **1479**, 1–19 (2017). https://doi.org/10.1016/j.chroma.2016.11.052
60. J. Weiss, Introduction, in *Handbook of Ion Chromatography*, vol. 3, (Wiley, 2016), pp. 1–14
61. R. Michalski, Ion chromatography applications in wastewater analysis. Separations **5**(1) (2018). https://doi.org/10.3390/separations5010016
62. P.N. Nesterenko, B. Paull, Ion Chromatography, in *Liquid chromatography: Fundamentals and Instrumentation*, ed. by S. Fanali, P.R. Haddad, C. Poole, P. Schoenmakers, D. Lloyd (2013)
63. J. Zhu et al., Colorectal cancer detection using targeted serum metabolic profiling. J. Proteome Res. **13**(9), 4120–4130 (2014). https://doi.org/10.1021/pr500494u
64. S. Hu, J. Wang, E.H. Ji, T. Christison, L. Lopez, Y. Huang, Targeted metabolomic analysis of head and neck cancer cells using high performance ion chromatography coupled with a Q exactive HF mass spectrometer. Anal. Chem. **87**(12), 6371–6379 (2015). https://doi.org/10.1021/acs.analchem.5b01350
65. J. Wang et al., Metabolomic profiling of anionic metabolites in head and neck cancer cells by capillary ion chromatography with orbitrap mass spectrometry. Anal. Chem **86**(10), 5124 (2014). https://doi.org/10.1021/ac500951v
66. C. Petucci et al., Use of ion chromatography/mass spectrometry for targeted metabolite profiling of polar organic acids. Anal. Chem. **88**(23), 11799–11803 (2016). https://doi.org/10.1021/acs.analchem.6b03435
67. S. Fekete, A. Beck, J.-L. Veuthey, D. Guillarme, Ion-exchange chromatography for the characterization of biopharmaceuticals. J. Pharm. Biomed. Anal. **113**, 43–55 (2015). https://doi.org/10.1016/j.jpba.2015.02.037
68. K.G. Moorhouse, W. Nashabeh, J. Deveney, N.S. Bjork, M.G. Mulkerrin, T. Ryskamp, Validation of an HPLC method for the analysis of the charge heterogeneity of the recombinant monoclonal antibody IDEC-C2B8 after papain digestion. J. Pharm. Biomed. Anal. **16**(4), 593–603 (1997). https://doi.org/10.1016/S0731-7085(97)00178-7
69. H. Lau, et al., Investigation of degradation processes in IgG1 monoclonal antibodies by limited proteolysis coupled with weak cation-exchange HPLC. J. Chromatogr. B Anal. Technol. Biomed. Life Sci. **878**(11–12), 868–876 (2010). https://doi.org/10.1016/j.jchromb.2010.02.003
70. G. Teshima et al., Separation of oxidized variants of a monoclonal antibody by anion-exchange. J. Chromatogr. A **1218**(15), 2091–2097 (2011). https://doi.org/10.1016/j.chroma.2010.10.107
71. B. Trathnigg, Size-exclusion chromatography of polymers. 1–26 (2006). https://doi.org/10.1002/9780470027318.a2032
72. A.M. Striegel, Size-exclusion chromatography, in *Liquid chromatography: Fundamentals and Instrumentation*, ed. by P.R. Haddad, C. Poole (Elsevier, 2013), pp. 193–220

73. P. Hong, S. Koza, E.S.P. Bouvier, A review size-exclusion chromatography for the analysis of protein biotherapeutics and their aggregates. J. Liq. Chromatogr. Relat. Technol. **35**(20), 2923–2950 (2012). https://doi.org/10.1080/10826076.2012.743724
74. H. Cho, S. Park, M. Ree, T. Chang, J.C. Jung, W.C. Zin, High temperature size exclusion chromatography. Macromol. Res. **14**(3), 383–386 (2006). https://doi.org/10.1007/BF03219098
75. D. Held, P. Montag, Tips & tricks GPC/SEC: high temperature GPC versus Ambient GPC. Column **15**, 23+ (2019)
76. Agilent Technologies, An introduction to gel permeation chromatography and size exclusion chromatography. https://www.agilent.com/cs/library/primers/Public/5990-6969ENGPCSECChromGuide.pdf
77. E.G. Malawer, L. Senac, Introduction to size exclusion chromatography, in *Handbook of size exclusion chromatography and related techniques*, ed. by C. Wu, 2nd edn. (Marcel Deker, 2004)
78. U. Just, H. Much, Characterization of polymers using supercritical fluid chromatography: application of adsorption chromatography, size exclusion chromatography and adsorption chromatography at critical conditions. Int. J. Polym. Anal. Charact. **2**(2), 173–184 (1996). https://doi.org/10.1080/10236669608233906
79. L. Plüschke, R. Mundil, A. Sokolohorskyj, J. Merna, J.U. Sommer, A. Lederer, High temperature quadruple-detector size exclusion chromatography for topological characterization of polyethylene. Anal. Chem. **90**(10), 6178–6186 (2018). https://doi.org/10.1021/acs.analchem.8b00619
80. O. Dragostin, L. Profire, Molecular weight of polymers used in biomedical applications, in *Characterization of Polymeric Biomaterials*, Elsevier Ltd. (2017), pp. 101–121
81. J. Engelke, J. Brandt, C. Barner-Kowollik, A. Lederer, Strengths and limitations of size exclusion chromatography for investigating single chain folding-current status and future perspectives. Polym. Chem. **10**(25), 3410–3425 (2019). https://doi.org/10.1039/c9py00336c
82. A. Revillon, Alternatives to size exclusion chromatography, **17**(14–15) (1994)
83. J. Skog et al., Glioblastoma microvesicles transport RNA and proteins that promote tumour growth and provide diagnostic biomarkers. Nat. Cell Biol. **10**(12), 1470–1476 (2008). https://doi.org/10.1038/ncb1800
84. J.Z. Nordin et al., Ultrafiltration with size-exclusion liquid chromatography for high yield isolation of extracellular vesicles preserving intact biophysical and functional properties. Nanomed. Nanotechnol. Biol. Med. **11**(4), 879–883 (2015). https://doi.org/10.1016/j.nano.2015.01.003
85. E.A. Mol, M.J. Goumans, P.A. Doevendans, J.P.G. Sluijter, P. Vader, Higher functionality of extracellular vesicles isolated using size-exclusion chromatography compared to ultracentrifugation. Nanomed. Nanotechnol. Biol. Med. **13**(6), 2061–2065 (2017). https://doi.org/10.1016/j.nano.2017.03.011
86. G. Brusotti, E. Calleri, R. Colombo, G. Massolini, F. Rinaldi, C. Temporini, Advances on size exclusion chromatography and applications on the analysis of protein biopharmaceuticals and protein aggregates: a mini review. Chromatographia **81**(1), 3–23 (2018). Friedr. Vieweg und Sohn Verlags GmbH. https://doi.org/10.1007/s10337-017-3380-5
87. M. Haberger et al., mAbs Rapid characterization of biotherapeutic proteins by size-exclusion chromatography coupled to native mass spectrometry. Taylor Fr. **8**(2), 331–339 (2016). https://doi.org/10.1080/19420862.2015.1122150
88. D. Gervais, A. Downer, D. King, P. Kanda, N. Foote, S. Smith, Robust quantitation of basic-protein higher-order aggregates using size-exclusion chromatography. J. Pharm. Biomed. Anal. **139**, 215–220 (2017). https://doi.org/10.1016/j.jpba.2017.03.004
89. D. Rood, Gas chromatography problem solving and troubleshooting. J. Chromatogr. Sci. **42**(9), 506–507 (2004). https://doi.org/10.1093/chromsci/42.1.54
90. W. Bertsch, Two-dimensional gas chromatography. concepts, instrumentation, and applications—part 1: fundamentals, conventional two-dimensional gas chromatography, selected applications. J. High Resol. Chromatogr **22**(12), 647–665 (1999)
91. P. Sunitha, K. Lavanya, A.A. Kumar, L.K. Kanthal, B. Kausik, Gas chromatography troubleshooting strategy for analytical chemistry. Int. Res. J. Pharm. **3**(8), 11–13 (2012)

92. K. Raval, H. Patel, Review on common observed HPLC troubleshooting problems. Int. J. Pharma Res. Heal. Sci. **8**(4), 3195–3202 (2020). https://doi.org/10.21276/ijprhs.2020.04.02
93. R. Slingsby, R. Kiser, Sample treatment techniques and methodologies for ion chromatography. TrAC Trends Anal. Chem. **20**(6–7), 288–295 (2001). https://doi.org/10.1016/S0165-9936(01)00069-3
94. R. Kadnar, Determination of amines used in the oil and gas industry (upstream section) by ion chromatography. J. Chromatogr. A **850**(1–2), 289–295 (1999). https://doi.org/10.1016/S0021-9673(99)00623-8
95. Thermo scientific, Ion Chromatography Troubleshooting (2020)
96. S.-C. Polymers, Problems associated with the gel permeation chromatogrphy of sulfur-containing polymers. J. Appl. Polym. Sci. **16**, 2425–2426 (1972)
97. V. Soneji Lafita, Gpc/Sec, *Encyclopedia of Chromatography,* 2nd edn (2005)
98. A.I. Lab, GPC/SEC Troubleshooting Guide (2019)

Chapter 9
Coupled Characterization Techniques

Andrea Rodríguez Vera, María José Rosales López, Euth Ortiz Ortega, Hamed Hosseinian, Ingrid Berenice Aguilar Meza, and Samira Hosseini

Abbreviations

ABS	Poly(acrylonitrile–butadiene–styrene)
AE	Auxiliary electrode
AFM	Atomic force microscope
APCI	Atmospheric pressure chemical ionization
Ca	Calcium
CaF_2	Calcium fluoride
$CaSO_4$	Calcium sulfate
CCD	Charge-coupled device
CNT	Carbon nanotubes
CRM	Confocal Raman microscope
CV	Cyclic voltammetry
CVD	Chemical vapor deposition
CYH	Cyclohexane
DMSO	Dimethyl sulfoxide
EDS	Energy-dispersive spectrometer
EDX	Energy-dispersive X-ray
EPR	Electron paramagnetic resonance
ESI	Electrospray ionization
E–T	Everhart–Thornley
GC	Gas chromatography
GFC	Gel filtration chromatography
GPC	Gel permeation chromatography
H	Hydrogen
He	Helium
HPGe	High-purity germanium
ITO	Indium tin oxide

Andrea Rodríguez Vera and María José Rosales López—share the first authorship.

E. O. Ortega et al., *Material Characterization Techniques and Applications*,
Progress in Optical Science and Photonics 19,
https://doi.org/10.1007/978-981-16-9569-8_9

LEDs	Light-emitting diodes
Li	Lithium
m/z	Mass/charge number of ions
MALDI	Matrix-assisted laser desorption ionization
MeOH	Methanol
MS	Mass spectrometry
nGPC-MS	Native GPC-MS
NS	Not specific
NT-MDT	Scanning probe optical unit
o-HVa	O-vanillin
OTEs	Optically transparent electrodes
PA6	Polyamide 6
PB	Polybutadiene rubber
PCA	Principal component analysis
PET	Poly(ethylene terephthalate)
PMMA	Poly(methyl methacrylate)
PTFE	Poly(tetrafluoroethylene)
REE	Rare earth element
RE	Reference electrode
ROS	Reactive oxygen species
RPLC-MS	Reversed-phase liquid chromatography-mass spectrometry
SAN-MA	Styrene acrylonitrile copolymer grafted with maleic anhydride
SEC	Size-exclusion chromatography
SEM	Scanning electron microscopy
SERS	Surface-enhanced Raman spectroscopy
Si(Li)	Lithium-drifted silicon
SMs	Secondary metabolites
SPEC	Spectroelectrochemistry
SWCNT	Single-wall carbon nanotube
$TBAPF_6$	Tetrabutylammonium hexafluorophosphate
TD	Thermal dissolution
TE	Transmitted electrons
TERS	Tip-enhanced Raman spectroscopy
THF	Tetrahydrofuran
Tm_2O_3	Thulium oxide
Tm^{3+}	Thulium
UV	Ultraviolet
UV–Vis	Ultraviolet–visible
UV–Vis-NIR	Ultraviolet–visible and near-infrared
WE	Working electrode

9.1 Raman and Atomic Force Microscopy for Material Characterization

9.1.1 History of Raman Coupled with Atomic Force Microscopy

In 1928, Sir Chandrasekhara Venkata Raman introduced the Raman scattering phenomenon for which he was awarded the Nobel Prize in Physics a few years later. Incorporating Raman spectroscopy into non-optical microscopes such as atomic force microscopy (AFM) had an immense impact on the quality of the spatial resolution. The coupling of these techniques also allowed a direct correlation between Raman spectra and topographical features [1].

For the development of a device that could combine both techniques, manufacturers overcame various problems inherent to each method such as the acquisition time and the spatial resolution. Modern manufacturers proposed two types of instruments: integrated and interfaced systems. The combined technique now correlates the chemical information from Raman, with image-based information from AFM as a full setup or as an add-on instrument to standard laboratory techniques [2]. The high-resolution AFM phase images have aided in the identification of molecules with diverse chain microstructures. AFM images demonstrate the topographic assembly with resolution as specific as a few nanometers. Also, the grouping of AFM with confocal Raman microscope (CRM) has been used for a wide range of studies including analysis of films of polymer blends, hybrid imaging, and chemical analysis instrumentation [2, 3].

Agriculture and biomass are areas where a combination of Raman and AFM has greatly served. Another field that has highly benefited from Raman-AFM is biochemistry, implying industries such as the pharmaceutical, chemical or nanotechnological. The AFM maps and Raman images demonstrate the correlation of mechanical and biochemical composition of cancer cells, which subsequently helps in the development of treatments for cancer. Additionally, the co-localized Raman-AFM is a dominant technique for many other areas as it serves to uncover topography and morphology, recognize the phase, and analyze the scattering on polymers and composite materials [4].

9.1.2 Mechanism of Operation of Raman Coupled with Atomic Force Microscopy

In AFM experiments, a sharp probe is positioned at a specific distance from the sample and brought near to the sample through a force-based feedback loop. A three-dimensional image of the nominated sample could be generated via a scanning tip moving above the sample. This allows accurate and non-destructive measurements of the surface topography, while the information provided by AFM is useful for industrial applications and scientific research [5]. In Raman experiments, monochromatic light is illuminated on the sample, and on the opposite side, the inelastically scattered light is detected. The scattering of light is on the basis of the Raman effect, which consists of both inelastic (Raman) scattering and elastic (Rayleigh) scattering. Due to molecular vibrations, the incident lights are scattered at different wavelengths. As Raman is less intense than Rayleigh scattering, it is essential to prevent Rayleigh scattering from overpowering the weaker Raman scattering to ensure the acquisition of Raman spectra. Raman and AFM, as subgroups of optical spectroscopies, are complementary methods. When Raman and AFM are coupled, improved results can be obtained than when utilized individually [5].

A specific principle does not exist for the integrated Raman-AFM technique; however, the individual components can be discussed. First, the Raman microscope is aligned via space optics in order to manage the collimated laser beam into the entry of the probe unit of scanning. The beam is then reflected through a set of mirrors, where a mirror is mounted on a closed-loop piezo-scanner. The piezoelectric scanner allows free movement and expansion on all three axes, thus obtaining a clear placement of the beam on the desired area of the sample. Finally, the beam strikes the sample via the scanning probe head [6]. The sample reflects Raman scattered light which is sent to the entry of the Raman microscope's spectrograph. An extra scanner pivots the sample with the AFM probe, hence activating the AFM measurements. During the AFM analysis, a laser detector reviews the probe-sample interface by recording the reflected beam from the cantilever. The reflected light and laser from the cantilever operate in the same optical path as the Raman probing laser. During the experiments, the operator directs the AFM laser onto the cantilever and leads the reflected beam to the photodetector. The Raman laser is placed adjacent to the AFM tip as the probe moves across the sample. The AFM and Raman data are gathered concurrently from the sample area (Fig. 9.1) [6, 7].

9.1.3 Advantages and Disadvantages of Raman Coupled with Atomic Force Microscopy

The most eminent advantage of coupling Raman spectroscopy and AFM is the acquisition of quantitative information of mechanical, chemical, and structural information about the samples. Among those data are parameters such as Young's modulus, viscosity, local adhesion forces, topographic information, chemical species, conformation order, crystalline and amorphous phases, and orientation [8–10]. Another beneficial feature is that the coupling with AFM enhances the spatial resolution of the Raman technique. This allows acquiring information within the micro- to nanoscale [8, 9]. As a result of the non-destructive nature of Raman and AFM individually, the coupled system does not damage the samples as well [9].

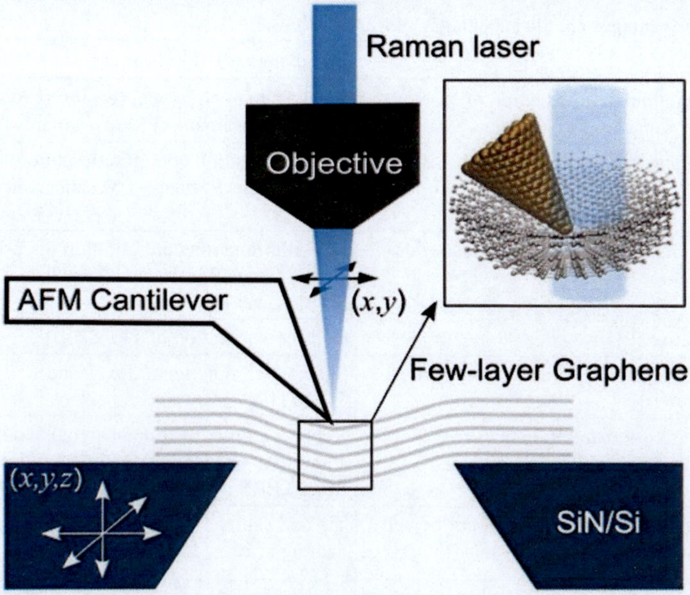

Fig. 9.1 An illustration of the AFM-Raman spectroscopy system. The image represents a concurrent AFM and hyperspectral Raman mapping on suspended two-dimensional membranes. Reproduced (or adapted) with permission [7], ©2016, Nature

Another beneficial aspect of Raman-AFM is the variety of instruments that are available in the market. Amidst the available instruments are tip-enhanced Raman spectroscopy (TERS) and co-located Raman-AFM instruments that perform Raman and AFM analyses simultaneously or in a one-after-another fashion [9, 10], each of which possesses advantages and limitations, as reviewed in Table 9.1. For instance, while most co-located Raman-AFM equipment operates in a one-after-another fashion, simultaneous co-located Raman-AFM instruments reduce analysis time. However, the top-illumination configuration of this form of Raman-AFM (which allows laser excitation of the sample on the top of an area free from the tip) causes a shadow effect. This effect influences the far-field signal of Raman. Nonetheless, the side-illumination configuration prevents such an effect. However, side-illumination yields a biased and lower power signal [9]. Also, since co-located Raman-AFM constitutes a simple joint of conventional Raman spectroscopy and AFM, weak Raman effect and sensitivity can be possible problems to encounter [10].

Meanwhile, TERS, which consists of the coupling of surface-enhanced Raman spectroscopy (SERS) and AFM, overcomes the limitations of co-located Raman-AFM [10, 11]. Despite this advantageous feature, the complex operation of TERS instrumentation represents its most important drawback [10]. Moreover, the need for specialized functional tips for this form of Raman-AFM, maintaining the reproducibility of Raman signal and stability, and preventing tip contamination are a few other obstacles to name [10, 12]. Additionally, special attention must be drawn toward

Table 9.1 Advantages and disadvantages of Raman-AFM

Advantages	Disadvantages
Provides quantitative information of mechanical properties [8–10]	Weak Raman effect and sensitivity of co-located Raman-AFM instruments [10]
Yields chemical and structural data [8–10]	Top-illumination configuration of co-located simultaneous Raman-AFM causes shadow effect [9]
High spatial resolution (micro- to nanoscale) [8, 9]	Side-illumination configuration of co-located simultaneous Raman-AFM yields biased and lower power signal [9]
Non-destructive [9]	Complex operation of TERS [10]
A variety of Raman-AFM instruments are in the market [9, 10]	Specialized functional tips are needed for TERS [10]
Simultaneous co-located Raman-AFM instruments reduce analysis time [9]	Reproducibility of Raman signal and its stability is challenging, and contamination of tip in TERS is possible [10, 12]
Side-illumination configuration of co-located simultaneous Raman-AFM prevents shadowing effect [9]	Tip and samples may be altered due to local heating effects [10, 13]
TERS overcomes limitations of co-located Raman-AFM [10, 11]	No universal wavelength and intensity of excitation laser illumination for samples in TERS is defined [10]

AFM atomic force microscopy; *TERS* tip-enhanced Raman spectroscopy

preventing tip and sample alteration on account of local heating effects due to the high power of the Raman laser [10, 13]. Finally, a universal or standard wavelength and intensity of the excitation laser illumination for samples do not exist; thus, the training of users for careful handling of the equipment is required [9].

9.1.4 Applications of Raman Coupled with Atomic Force Microscopy

Enhanced analysis of the chemical and physical characteristics of a sample can be obtained by combining the Raman and AFM techniques. For the past two decades, commercially available equipment has been able to analyze a sample using Raman and AFM simultaneously. However, one of the limitations is that the sample must be transparent in order to be analyzed by Raman considering the fact that the illumination is done underneath the sample [9]. Moreover, if an illumination above the sample was utilized to obtain the Raman analysis, the light with a wavelength close to 785 nm heats the cantilever and affects the AFM measurement. The most efficient solution to address these problems when operating the two techniques simultaneously is to perform the techniques in series: one after the other. In biology, Raman microscopy allows the analysis of samples regardless of their size or presentation (solution or attached to a surface), while AFM provides information about the structure and mechanical properties of biological systems. For example, it is possible to quantify proteins and DNA by correlating topographic images obtained by AFM with Raman maps [9].

In polymer science, the Raman-AFM combination serves to differentiate polymers that exhibit similar mechanical properties which would be difficult to identify using AFM alone. The information about the polymer provided by Raman microscopy can be qualitative or quantitative. This includes characterizing the functional groups present in the sample, whether the polymer is crystalline or not, and the orientation of the polymer chain, among others. In this sense, the primary technique in determining the chemical properties of the sample is the confocal Raman microscope (CRM). Two polymers with similar mechanical properties such as poly(methyl methacrylate) (PMMA) and poly(ethylene terephthalate) (PET) can be analyzed and differentiated using a combination of AFM and CRM techniques with a resolution below 200 nm. This served as a means for associating the topography of the polymers to their chemical composition while differentiating them [3].

In polymer engineering, polyamides are polymeric materials that stand out for their physical and chemical properties, such as chemical resistance and resistance at room temperature. However, when subjected to stress (e.g., molding flows) they are brittle. In order to offer polyamides as commercially attractive materials, polyamide 6 (PA6) was blended with other polymers such as poly(acrylonitrile–butadiene–styrene) (PABS) to improve the stiffness/toughness balance. In this case, the combination of the Raman-AFM techniques provides valuable information on multiphase polymeric systems. Raman spectroscopy provides information on the physical arrangement of the chains in the polymer, the distribution of the crystalline phases, and the end groups, while AFM provides data such as energy dissipation, modulus, and adhesion. J. P. Cosas Fernandes et al. employed Raman-AFM to analyze mixtures of PA6/ABS in combination with styrene acrylonitrile copolymer grafted with maleic anhydride (SAN-MA) [14]. The Raman-AFM results determined the ϒ-phase content, crystallinity, and polymer distribution of (PA6/ABS/SAN-MA), (ABS/SAN-MA) + PA6, and (PA6/SAN-MA) + ABS samples as shown in Fig. 9.2. The highest amount of ϒ-phase was presented by the sample (PA6/SAN-MA) + ABS with about 51%. It also obtained the highest value of the I_{1080}/I_{1125} intensity ratio, which indicates a higher crystallinity compared to the other polymers [14].

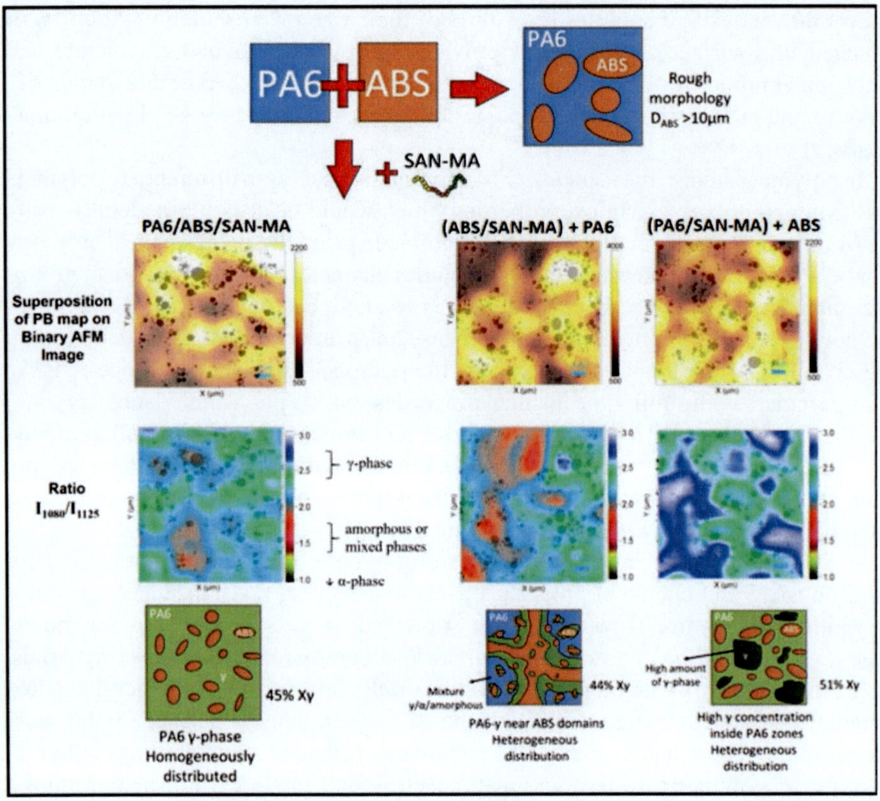

Fig. 9.2 Superposition of Raman mapping on the binary AFM modulus image, presenting the polybutadiene rubber (PB) phase in black. The greater the concentration of PB, the brighter are the pixel. The ratio of the peak intensities I_{1080}/I_{1125}. Reproduced (or adapted) with permission [14], © 2018, Elsevier

9.2 Gas Chromatography Coupled with Mass Spectrometry for Material Characterization

9.2.1 History of Gas Chromatography Coupled with Mass Spectrometry

The beginning of mass spectrometry (MS) dates to the early twentieth century with Sir Joseph John Thomson's experiments on the behavior of molecules. Years later, in 1952, Archer JP Martin and Allen T. James invented gas chromatography (GC), which today is among the extensively used analytical techniques in modern chemistry. MS is a tool that provides a means to identify pure materials, whereas GC separates components of a mixture. Thus, there was a need to combine their characteristics to improve their analytical power. In this way, each element separated by GC could be individually analyzed by MS [15–18].

In the earliest versions, the analyzed sample by MS had to be vaporized in a heater glass container, which was linked to the MS's ion source through a pinhole leak. Therefore, the sample had to have a substantial vapor pressure at temperatures below 200–250 °C. Similarly, GC separates mixtures that could be volatilized at temperatures below 250 °C. Both, MS and GC, required volatilized samples; however, to combine these techniques, some incompatibilities had to be overcome [19].

Reports to couple the gas chromatograph with the mass spectrometer date back to after the development of GC. In 1955, Fred McLafferty and Roland Gohlke started to conduct experiments with GC–MS, and their work was published in 1959. However, early GC–MS was not sustainable due to the advancements of the time, where the adaptation of these technologies was rather poor [19]. It was in 1964 when, independently but simultaneously, Ragnar Ryhage and J. Throck Watson introduced their devices: a jet carrier gas separator and a fritted gas carrier gas separator. These devices overcame the drawbacks presented by early GC–MS devices such as the absence of a rapid transient recorder, the loss of mass spectrometer's sensitivity, or the needed overloading of the GC column. Thus, they marked a milestone in the development of GC–MS [19].

Advances in technology have allowed the progress of GC–MS instruments from analog output devices to computerized automatic devices. In addition, a considerable factor in the development of GC–MS was the expansion of data systems. Nowadays, devices can carry out dozens of samples without human interference, producing a flood of data that can be easily interpreted [19]. Moreover, advances in capillary column injectors, improvement of fused silica capillary columns, electronic flow, and pressure control have steered to easy-to-operate GC–MS setups with lower detection limits [20].

9.2.2 Mechanism of Operation of Gas Chromatography Coupled with Mass Spectrometry

As a powerful method, GC–MS is extensively applied for the analysis of small and volatile molecules. As the name implies, GC is coupled with MS to form a single method of analyzing unknown compounds and multicomponent quantification. Principally, GC differentiates the components of a mixture and MS characterizes each of the components. On the one hand, MS can detect many compounds in a selective manner but not necessarily separate them. On the other hand, GC can separate several volatile and semivolatile mixtures but offers a limited detection. By combining both methods, a solution containing several chemicals can be evaluated in both qualitatively and quantitatively measures [21].

The instrumentation fundamentally consists of three components: GC, MS, and a data system. The data system facilitates both an instrumental control unit and data

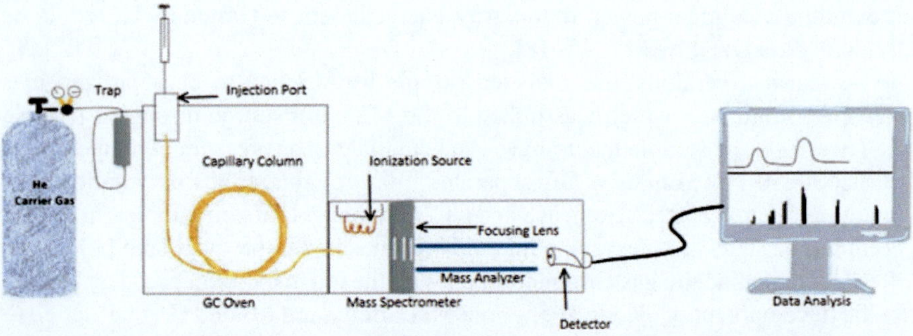

Fig. 9.3 Schematic illustration of a GC–MS setup. The sample enters from the injection port and after passing the GC oven, the MS detector will analyze the data and send the information to a computer. Reproduced (or adapted) with permission [24], © 2012, InTech

gain and handling unit. In earlier generations of GC–MS systems that were packed with GC columns, the mass spectrometer's vacuum system was built in amid the column and the spectrometer to decrease the carrier gas flow from about 30 mL/min to 1.5–2 mL/min. Nonetheless, modern systems occupy capillary columns which are formed of polyimide-coated fused silica and are linked with the mass spectrometer. In some versions of the device made by different companies, there could be some additional components to the above-mentioned segments including an autosampler, an infrared spectrometer, or a flame-ionization detector (Fig. 9.3) [22].

To overcome the significant differences in pressure, the most widely accepted solution is to use a jet separator for packed column GC. However, the main disadvantage is that it can be easily blocked and needs routine service. The principal approach for capillary column GC is that all the GC effluent can be delivered into the MS when the GC gas flow is a few ml/min (low) and the pumping pace of the MS vacuum system is some hundred liters per second (high). Nowadays, for capillary GC–MS, the capillary column is directly fixed within the ion source. The temperature of the effluent traveling the GC reaches up to around 300–400 °C. The next step for each individual compound that is eluted from the GC is to insert the MS and be subsequently identified by the electron ionization detector [23]. In order to generate ions, an ion trap or time-of-flight could be applied. Afterward, each individual compound is flooded by a stream of electrons producing fragmentation. At this stage, the purpose of the MS is to measure the mass-to-charge ratio of ions (m/z) which represents the molecular weight of the fragment where the charge is usually $+1$. Each fragment will be focused through a slit into the detector by a group of four electromagnets and a quadrupole. Only certain fragments are directed via these quadrupoles that are automated by a computer. Subsequently, when the computer has the quadrupoles, it begins scanning the fragment individually until the full recovery of m/z is in range. Finally, the mass spectrum, which is a graphical representation of the ion abundance versus the mass-to-charge ratio of the ions separated in a mass spectrometer, would be produced. It is relevant to mention that researchers can measure the concentrations of each of the components after careful calibration [23].

9.2.3 *Advantages and Disadvantages of Gas Chromatography Coupled with Mass Spectrometry*

The coupled technique of GC and MS constitutes a highly selective and sensitive method. Due to its ease of molecular identification and quantitation, it is considered a universal detection technique, thus allowing an extensive number of applications. Additional to its simple construction, the equipment possesses a relatively low cost [25, 26]. Also, the simple operation of GC–MS instrumentation prevents specialized training for users [25]. Moreover, this technique permits additional coupling with separation techniques [26].

As in individual GC, GC–MS is also limited to only analyzing volatile components of samples. Therefore, samples with involatile components must be processed prior to analysis [25]. Although this technique involves multiple alternatives for sample preparation, sample preparation is typically time-consuming [26]. Among other disadvantages of GC–MS are the prolonged analysis time and the inability to perform real-time analysis [27]. This coupled technique is constrained to compounds that do not surpass 1000 Da and are thermally stable. Finally, the destructive nature of GC–MS to samples is listed in Table 9.2 as another drawback [26].

Table 9.2 Advantages and disadvantages of GC–MS

Advantages	Disadvantages
High selectivity [25]	Limited to analyzing volatile components of samples [25]
High sensitivity [25, 26]	Time-consuming sample preparation [26]
Ease of molecular identification and quantification [25, 26]	Prolonged analysis time [27]
Simple construction and relatively low cost [25, 26]	Not able to perform real-time analysis [27]
Simple operation [25]	Not appropriate for compounds surpassing 1000 Da [26]
Able to couple with other separation techniques [26]	Destructive to samples [26]
Allows multiple alternatives for sample preparation [26]	Limited to thermally stable compounds [26]

9.2.4 *Applications of Gas Chromatography Coupled with Mass Spectrometry*

The identification of organic compounds in complex mixtures has been facilitated by GC–MS. A small quantity of sample is necessary to perform a complete analysis where GC separates the sample components and MS identifies the compounds once separated [18]. The chemical reactions involved in metabolites, as well as the characterization and quantification of biological molecules, are analyzed by means of omics sciences, e.g., metabolomics and proteomics. In metabolomic analyses, GC–MS obtains biological data involved in metabolomics processes. This coupled technique allows the user to identify a wide variety of metabolites due to its broad dynamic range, reproducible results, accuracy, good precision, and high sensitivity. Additionally, complex samples can be studied as GC supplies high resolving power, whereas high sensitivity is provided by MS. The chemometric methods used to analyze GC–MS data depend on the data structure utilized during the sample analysis. For example, Fig. 9.4 presents hyphenated chromatographic data that can be

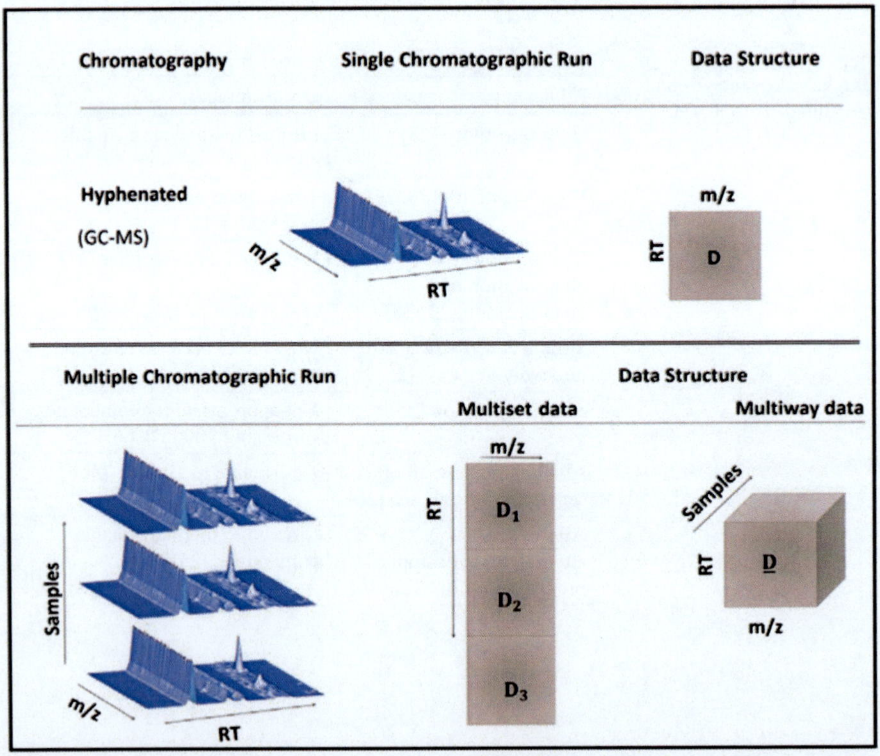

Fig. 9.4 Data structure for GC–MS chromatographic run. Reproduced (or adapted) with permission [28], © 2021, Elsevier

constructed with a second-order matrix or three-way data cube where the mass/charge number of ions (m/z), number of samples, and elution time determine the magnitude of the cube [28].

The demand for tea consumption is high across the world, so the quality of tea depends on its country of origin. In this sense, experts in the black tea industry have shown interest in an analytical method that permits the identification of the geographical origin of the existing black tea varieties in a reliable way. Jing Yun et al. analyzed volatile components of black tea samples by GC–MS from 10 different geographical areas corresponding to China and India. The study revealed that variation of the concentration of volatile compounds including aldehydes, alcohols, esters, and ketones allowed the geographical origin of black tea to be identified [29].

Normal cellular metabolism present in aerobic organisms can generate reactive oxygen species (ROS), which are involved in biological processes such as cellular damage [30]. Hydrogen peroxide is an example of ROS that can cause physiological or pathological alterations through its homeostasis. When ROS levels become too high or too low, they can negatively alter the biological functions of an organism and create damage. For instance, the mutation or death of cells can occur due to cellular damage, specifically DNA damage. Consequently, humans can manifest cancer, cardiovascular or neurodegenerative diseases. To reliably identify the compounds that are produced when there is DNA damage caused by ROS, the GC–MS technique can serve as a powerful tool [30].

Secondary metabolites (SMs) from plants present pharmacological activity and can therefore be considered as a renewable source to produce medicines. The scientific community has been concerned with identifying, isolating, and purifying SMs from plants to compare them with artificial ones and to avoid the adverse effects that occur when using synthetic materials in the human body. One of the techniques used for the analysis of SMs from natural sources is GC–MS. This technique also serves the study of essential oils extracted from the plants' roots [31].

In the coal industry, identifying the structure and composition of the compounds present in coal is important to determine which chemical process should be used to convert it into fuels. The first step is to obtain soluble compounds from the complex matrix present in the coal, where the most used method is thermal dissolution (TD). Subsequently, GC–MS identifies relatively volatile compounds, low polar species, and thermally stable compounds. The data is statistically analyzed with a tool called principal component analysis (PCA) to obtain a rapid classification of the components present within the sample. Guo-Sheng Li et al. used GC–MS to study the Shenfu sub-bituminous coal, Shengli lignite, and Zhaotong lignite from China mines. An evaluation was performed regarding the influence of the type of solvent (methanol and cyclohexane as polar and nonpolar solvents, respectively) used in the TD extraction [32]. The results showed that, for the 3 carbon samples, the extraction efficiency was higher in methanol than that of cyclohexane. However, more compounds were detected in the extraction that used the nonpolar solvent (cyclohexane). Figure 9.5 presents the compounds that were identified after the analysis and were classified as alcohols, alkenes, ketones, aldehydes, arenes, esters, phenols, alkanes, and ethers, among others [32].

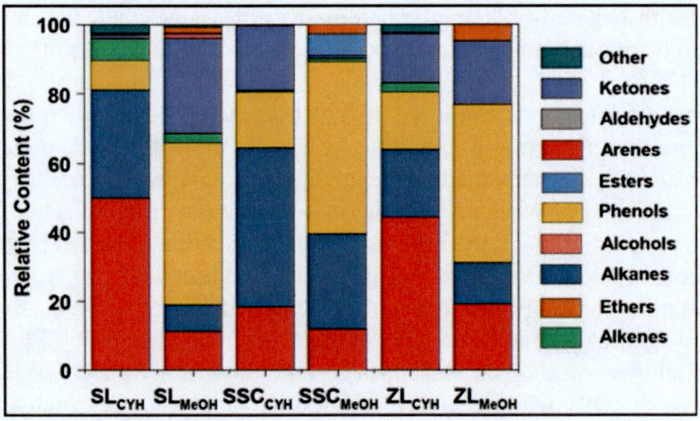

Fig. 9.5 Distribution of the six TD isolated compounds by GC–MS. Reproduced (or adapted) with permission [32], © 2019, Elsevier

9.3 Gel Permeation Chromatography Coupled with Mass Spectrometry for Material Characterization

9.3.1 History of Gel Permeation Chromatography Coupled with Mass Spectrometry

The need for a more complete sample identification has led to the integration of mass spectrometry (MS) and chromatographic analysis for compound identification and structure elucidation. The sample can be separated first through gel permeation chromatography (GPC), size-exclusion chromatography (SEC), or gel filtration chromatography (GFC), and then structurally characterized by MS. The combination of mass spectrometry and chromatographic separation is a feasible way to diminish the sample complexity and obtain more versatile data [33, 34].

J. C. Moore conceived the term "gel permeation chromatography" in 1964 at the Dow Chemical Company. This technique describes the separation of polymer molecules according to the size of the polymer chains. Water Associates licensed the technology and later commercialized it. Although it was well known that reducing the size of the column particles would provide a higher efficiency separation, it was only in 1972 that a 10 μm porous silica particle was introduced to the market. In order to apply the MS detection, SEC methods have been modified by changing the buffers or the mobile phases [35, 36].

There was a need to evaporate the mobile phase of GPC in order to be analyzed by MS. However, the problem that arose from the union of GPC and MS was the undesired liquid-corresponding peaks in the mass spectrometer. Therefore, developments in ionization systems such as matrix-assisted laser desorption/ionization (MALDI) and electrospray ionization (ESI) in the late 1980s had a significant impact on the coupling of SEC and MS. Consequently, the first approach of joining MALDI-MS to an SEC system was done by Giorgio Montaudo's group in 1995. Two years prior, in 1993, Laszlo Prokai and William J. Simonsick Jr. had pioneered the joint use of SEC-ESI–MS. Since then, MALDI and ESI–MS have evolved into powerful tools for synthetic polymer analysis [37–39].

The trend for the future progress of GPC-MS analysis of synthetic polymers points toward the development of more advanced chromatographic separation and enhanced resolution in mass spectrometry systems. Quicker scanning is also needed to record as much data as possible with respect to the separation prior to sample loss. Thus, further developments could be administered by new interfaces and separation methods in addition to instrumental advancements [33, 34, 37, 38].

9.3.2 Mechanism of Operation of Gel Permeation Chromatography Coupled with Mass Spectrometry

One of the leading techniques utilized for molecular weight prediction of proteins in their native state is GPC. This characterization technique has been employed in various applications and different fields of study. Till recent times, the direct coupling of GPC and MS was unsuccessful, mostly due to the oldest generations of MS systems which were not necessarily compatible with high salt contents (even volatile salts such as ammonium acetate). Moreover, GPC columns were not inert enough (weak peaks were recorded for volatile salts). The reports of the literature suggest the offline integration of GPC with a conventional denaturing MS or a native MS, mainly via the reversed-phase liquid chromatography coupled with MS (RPLC-MS) [40, 41]. In modern configurations, GPC fractions are collected and again injected into the system manually and the mechanism of operation is the same as the individual working principle. Processes including the direct online coupling of GPC and denaturing MS, employing ammonium acetate with a mobile phase comprised of organic solvent or modifiers for enhanced chromatographic and MS performance, were introduced for the characterization of proteins and biotherapeutics [42, 43].

To overcome the barriers, a GPC-MS method was established for the study of proteins and their interactions. The method employs an MS-compatible mobile phase containing ammonium formate with a post-column integration of acetonitrile and formic acid. GPC-MS combines the high mass accuracy of electrospray ionization MS with the ability of GPC to resolve protein complexes. Another version of GPC-MS is native GPC-MS (nGPC-MS), which is particularly desirable from a chromatographic perspective as it is based on a simple isocratic separation principle using a single solvent [44].

Gel permeation chromatography can be linked with mass spectrometry by means of an atmospheric pressure chemical ionization (APCI) configuration [45]. This technique allows clear identification of the chemical structures in compounds with a molecular mass lower than 2000 g/ml. During the GPC, chloroform serves as the solvent in the ionization mechanism of the injected sample. The ionization occurs due to the electrons generated by a corona needle. The even electron anions [$CHCl^3$ + $Cl]^-$ resulting from thermal electrons in the negative APCI mode further ionize the solutes as $[M + Cl]^-$ adducts. This stable even electron anion is due to chloroform being present in the APCI interface. However, if odd electron anions surge, a mass spectrum with high background noise is gained for low potentials. If the corona needle is triggered around 2000 V, there is sufficient energy to rupture the background anions. It is important to note that the maximum ionization occurs at zero capillary potential. Furthermore, in the positive APCI mode, high background noise is produced from the contaminants in the chloroform such as the ionization of solvent impurities by charge or photon transfers. To counteract this effect, 10% of tetrahydrofuran (THF) may be added [45].

9.3.3 Advantages and Disadvantages of Gel Permeation Chromatography Coupled with Mass Spectrometry

Studies have reported GPC-MS as a robust technique capable of clarifying structures of low molecular mass compounds below 2000 g mol^{-1}, such as oligomers and additives [45]. Also, this characterization technique delivers data that aids the construction of calibration curves for low molecular masses where Mark–Houwink–Sakurada relationships (of intrinsic viscosity and molecular weight) are unsuitable [45–49]. A rapid analysis time, within minutes, is another advantageous attribute of modern GPC-MS [49].

Several disadvantages are associated with GPC-MS which are summarized in Table 9.3. For instance, GPC-MS is not considered an efficient method for examining molecular weight distribution [46, 50]. Additionally, its combination with a concentration detector (including the differential refractive index or ultraviolet detectors) is highly recommended as it is considered to be a poor method for weight distribution independently [46, 50]. Likewise, the complicated prediction of ionization of chemical compounds is a known limitation of this coupled technique [45–49].

Table 9.3 Advantages and disadvantages of GPC-MS

Advantages	Disadvantages
Robust in clarifying structures of low molecular mass compounds (<2000 g mol^{-1}) [45–49]	Not efficient for examining molecular weight distribution [46, 50]
Delivers crucial data that aids the construction of calibration curves [46–49]	Combination with concentration detector is highly recommended [46, 50]
Analysis within minutes [49]	Complicated prediction of ionization of chemical compounds [45–49]

9.3.4 Applications of Gel Permeation Chromatography Coupled with Mass Spectrometry

In the food industry, the analysis of mycotoxins is essential to remove the fat matrices and edible oils (e.g., aflatoxins, trichothecenes, and zearalenone). From the oil fraction, it is possible to separate the analytes with a yield of up to 0.1 g/ml in tetrahydrofuran (THF) using GPC-MS. Parameters such as temperature, pH, column composition, and column capacity greatly improve the separation performance of this method [51].

Another application of GPC-MS is in determining the molecular weight of a sample. This is an essential attribute in the engineering of materials such as asphalt. The parameter called "fragmentor" controls the ionization efficiency in a GPC-MS system. In a high fragmentor configuration, it is possible to ionize heavy molecules and simultaneously fragment them. The advantage offered by MS is that no solvents are required for its analysis, while GPC is an inexpensive and easy-to-maintain technique [52].

Essential oils from plants have various fields of application such as medicine, pharmaceuticals, food, and cosmetics, among others. There are different extraction methods that are friendly with the environment, thus making these products even more attractive. The quality and efficiency of essential oils extracted from plants depend on their geographical origin due to environmental conditions involved in the plants' growth such as altitude and climate, and the composition of the oils. GPC-MS acts as a reliable tool to analyze essential oils and identify their geographic origin [53].

The way cocoa is processed determines the characteristic flavor of chocolate. The key step in granting chocolate a nice taste is fermentation, where enzymatic and microbiological reactions are involved. Peptides are a product of cocoa fermentation and are considered to be the most important flavor precursors [54]. One way to identify peptide structures within complex matrices that are formed during fermentation is through GPC-MS. For example, from the cocoa vicilin protein, a hexapeptide and a nonapeptide are produced. The hydrolysis of polypeptides that takes place during cocoa fermentation generates small peptides and amino acids. This can be detected by the chromatogram (around 15 and 21 min of elution time) at different days of fermentation corresponding to the region where the peptides eluted (Fig. 9.6a). After 5 days, the fermented extract presented ions that can be eluted between 12 and 25 min (Fig. 9.6b) [54].

Additives and oligomers presenting molecular masses below 2000 g mol^{-1} are of interest to the chemical and food industry. The chemical structure of polystyrene oligomers (from $n = 2$ to $n = 18$) can be studied by GPC-MS using chloroform as a solvent. In addition, the background noise caused by some chemical compounds of complex matrices can be suppressed by adding 10% of THF. This method consumes a low amount of solvent while ionization can be controlled, the sample preparation is minimal, and the obtained data is easy to interpret. The GPC-MS linked with

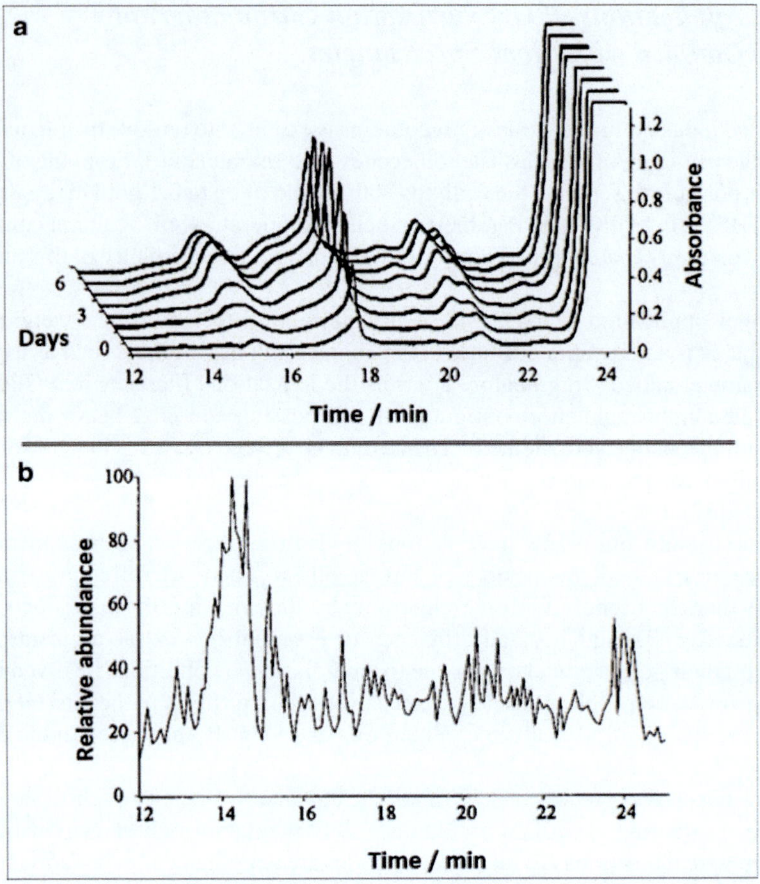

Fig. 9.6 **a** Incomplete GPC chromatograms of cocoa extracts at different fermentation stages. **b** Complete GPC-MS scan of cocoa extract after 5 days of fermentation. Reproduced (or adapted) with permission [54], © 2001, American Chemical Society

atmospheric pressure chemical ionization (APCI) makes a great union for obtaining essential information on additives and evaluating the composition of oligomers [45].

Tea is a widely consumed beverage across the globe. During its production, pesticides are used to ensure a high-quality harvest. However, these chemical compounds can affect human health. The recognition of pesticides in tea is complex since, in the pre-treatment procedure of the sample, several unwanted extracts such as sugars or alkaloids are obtained. The GPC-MS system (Fig. 9.7a) is a valuable method for pesticide identification in an efficient and fast manner compared to other techniques that require long processing time, sample pre-treatment, and high reagent consumption. Pesticides can be identified by the retention time depending on the size of the molecule. Heavier compounds will have shorter retention times in comparison with lighter compounds. For example, chinomethionat has a molecular weight of

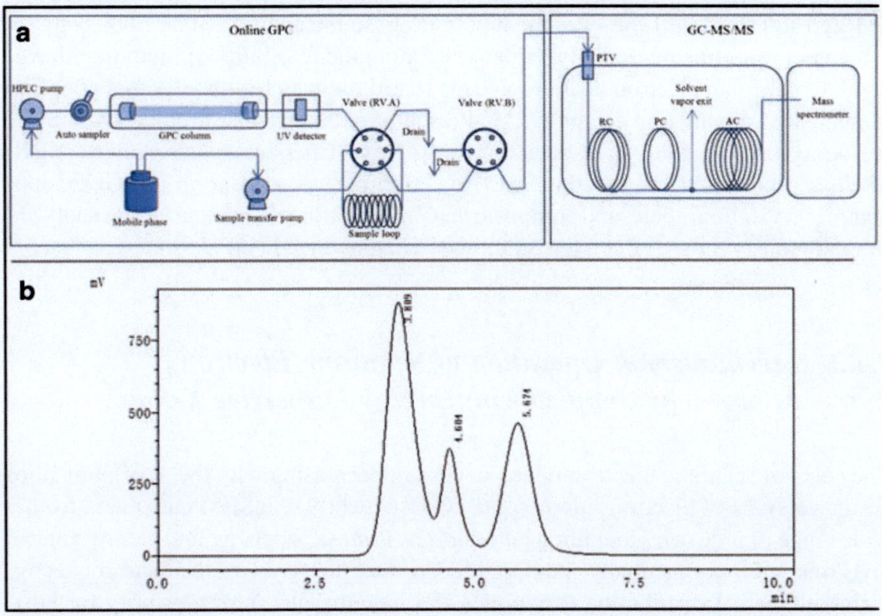

Fig. 9.7 **a** Schematic representation of GPC-GC–MS/MS. **b** GPC chromatogram of fluvalinate at 3.8 min as well as chinomethionate at 5.6 min using a UV wavelength of 210 nm. Reproduced (or adapted) with permission [55], © 2019, Elsevier

234 g mol^{-1} while fluvalinate has a molecular weight of 502 g mol^{-1}. Therefore, the GPC chromatogram reveals a retention time of 5.6 min for the lighter compound (chinomethionat) and 3.8 min for the heavier compound (fluvalinate) (Fig. 9.7b) [55].

9.4 Scanning Electron Microscopy Coupled with Energy-Dispersive X-Ray for Material Characterization

9.4.1 History of Scanning Electron Microscopy Coupled with Energy-Dispersive X-Ray

The combination of scanning electron microscopy (SEM) with energy-dispersive X-ray analysis (EDX) is an example of a widely known hyphenated technique used for characterization [56]. SEM was first developed by Max Knoll in 1935, while it was in 1965 that the commercialization of the instrument began. Over time, this technology has had different developments which include three classifications: its instrumentation, multisignal capacity, and computerization [57, 58].

Ray Fitzgerald, in 1968, was the first to propose the addition of an EDX detector to an electron probe microanalyzer demonstrating the feasibility of mounting the X-ray detector to an electron beam tool. This paved the way for the eventual coupling of such instrumentation to the SEM. Present-day X-ray measurements for SEM are done through solid-state detectors or energy-dispersive spectrometers (EDS) of either lithium-drifted silicon (Si(Li)) or high-purity germanium (HPGe). Topographic, crystallographic, and compositional information of a sample can be analyzed simultaneously due to effective coupling of SEM and EDX [57, 59].

9.4.2 Mechanism of Operation of Scanning Electron Microscopy Coupled with Energy-Dispersive X-Ray

The costs of scanning electron microscopes are decreasing with the development of newer versions of electron microscopes. Consequently, the SEM customers from a wide range of industries including laboratories, forensic sciences, and quality control divisions have exponentially increased. Also, with the growing demand for recognizing elemental distribution in samples and imaging, it is now a priority for businesses and organizations to introduce EDX paired with SEM. Analytical technology specialists have contributed to improving the functionality of these devices. Additionally, with advancing electronics, the operation of SEM–EDX has transformed from an analog to a digital operation, with more control over the equipment and data processing via computers. Previously, an individual EDX equipment was combined with SEM instrumentation, so each system had to operate separately, and the user had to learn how to operate both systems. Nowadays, SEM and EDX are coupled into a single device to facilitate complex operations. The SEM unit includes the EDX detector, and the operation panel contains two monitors [60].

As illustrated in Fig. 9.8, the primary electrons transmitted by the sample impact a gold plate under the bright and dark-field aperture. In the case of applying the SEM, the detector is placed directly underneath the thin sample in order to receive the transmitted electrons (TEs). The purpose of using the gold plate is to multiply and convert the transmitted electrons into secondary electrons. These are then gathered by an Everhart–Thornley (E–T) detector. Subsequently, a dedicated screening ring blocks the secondary electrons emitted by the sample surface toward the E–T detector. Finally, EDX detectors capture the X-ray photon fluxes [61].

9.4.3 Advantages and Disadvantages of Scanning Electron Microscopy Coupled with Energy-Dispersive X-Ray

The integration of SEM and EDX into a single equipment yields a characterization technique capable of providing information on surface morphology, including the

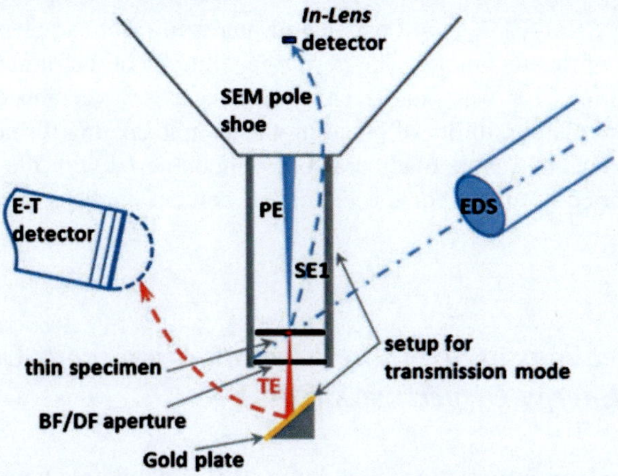

Fig. 9.8 A representation of a transmission mode SEM–EDX system [primary electrons (PEs), secondary electrons (SE1), transmitted electrons (TEs), bright field (BF), dark field (DF), Everhart–Thornley detector (E–T)]. Reproduced (or adapted) with permission [62], © 2014, RSC Advances

shape size and surface constructs [63, 64]. Moreover, SEM–EDX offers compositional data of the samples [63, 65]. This method identifies and quantifies most elements in the periodic table except for hydrogen, helium, and lithium [65]. Other benefits of SEM–EDX are listed in Table 9.4 and include its rapid analysis, high spatial resolution (in the nanometer scale), and the ability to perform high magnifications [63, 64, 66]. Additionally, SEM–EDX equipment is suitable to work in low vacuum environmental mode [66, 67]. This has contributed to the versatility of the technique, thus enabling its application in a range of different areas.

Table 9.4 Advantages and disadvantages of SEM–EDX

Advantages	Disadvantages
Provides information about surface morphology [63, 64]	No sensitivity to H, He, and Li [65]
Offers compositional data of samples [63, 65], and a new generation of the tool can perform wet sample analysis [66, 67]	Not capable of analyzing wet samples [67]
Identifies and quantifies most elements in the periodic table [65]; high spatial resolution (in the nanometer scale) [63, 64, 66]	Non-electrically conductive and grounded samples may need a coating of electroconductive material to be suitable for analysis [63, 67]
Quick analysis [63]; high magnification range [66]	Limited accuracy for the elemental content above 1000 ppm [67]
Suitable to work in low vacuum [66, 67]	

H hydrogen; *He* helium; *Li* lithium

The fact that SEM–EDX is not capable of analyzing wet samples is among the main limitation of this technique. The newer generations of instrumentation supply improved conditions for wet sample analysis, where low vacuum conditions are implemented and the possibility of avoiding the sample coating for non-conductive materials nowadays exists. Another drawback encountered with the use of SEM–EDX is the limited accuracy for the elemental content analysis above 1000 ppm [67].

9.4.4 Applications of Scanning Electron Microscopy Coupled with Energy-Dispersive X-Ray

Detailed morphology information in addition to spatial and elemental analyses of a sample can be performed using SEM coupled with EDX. SEM provides detailed morphological information of high magnification that is not perceptible using optical microscopy, while EDX can identify and provide localization of the elements present in the sample. SEM–EDX allows users to obtain a fast, automated analysis and reliable statistical data from nano- and microsized materials. Some important aspects of SEM–EDX are the sample preparation (substrate selection), the possibility of damage to the sample by the beam, and the vacuum that is exerted during the analysis. The identification of environmental particles, such as ammonium sulfate and nitrate, is one of the applications where the SEM–EDX has widely served the field of environmental monitoring [68]. Other applications include life science, materials science, geomaterials, and forensic research, among others [69].

The archaeological record allows researchers to rebuild the life of ancient civilizations. The data included in organic residues from food significantly helps archaeologists to reconstruct the identity of the individuals who formed these civilizations. Tanasi et al. analyzed samples obtained from Monte Kronio located northeast of Sciacca from the Copper Age. SEM–EDX was employed to obtain data about organic residues that would identify the type of food consumed at that time. The results revealed the presence of bone microfragments which could be related to animal residues. In the EDX spectrum, they were related to the presence of potassium, calcium, and phosphorus. The elevated levels of carbon identified around the bone fragments could be attributed to the consumption of protein and organic residues. In addition, vegetable fibers were also detected in the samples (Fig. 9.9) [70].

Rare earth elements (REEs) have intriguing properties for the scientific community that makes their applications rather diverse. Some of such elements are luminescence in their nature. The industrial and healthcare sectors use light-emitting diodes (LEDs) or lasers made from REE-based phosphor materials for which luminescent elements can serve a purpose. The synthesis of luminescent materials can be done by introducing trivalent REE ions such as thulium (Tm^{3+}) into calcium (Ca)-based compounds such as calcium fluoride (CaF_2) or calcium sulfate ($CaSO_4$). Rodríguez et al. synthesized luminescent materials by homogeneously mixing thulium oxide

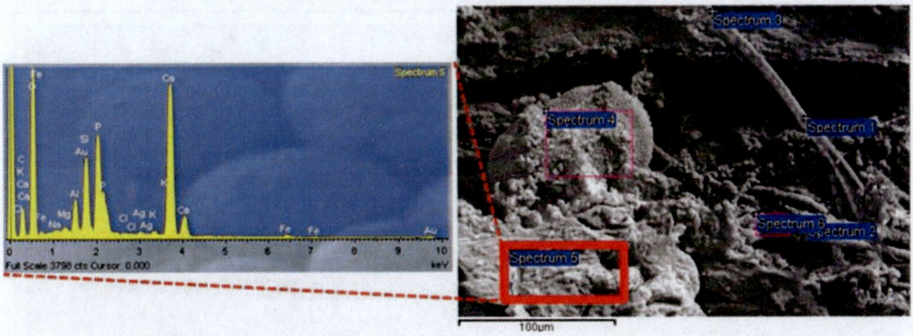

Fig. 9.9 SEM recording of the sample in addition to the EDX spectrum. Reproduced (or adapted) with permission [70], © 2017, Elsevier

(Tm_2O_3) and CaF_2 powders. The composition and structure of these materials were analyzed by SEM, while EDX determined the elemental composition. The images and data presented a micrometric-grained structure as well as the presence of Tm (red circle), Ca (white circle), and a mixture of Ca-Tm (yellow circle). The mapping indicated that Ca is present in areas illuminated with yellow, while Tm was found in green areas, which are also identified in the composition spectrum as shown in Fig. 9.10 [71].

9.5 Ultraviolet–Visible Coupled with Cyclic Voltammetry for Material Characterization

9.5.1 History of Ultraviolet–Visible Coupled with Cyclic Voltammetry

Ultraviolet–visible (UV–Vis) spectroscopy is a technique that was originated in 1801 when Johann Wilhelm Ritter detected ultraviolet (UV) radiation by extending the electromagnetic spectrum beyond the visible region. In 1834, William Henry Fox Talbot discovered analytical optical spectroscopy. However, it was in the 1900s when the relevance of the UV spectral range was recognized. The quantitative determination of vitamin A was the key event in the development of the UV–Vis spectrometer. In 1941, the Zeiss Company launched the first spectrophotometer that included the UV range with a nanometric resolution [72].

Moreover, cyclic voltammetry (CV) is an electrochemical technique established in 1940 when Izaak Maurits Kolthoff and Herbert August Laitinen introduced the term voltammetry to describe the current measurement in correlation with the potential on a solid working electrode [73]. However, it was in 1965 when Richard S. Nicholson formulated the theory of CV rapidly gaining popularity and relevance [74].

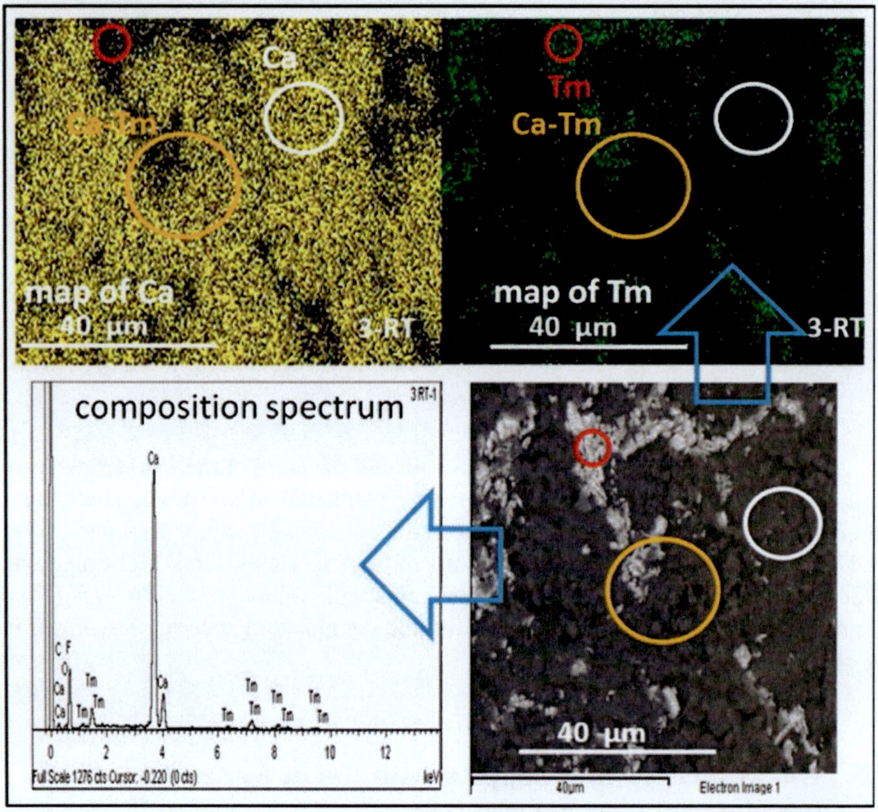

Fig. 9.10 Analysis of a blended sample with ESEM-EDX shows traces of Ca and Tm. Reproduced (or adapted) with permission [71], © 2021, Elsevier

In 1969, Theodore Kuwana introduced the term "spectroelectrochemistry" (SPEC) which is comprised of techniques that allow the concurrent collection of electrochemical and spectroscopic data of an electrochemical cell in situ which was followed by a considerable wave of studies on this field in 1958. Moreover, this field was broadened by the introduction of optically transparent electrodes (OTEs) in 1964 by Kuwana allowing the parallel measurement of electrochemical and spectroscopic responses [75, 76]. The evolution of UV–Vis-SPEC was initially focused on the development of new devices. Various sensitive and selective sensing systems were proposed in which the configurations of the light beam allowed a further enhancement in sensitivity of the platform [77].

The evolution of manufacturing techniques has also had an impact on the development of SPEC. In 2015, carbon nanotube (CNT) films were synthesized through chemical vapor deposition (CVD) and were employed as OTEs for thin layer SPEC. Other microelectrodes, including gold ultramicroelectrode, and screen-printed electrodes have also been explored for OTEs [78]. Moreover, 3D printing technologies have been utilized to fabricate conductive electrodes. However, this technology has yet to be perfected since it contributes to the building of substantial intrinsic kinetic barriers for electron transfer via the electrode–electrolyte interface [79].

9.5.2 Mechanism of Operation of Ultraviolet–Visible Coupled with Cyclic Voltammetry

Spectroelectrochemistry (SPEC) is a technique that brings together spectroscopy and electrochemistry in order to produce a series of chemical and electrochemical information about the sample. Unlike other hybrid models, the two measurements in UV–Vis-CV are obtained simultaneously. This prevents misinterpretation of electrochemical results due to the difference in composition between the surface of the electrode and the bulk solution [80].

It is important to note that spectroscopy is rather structure-oriented, as opposed to electrochemistry which can be classified as reaction-oriented. The integration of the two allows the electrochemical reactions and subsequent progressions to be concurrently recorded by the coupled device. Additionally, detailed information can be obtained regarding redox chemistry, reaction mechanisms, and kinetics, in addition to the electrode surface phenomena of organic and inorganic natures. SPEC analyses are implemented by the UV–Vis and cyclic voltammetry methods individually or together (Fig. 9.11) [81].

For a CV analysis, all three electrodes (reference electrode (RE), working electrode (WE), and auxiliary electrode (AE)) are placed in a cell where the WE and RE are positioned as close as possible and connected to a potentiostat. The cap of the cell includes a poly(tetrafluoroethylene) (PTFE) electrode holder; thus, the electrodes are fixed within the holes of the holder. The argon or nitrogen pipe should be immobilized through another hole on the holder in order to bubble the solution. Once completed, the pipe can be relocated above the solution to allow the gas flow for the measurements. The excess gas must escape; thus, the cell should not be entirely leak-proof [82].

In a UV–Vis-NIR SPEC analysis, the PTFE seal is on one side of the cell with one face in contact with a conductive border of the indium tin oxide (ITO) electrode. The upper side of the ITO electrode should be wrapped in copper foil to enhance its conductivity. The RE and AE should be fixed through the PTFE electrode holder. This assembled cell is placed in the spectrometer holder, and the electrodes are connected to a potentiostat (Fig. 9.11) [82]. The analysis requires the WE to be placed near the bottom of the cell, while the RE is on the top of the active site of the WE. The cell containing an electrode is positioned in the sample area of the spectrometer, and all electrodes are connected to a potentiostat. This setup is shown in Fig. 9.11 [82].

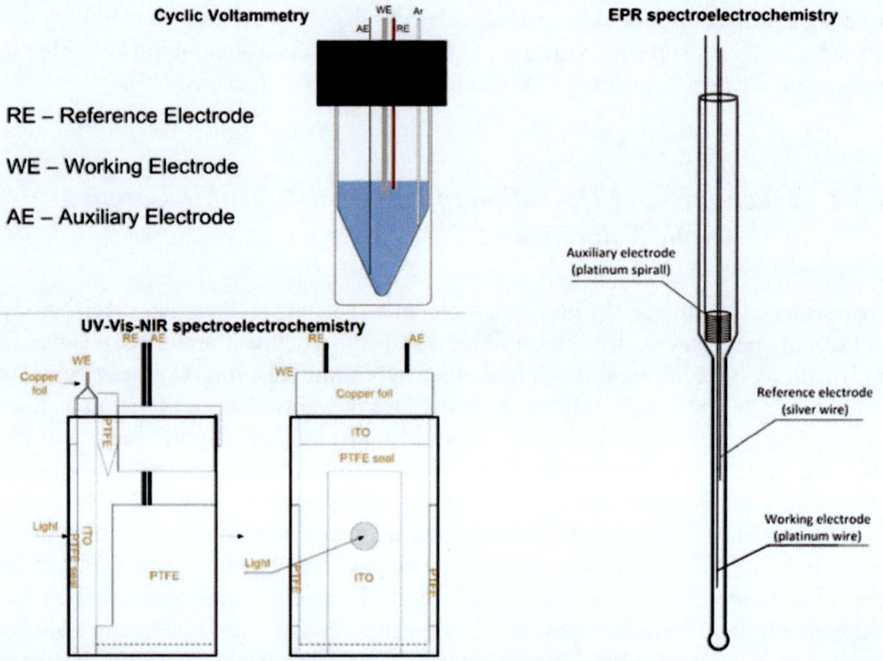

Fig. 9.11 An electrochemical and spectroelectrochemical setup for UV–Vis-CV measurements. Reproduced (or adapted) with permission [82], © 2018, Journal of Visualized Experiments

9.5.3 Advantages and Disadvantages of Ultraviolet–Visible Coupled with Cyclic Voltammetry

The UV–Vis-CV hybrid technique presents descriptive data that defines the kinetics and dynamics of electrochemical processes on electrode surfaces [78, 83, 84]. Additionally, it provides information regarding molecular vibrational processes [78, 85]. Moreover, UV–Vis-CV cross-correlates data extracted from UV–Vis variations with synchronous electrochemical scans, which constitutes its principal advantageous feature [78, 86]. Moreover, it is considered to be a rapid and low-cost technique [83].

The spectrum of UV–Vis light is between 100 and 800 nm. However, UV–Vis spectrometers operate along with an extended range of wavelengths, normally from 200 to 1100 nm; thus, higher-quality results are obtained inside a narrower interval (220 to 780 nm). Another limitation is the interference of by-products in the interest regions of the spectra due to solute–solvent interactions [87]. Additionally, the system requires calibration for reliable data acquisition [88] as presented in Table 9.5.

Table 9.5 Advantages and disadvantages of UV–Vis-CV

Advantages	Disadvantages
Describes the kinetics and dynamics of electrochemical processes on electrode surfaces [78, 83, 84]	Distorted results outside the narrow range of wavelength [87]
Provides information regarding molecular vibrational processes [78, 85]	Interference of by-products from the reaction in the spectrum [87]
Cross-correlates data extracted from UV–Vis variations with synchronous electrochemical scans [78, 86]	
Rapid method [83]	Requires frequent calibration [88]
Low cost [83]	

UV–Vis ultraviolet–visible

9.5.4 Applications of Ultraviolet–Visible Coupled with Cyclic Voltammetry

Understanding reaction mechanisms is a fundamental step in the understanding of chemical and electrochemical processes. This involves knowing the potential at which the reaction occurs, the structures of the products and by-products of the reaction, in addition to the processes involved in the electron transfer. SPEC is a novel tool that merges spectroscopic techniques such as UV–Vis with electrochemical characterization techniques such as CV. This characterization tool allows users to obtain in situ information on the products generated during a chemical reaction, the electrochemical behavior of chemical species, the chemical changes presented by the species involved in a reaction (ionic species mainly), and the kinetic study of the reaction. An effective technique for the synthesis of conductive polymers such as polypyrrole is electro-polymerization. This procedure can be studied by SPEC, which provides information about the structure, morphology, and thickness, among other characteristics of the synthesized polymer [89]. Additionally, SPEC is a technique capable of providing kinetic and thermodynamic data about processes involving charge transfer at the electrode interface. This indicates that in electro-polymerization, the user has control of the applied voltage to the system. It is possible to perform a quantitative study of the dynamics of homogeneous or heterogeneous physicochemical phenomena occurring at either liquid–liquid or solid–liquid interfaces using SPEC, for instance, an in-depth analysis of the electrochemical mechanism of the systems involving reversible reactions with complex stoichiometries [90].

Antioxidants are molecules capable of reducing the risk of disease through prevention. These compounds drastically reduce or prevent other species from oxidizing, thus acting against reactive oxygen species. O-vanillin (o-HVa) is a natural product extracted from plants and employed as a flavoring agent in therapeutic activities, in addition to presenting antioxidant properties at low concentrations. The generated products from o-vanillin act as an antioxidant and can be studied using SPEC as it allows the study of the phenomenon in situ and the real-time recording of spectroscopic and electrochemical data throughout the chemical transformation of o-vanillin. In this case, the chemical reaction is electrochemically manipulated while the information of the generated products is obtained by means of the optical signal. Gonzalez-Baro et al. studied the antioxidant properties of o-HVa by SPEC in the absence and presence of oxygen [91]. The results demonstrated that both in the absence and presence of oxygen, an anodic peak appeared that is attributed to the oxidation of o-HVa (around +1.17 V) into o-quinoid and dimerization products. On the other hand, an irreversible reduction peak (of approximately −0.75 V) only appears in the presence of oxygen, which is attributed to the oxygen reduction reaction generating superoxide ions (Fig. 9.12a). Absorption spectra confirm the presence of compounds generated by the oxidation of o-HVa as a result of the presence of an absorption band at 507 nm (Fig. 9.12b). Meanwhile, an absorption band at 410 nm appeared in the presence of oxygen which could be attributed to the superoxide ion generated during the oxygen reduction reaction chemically reacting with o-HVa (Fig. 9.12c) [91].

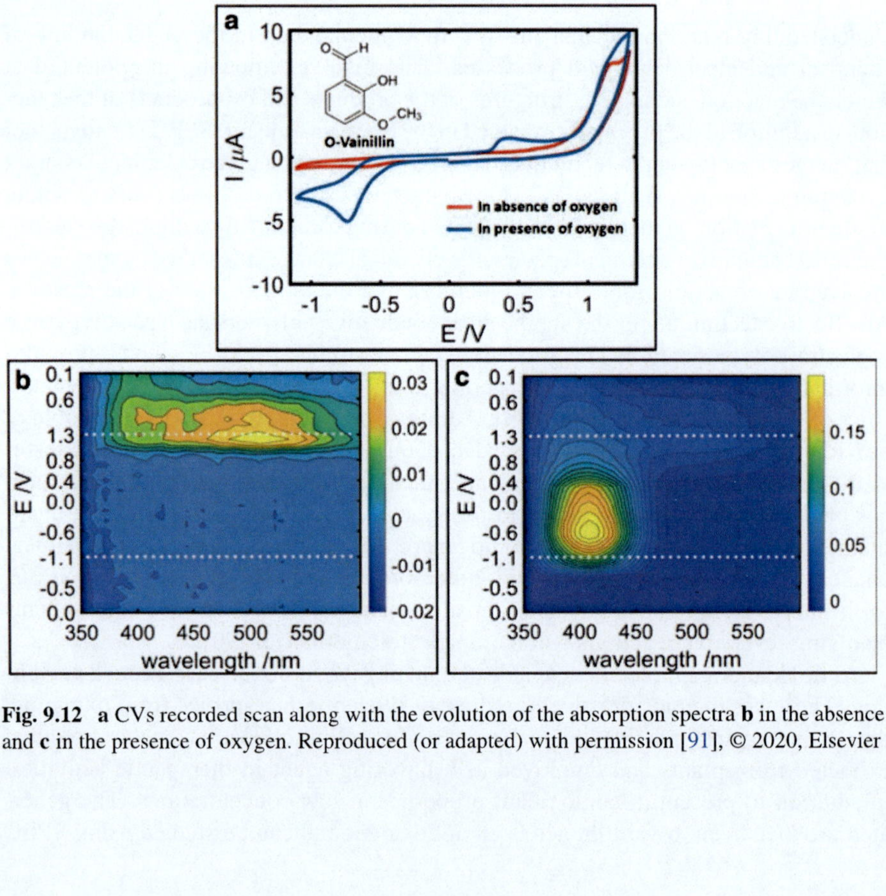

Fig. 9.12 **a** CVs recorded scan along with the evolution of the absorption spectra **b** in the absence and **c** in the presence of oxygen. Reproduced (or adapted) with permission [91], © 2020, Elsevier

9.6 Troubleshooting of the Coupled Characterization Techniques

Coupling Raman and AFM techniques allow the user to obtain extensive and accurate data on the surface topology and structural composition of the sample. Obtaining superior quality results depends on both the skill of the person operating the equipment and the correct operating conditions. Table 9.6 presents the common problems that may occur when dealing with this coupled technique, the potential causes of such errors, and suggested solutions. For instance, it is important to choose the right type of stimulation (buffer or acoustic) for the cantilever beam according to the AFM operating conditions (air or liquid). In the case of Raman, regulating laser power is one of the most common challenges to face. For instance, in the case of excessive power, the sample can be burnt, and if the power is too low, the image resolution is compromised. Moreover, to improve the quality of the data, it is important to select the appropriate specimen holder and objectives based on the sample. Furthermore, the user should avoid sample contamination and select the correct integration time to decrease the signal-to-noise ratio. When the software fails, the recommendation is to restart the computer and check that all connections between the computer system and the equipment are correct. Finally, it is necessary to dampen the sample or shorten the data acquisition time to prevent fluorescence from reducing the image quality during the analysis [92–97].

The most universal analytical method for the identification and quantitation of organic compounds in complex matrices is GC–MS. Table 9.6 lists problems that commonly originate during the use of GC–MS and suggests viable solutions. To avoid errors with the gas chromatography, the first step is to have a correct installation of the instrument followed by its right maintenance, which includes the timely change of filters. It is also important to determine if the problem is specific to the gas chromatograph or the mass spectrometer [20, 98]. Moreover, to eliminate interference, it is necessary to obtain a mass resolution of the sample ions or by the aim of alternative ions that would appear at the same nominal mass. If adequate chromatography protocols are used, such interferences are removed. Since an increase in resolving power may result in a decrease in sensitivity in dual-focused setups, it is recommended to use the system at the lowest resolving power which would correspond with satisfactory results [20].

When it comes to coupled GPC-MS technology, since not all samples can ionize without significant fragmentation, a universal GPC-MS setup cannot be developed and each set of experiments requires its own specifications. Table 9.6 provides some parameters that may be adjusted for the improvement of the operating conditions in the identification of samples [99]. Both the MALDI and ESI ionization techniques used in mass spectrometry present multiple factors in the GPC-MS setup that have a great impact on the ionization procedure. The flow rate in addition to the split ratio, the type, and the amount of the salt in methanol can vary to amplify the ionization effectiveness while maintaining the salt clustering to a minimum [99].

Table 9.6 Summarized guide of common problems, possible causes, and solutions for Raman-AFM, GC–MS, GPC-MS, SEM–EDX, and UV–Vis CV

Technique	Problem	Cause	Solution
Raman-AFM	Low quality and resolution of image [92]	The sample could be contaminated [92]	The sample should be exposed to UV light prior to analysis [92]
	Low contrast [95]	Image parameters are likely to be incorrect [95]	Reduce the signal-to-noise ratio [95]
	Low image resolution [93, 95]	NS	Use a material that increases the conductivity at the tip, e.g., carbon nanotubes [93]
		The sample is damaged [95]	A sharp tip with a high aspect ratio is necessary [95]
	Failure of software or any unit of the equipment [94, 96]	NS	Restart the software and the computer equipment and check the connections [94, 96]
	Poor signal intensity [97]	The laser power is low [97]	Increase the laser power [97]
	Burned specimen [97]	The laser power is high [97]	Decrease the laser power [97]
	Changes in the signal-to-noise ratio [96, 97]	The focus is not correct [96, 97]	Refocus the setup [96, 97]
		The integration time is short [96, 97]	Adjust the integration time [96, 97]
	Excess of fluorescence in the image [97]	The sample is too dry or bleached [97]	Make the sample wet or reduce the integration time [97]
GC–MS	Peaks have a rear tail [20, 102]	Incorrect column installation or injector [20]	Verify that nonpolar compounds such as hydrocarbons do not have a tail and verify the installation [20]
	No peaks were obtained [20]	No mobile phase flow [98]	Verify that all the flows are correct [20]
		Blocking on the line [98]	Verify if the injector liner is healthy [20]
	Leak on the injectors [98]	Bad installation or improper condition of the connections [98]	Check the installation points [98]
	Low or no signal [103]	Insufficient ion beam [103]	Verify the isolation valve, filament continuity, and ion lens voltages [103]
	Poor sensitivity [104]	Degradation of the electron multiplier [104]	Sonicate the electron multiplier to clean and dry it [104]

(continued)

Table 9.6 (continued)

Technique	Problem	Cause	Solution
		Contamination at the source [104]	Replace the filament [104]
GPC-MS	Unpredictable flow rates [105]	Air in the system [105]	Prime the pump to remove air, remove and replace the solvent bottle, degas the mobile phase [105]
	Increase or decrease in the retention times [105]	Column contamination [105]	Inspect the column and equilibrium of the system, and improve the solvent blending [105]
	Split and distorted peaks [105]	Poor tubing connections or a blocked in-line filter [105]	Review all fittings and tubing on the outlet of valve to the detector [105]
	Salt formation [99]	Buffer is inappropriate [99]	Change the solvent [99]
	Unclear image [106]	NS	Slow the scan rate, increase the beam intensity on the sample, or try a higher current [106]
SEM–EDX	The inappropriate analytical signal from the sample [107]	NS	Modify the parameter of the voltage and the beam current [107]
	No image on the screen [108]	NS	Reduce magnification to the lowest value [108]
	Concentration measurement failure [87]	Solvent dilatation with temperature [87]	Correct the factor of the coefficient of thermal expansion [87]
	Insufficient band intensity [109]	NS	Increase analyte concentration [109]
UV–Vis-CV	Increased peak-to-peak separation in the voltammogram [110]	Ohmic drop [110]	Reduce the size of the WE to decrease current or use low scan rates [110]
			Increase electrolyte concentration [110]
			Place the working and reference electrode closer [110]
	Background scan [101]	Electrochemically active solution [101]	Verify that the range of the scan window is not electrochemically active for the solution [101]

AFM atomic force microscopy; *NS* not specific; *GC* gas chromatography-mass spectrometry; *GPC-MS* gel permeation chromatography-mass spectrometry; *SEM–EDX* scanning electron microscopy-electron-dispersive X-ray; *UV–Vis-CV* ultraviolet–visible spectroscopy-cyclic voltammetry; *WE* working electrode

In the case of modern SEM–EDX, the images are rather simple to obtain. However, it necessitates the knowledge of the major modes of microscopy and electron beam variables. The reliability of the image depends on the beam voltage. With a smaller electron probe, the probe current available would be lower; hence, the image features visibility would turn poorer as well [59, 100].

Finally, with respect to UV–Vis-CV, a drawback is the sensitivity of the system. However, the right choice of optical configuration can greatly improve the sensitivity. A correct cyclic voltammetry experiment must be done in a solvent that would not react with the analyte and must not be redox-active within the predetermined scan window. Another critical point is to ensure that the analyte does not react with the electrode surface. Therefore, their homogeneity in solution should be assessed prior to analysis [77, 101]. Some additional considerations to avoid errors in running UV–Vis-CV are itemized in Table 9.6.

References

1. V. Stancovski, S. Badilescu, In situ Raman spectroscopic–electrochemical studies of lithium-ion battery materials: a historical overview. J. Appl. Electrochem. **44**(1), 23–43 (2014)
2. B. Foster, AFM/Raman opens new horizons for research and industrial characterization **39**(6), 13–14 (2007)
3. U. Schmidt, S. Hild, W. Ibach, O. Hollricher, Characterization of thin polymer films on the nanometer scale with confocal Raman AFM. Macromol. Symp. **230**(1), 133–143 (2005)
4. H. Mitsutake, R.J. Poppi, M.C. Breitkreitz, Raman imaging spectroscopy: history, fundamentals and current scenario of the technique. J. Braz. Chem. Soc. **30**, 2243–2258 (2019)
5. H. Kuzmany, Light scattering spectroscopy, in *Solid-State Spectroscopy*, 2nd edn. (2009), pp. 183–215
6. M. Surtchev, S. Magonov, M. Wall, Characterization of materials with a combined AFM/Raman microscope, in *NT-MDT Application Note*, vol. 89 (2015)
7. K. Elibol, B. Bayer, S. Hummel, J. K.-S. Reports, Visualising the strain distribution in suspended two-dimensional materials under local deformation. nature.com (2016)
8. J.P.C. Fernandes, V.H. Mareau, L. Gonon, AFM-Raman colocalization setup: advanced characterization technique for polymers. Int. J. Polym. Anal. Charact. **23**(2), 113–119 (2018). https://doi.org/10.1080/1023666X.2017.1391740
9. B. Prats-Mateu, N. Gierlinger, Tip in–light on: advantages, challenges, and applications of combining AFM and Raman microscopy on biological samples. Microsc. Res. Tech. **80**(1), 30–40 (2017). https://doi.org/10.1002/jemt.22744
10. W. Fu, W. Zhang, Hybrid AFM for nanoscale physicochemical characterization: recent development and emerging applications. Small **13**(11) (2017). https://doi.org/10.1002/smll.201603525
11. K.F. Domke, B. Pettinger, Studying surface chemistry beyond the diffraction limit: 10 years of TERS. ChemPhysChem **11**(7), 1365–1373 (2010). https://doi.org/10.1002/cphc.200900975
12. T. Schmid, L. Opilik, C. Blum, R. Zenobi, Nanoscale chemical imaging using tip-enhanced Raman spectroscopy: a critical review. Angew. Chem. Int. Ed. **52**(23), 5940–5954 (2013). https://doi.org/10.1002/anie.201203849
13. A. Downes, D. Salter, A. Elfick, Heating effects in tip-enhanced optical microscopy. Opt. Express **14**(12), 5216 (2006). https://doi.org/10.1364/oe.14.005216

14. J.P.C. Fernandes, L.D.C. Castro, V.H. Mareau, L.A. Pessan, L. Gonon, New insights on the compatibilization of PA6/ABS blends: a co-localized AFM-Raman study. Polymer **146**, 151–160 (2018). https://doi.org/10.1016/j.polymer.2018.05.012
15. J.R. Yates III, A century of mass spectrometry: from atoms to proteomes. Nat. Methods **8**(8), 633–637 (2011). https://doi.org/10.1038/nmeth.1659
16. I.G. Kolomnikov, A.M. Efremov, T.I. Tikhomirova, N.M. Sorokina, Y.A. Zolotov, Early stages in the history of gas chromatography. J. Chromatogr. A **1537**, 109–117 (2018)
17. K.D. Bartle, P. Myers, History of gas chromatography. TrAC Trends Anal. Chem. **21**(9–10), 547–557 (2002)
18. M.A. Grayson, A history of gas chromatography mass spectrometry (GC/MS), in *The Encyclopedia of Mass Spectrometry* (2016), pp. 152–158. https://doi.org/10.1016/b978-0-08-043848-1.00020-1
19. R.A. Hites, Development of gas chromatographic mass spectrometry. Anal. Chem. **88**(14), 6955–6961 (2016). https://doi.org/10.1021/acs.analchem.6b01628
20. D.O. Sparkman, Z. Penton, F. Kitson, *Gas chromatography and mass spectrometry. A practical guide* (2011)
21. O.D. Sparkman, Z.E. Penton, F.G. Kitson, The fundamentals of GC/MS, in *Gas Chromatography and Mass Spectrometry: A Practical Guide* (2011), pp. 2–13. https://doi.org/10.1016/B978-0-12-373628-4.00001-0
22. D.J. Harvey, Gas chromatography|Gas chromatography/mass spectrometry, in *Encyclopedia of Analytical Science* (2019), pp. 169–179. https://doi.org/10.1016/B978-0-12-409547-2.14103-4
23. J. Sneddon, S. Masuram, J.C. Richert, Gas chromatography-mass spectrometry-basic principles, instrumentation and selected applications for detection of organic compounds. Anal. Lett. **40**(6), 1003–1012 (2007). https://doi.org/10.1080/00032710701300648
24. R. Davarnejad, M. Jafarkhani, *Applications of Gas Chromatography* (2012)
25. H.-J. Hübschmann, Introduction, in *Handbook of GC-MS: Fundamentals and Applications*, 3rd edn. (Wiley-VCH Verlag, Singapore, 2015), pp. 1–6
26. G. Lubes, M. Goodarzi, GC–MS based metabolomics used for the identification of cancer volatile organic compounds as biomarkers. J. Pharm. Biomed. Anal. **147**, 313–322 (2018). https://doi.org/10.1016/j.jpba.2017.07.013
27. T. Majchrzak, W. Wojnowski, M. Lubinska-Szczygeł, A. Różańska, J. Namieśnik, T. Dymerski, PTR-MS and GC-MS as complementary techniques for analysis of volatiles: a tutorial review. Anal. Chim. Acta **1035**, 1–13 (2018). https://doi.org/10.1016/j.aca.2018.06.056
28. N. Feizi, F.S. Hashemi-Nasab, F. Golpelichi, N. Sabouruh, H. Parastar, Recent trends in application of chemometric methods for GC-MS and GC×GC-MS-based metabolomic studies. TrAC Trends Anal. Chem. **138**, 116239 (2021). https://doi.org/10.1016/j.trac.2021.116239
29. J. Yun et al., Use of headspace GC/MS combined with chemometric analysis to identify the geographic origins of black tea. Food Chem. **360**, 130033 (2021). https://doi.org/10.1016/j.foodchem.2021.130033
30. Ö. Aybastıer, C. Demir, Optimization and validation of ultrasensitive GC–MS/MS method to measure oxidatively induced DNA damage products and role of antioxidants in oxidation mechanism. J. Pharm. Biomed. Anal. **200** (2021). https://doi.org/10.1016/j.jpba.2021.114068
31. A. Sharma, G. Bhardwaj, D.S. Canno, Antioxidant potential, GC/MS and headspace GC/MS analysis of essential oils isolated from the roots, stems and aerial parts of Nepeta leucophylla. Biocatal. Agric. Biotechnol. **32**, 101950 (2021). https://doi.org/10.1016/j.bcab.2021.101950
32. G.S. Li et al., Molecular characteristics of the soluble components from three low-rank coals based on the analyses using GC/MS and GC/Q-TOF MS. Fuel **254**(Oct 2018), 115602 (2019). https://doi.org/10.1016/j.fuel.2019.06.010
33. L. Molnárné Guricza, W. Schrader, Optimized asphaltene separation by online coupling of size exclusion chromatography and ultrahigh resolution mass spectrometry. Fuel **215**, 631–637 (2018). https://doi.org/10.1016/j.fuel.2017.11.054

34. A. Viodé et al., Coupling of size-exclusion chromatography with electrospray ionization charge-detection mass spectrometry for the characterization of synthetic polymers of ultra-high molar mass. Rap. Commun. Mass Spectrom. **30**(1), 132–136 (2016). https://doi.org/10.1002/rcm.7426
35. A.M. Striegel, Size-exclusion chromatography, in *Liquid Chromatography: Fundamentals and Instrumentation*, ed. by P.R. Haddad, C. Poole (Elsevier, 2013), pp. 193–220
36. E.S.P. Bouvier, S.M. Koza, Advances in size-exclusion separations of proteins and polymers by UHPLC. TrAC, Trends Anal. Chem. **63**, 85–94 (2014). https://doi.org/10.1016/j.trac.2014.08.002
37. R. Murgasova, D.M. Hercules, Polymer characterization by combining liquid chromatography with MALDI and ESI mass spectrometry, pp. 481–489 (2002). https://doi.org/10.1007/s00216-002-1332-9
38. G. de Koster, P.J. Schoenmakers, History of liquid chromatography—mass spectrometry couplings, in *Hyphenations of Capillary Chromatography with Mass Spectrometry* (Elsevier Inc., 2020), pp. 279–295. https://doi.org/10.1016/B978-0-12-809638-3.00007-7
39. H. Pasch, Hyphenated techniques in liquid chromatography of polymers. Adv. Polym. Sci. **150**, 48–55 (2000)
40. W. Cai et al., Top-down proteomics of large proteins up to 223 kDa enabled by serial size exclusion chromatography strategy. Anal. Chem. **89**(10), 5467–5475 (2017). https://doi.org/10.1021/ACS.ANALCHEM.7B00380
41. S. Zheng, D. Qiu, M. Adams, J. Li, R.V. Mantri, R. Gandhi, Investigating the degradation behaviors of a therapeutic monoclonal antibody associated with pH and buffer species. AAPS Pharm. Sci. Tech. **18**(1), 42–48 (2017). https://doi.org/10.1208/S12249-015-0403-0
42. M. Smoluch, P. Mak, J. Chervet, G. Höhne, J. Silberring, Size-exclusion chromatography performed in capillaries: studies by liquid chromatography–mass spectrometry (Elsevier, 1999)
43. B. Bobály, S. Fleury-Souverain, A. Beck, J.-L. Veuthey, D. Guillarme, S. Fekete, Current possibilities of liquid chromatography for the characterization of antibody-drug conjugates (Elsevier, 2018)
44. T.M. Dan Bach Kristensen, P.F.J. Sloth, M. Ørgaard, Increasing throughput of native SEC-MS in biopharmaceutical development using a tandem UHPLC setup. Thermofisher, 5–8 (2020)
45. A. Boborodea, M. Haex, Ionization mechanism in gel permeation chromatography mass spectrometry (GPC-MS) with atmospheric pressure chemical ionization (APCI) interface. Int. J. Polym. Anal. Charact. **24**(6), 496–503 (2019). https://doi.org/10.1080/1023666X.2019.1603589
46. A. Boborodea, G. Cleaver, Gel permeation chromatography–atmospheric pressure chemical ionization–mass spectrometry for oligomer characterization. Int. J. Polym. Anal. Charact. **22**(2), 180–186 (2017). https://doi.org/10.1080/1023666X.2016.1267482
47. A. Boborodea, A. Brookes, Gel permeation chromatography–atmospheric pressure chemical ionization–mass spectrometry for characterization of polymer additives. Int. J. Polym. Anal. Charact. **22**(3), 210–214 (2017). https://doi.org/10.1080/1023666X.2016.1276740
48. A. Boborodea, S. O'Donohue, Assessing the suitability of a green solvent for gel permeation chromatography–mass spectrometry analysis. Int. J. Polym. Anal. Charact. **22**(4), 305–309 (2017). https://doi.org/10.1080/1023666X.2017.1288348
49. A. Boborodea, G. Cleaver, Fast gel permeation chromatography–mass spectrometry method for polymer and additive analysis. Int. J. Polym. Anal. Charact. **22**(6), 490–496 (2017). https://doi.org/10.1080/1023666X.2017.1330739
50. L. Prokai, D.J. Aaserud, W.J. Simonsick, Microcolumn size-exclusion chromatography coupled with electrospray ionization mass spectrometry. J. Chromatogr. A **835**(1–2), 121–126 (1999). https://doi.org/10.1016/S0021-9673(98)01082-6
51. C. Gottschalk, J. Barthel, U. Aulwum, G. Engelhardt, J. Bauer, K. Meyer, Application of a GPC-LC-MS/MS method for the determination of 31 mycotoxins in edible oils. LC GC North Am. **1**(FEV), 38 (2009)

52. S. Sato, T. Takanohashi, R. Tanaka, Molecular weight calibration of asphaltenes using gel permeation chromatography/mass spectrometry. Energy Fuels **19**(5), 1991–1994 (2005). https://doi.org/10.1021/ef049672r
53. M. Chikri et al., Discrimination of chemical compounds of the oil of Menthasuaveolens (L.) of Eastern Morocco by the GPC-MS and chemometric methods. J. Mater. Environ. Sci. **9**(3), 909–917 (2018). https://doi.org/10.26872/jmes.2018.9.3.101
54. E. Buyukpamukcu, D.M. Goodall, C.E. Hansen, B.J. Keely, S. Kochhar, H. Wille, Characterization of peptides formed during fermentation of cocoa bean. J. Agric. Food Chem. **49**(12), 5822–5827 (2001). https://doi.org/10.1021/jf0104127
55. B. Zhu et al., Simultaneous determination of 131 pesticides in tea by on-line GPC-GC–MS/MS using graphitized multi-walled carbon nanotubes as dispersive solid phase extraction sorbent. Food Chem. **276**, 202–208 (2019)
56. G. Armstrong, L. Kailas, Hyphenated analytical techniques for materials characterisation. Eur. J. Phys. **38**(5), 53001 (2017). https://doi.org/10.1088/1361-6404/aa7e93
57. J.I. Goldstein et al., Introduction, in *Scanning Electron Microscopy and X-Ray Microanalysis* (Springer US, Boston, MA, 1992), pp. 1–19. https://doi.org/10.1007/978-1-4613-0491-3_1
58. P.J. Breton, From microns to nanometres: early landmarks in the science of scanning electron microscope imaging. Scan. Microsc. **13**(1), 1–6 (1999)
59. J.I. Goldstein et al., X-ray spectral measurement: EDS and WDS, in *Scanning Electron Microscopy and X-Ray Microanalysis* (Springer US, Boston, MA, 2003), pp. 297–353. https://doi.org/10.1007/978-1-4615-0215-9_7
60. T. Yurugi, S. Ito, Y. Numata, K. Sykes, *SEM/EDX-Integrated Analysis System SEMEDX Series*, vol. 22. horiba.com (2001)
61. V. Hodoroaba, C. Motzkus, T. Macé, S. Vaslin-Reimann, Performance of high-resolution SEM/EDX systems equipped with transmission mode (TSEM) for imaging and measurement of size and size distribution of spherical. cambridge.org (2014). https://doi.org/10.1017/S1431927614000014
62. S. Rades, V. Hodoroaba, T. Salge, T. Wirth, M.P. Lobera, R.H. Labrador, K. Natte, T. Behnke, T. Gross, W.E.S. Unger, High-resolution imaging with SEM/T-SEM, EDX and SAM as a combined methodical approach for morphological and elemental analyses of single. pubs.rsc.org
63. A. Barhoum, M.L. García-Betancourt, Physicochemical characterization of nanomaterials: size, morphology, optical, magnetic, and electrical properties, in *Emerging Applications of Nanoparticles and Architectural Nanostructures: Current Prospects and Future Trends* (2018), pp. 279–304. https://doi.org/10.1016/B978-0-323-51254-1.00010-5
64. M. Gniadek, A. Dąbrowska, The marine nano- and microplastics characterisation by SEM-EDX: the potential of the method in comparison with various physical and chemical approaches. Mar. Pollut. Bull. **148**, 210–216 (2019). https://doi.org/10.1016/j.marpolbul.2019.07.067
65. D.E. Newbury, N.W.M. Ritchie, Is scanning electron microscopy/energy dispersive X-ray spectrometry (SEM/EDS) quantitative? Scanning **35**(3), 141–168 (2013). https://doi.org/10.1002/sca.21041
66. E. Hayes, D. Cnuts, V. Rots, Integrating SEM-EDS in a sequential residue analysis protocol: benefits and challenges. J. Archaeol. Sci. Rep. **23**, 116–126 (2019). https://doi.org/10.1016/j.jasrep.2018.10.029
67. S. Erić, The application and limitations of the SEM-EDS method in food and textile technologies. Adv. Technol. **6**(2), 5–10 (2017)
68. A. Worobiec, S. Potgieter-Vermaak, A. Brooker, L. Darchuk, E. Stefaniak, R. van Grieken, Interfaced SEM/EDX and micro-Raman spectrometry for characterisation of heterogeneous environmental particles—fundamental and practical challenges. Microchem. J. **94**(1), 65–72 (2010)
69. C. Cardell, I. Guerra, An overview of emerging hyphenated SEM-EDX and Raman spectroscopy systems: applications in life, environmental and materials sciences. TrAC Trends Anal. Chem. **77**, 156–166 (2016). https://doi.org/10.1016/j.trac.2015.12.001

70. D. Tanasi, E. Greco, V. di Tullio, D. Capitani, D. Gullì, E. Ciliberto, ^{1}H-^{1}H NMR 2D-TOCSY, ATR FT-IR and SEM-EDX for the identification of organic residues on sicilian prehistoric pottery. Microchem. J. **135**, 140–147 (2017)
71. R. Rodríguez, V. Correcher, J.M. Gómez-Ros, J.L. Plaza, J. García-Guinea, Cathodoluminescence, SEM and EDX analysis of CaF_2 and Tm_2O_3 pellets for radiation dosimetry applications. Radiat. Phys. Chem. **188**, 109621 (2021)
72. W. Schmidt, Introduction to optical spectroscopy, in *Optical Spectroscopy in Chemistry and Life Sciences* (Wiley-VCH Verlag, 2005), pp. 1–11
73. K. Lubert, K. Kalcher, History of electroanalytical methods. Electroanalysis **22**(17–18), 1937–1946 (2010)
74. A.J. Bard, C.G. Zoski, *Voltammetry Retrospective* (ACS Publications, 2000)
75. L. Dunsch, Recent Advances in in situ multi-spectroelectrochemistry. J. Solid State Electrochem. **15**(7–8), 1631–1646 (2011). https://doi.org/10.1007/s10008-011-1453-1
76. R.J. Mortimer, Spectroelectrochemistry, applications, in *Encyclopedia of Spectroscopy and Spectrometry* (Elsevier, 2017), pp. 160–171. https://doi.org/10.1016/B978-0-12-803224-4.00288-0
77. J. Garoz-Ruiz, J.V. Perales-Rondon, A. Heras, A. Colina, Spectroelectrochemical sensing: current trends and challenges. Electroanalysis **31**(7), 1254–1278 (2019). https://doi.org/10.1002/elan.201900075
78. Y. Zhai, Z. Zhu, S. Zhou, C. Zhu, S. Dong, Recent advances in spectroelectrochemistry. Nanoscale (2018). https://doi.org/10.1039/C7NR07803J
79. E. Vaněčková et al., UV/VIS spectroelectrochemistry with 3D printed electrodes. J. Electroanal. Chem. **857**, 113760 (2020). https://doi.org/10.1016/j.jelechem.2019.113760
80. L. León, J.D. Mozo, Designing spectroelectrochemical cells: a review. TrAC Trends Anal. Chem. **102**, 147–169 (2018)
81. K. Karon, M. Łapkowski, J.C. Dobrowolski, ECD spectroelectrochemistry: a review. Spectrochim. Acta Part A Mol. Biomol. Spectrosc. **250**, 119349 (2021)
82. S. Pluczyk, M. Vasylieva, P. Data, Using cyclic voltammetry, UV–Vis-NIR, and EPR spectroelectrochemistry to analyze organic compounds. JoVE **140**, e56656 (2018). https://doi.org/10.3791/56656
83. C. Fernández-Blanco, Á. Colina, A. Heras, UV/Vis spectroelectrochemistry as a tool for monitoring the fabrication of sensors based on silver nanoparticle modified electrodes. Sensors **13**(5), 5700–5711 (2013)
84. C.P. Byers et al., From tunable core-shell nanoparticles to plasmonic drawbridges: active control of nanoparticle optical properties. Sci. Adv. **1**(11), e1500988 (2015)
85. A.M. Lines, Z. Wang, S.B. Clark, S.A. Bryan, Electrochemistry and spectroelectrochemistry of luminescent europium complexes. Electroanalysis **28**(9), 2109–2117 (2016)
86. P. Zassowski et al., Spectroelectrochemistry of alternating ambipolar copolymers of 4, 4′-and 2, 2′-bipyridine isomers and quaterthiophene. Electrochim. Acta **231**, 437–452 (2017)
87. F.S. Rocha, A.J. Gomes, C.N. Lunardi, S. Kaliaguine, G.S. Patience, Experimental methods in chemical engineering: ultraviolet visible spectroscopy-UV–Vis. Can. J. Chem. Eng. **96**(12), 2512–2517 (2018). https://doi.org/10.1002/cjce.23344
88. J. Sworakowski, How accurate are energies of HOMO and LUMO levels in small-molecule organic semiconductors determined from cyclic voltammetry or optical spectroscopy? Synth. Met. **235**, 125–130 (2018)
89. M. Zhao, X. Tang, H. Zhang, C. Gu, Y. Ma, Characterization of complicated electropolymerization using UV–vis spectroelectrochemistry and an electrochemical quartz-crystal microbalance with dissipation: a case study of tricarbazole derivatives. Electrochem. Commun. **123**, 106913 (2021)
90. E. Laborda, J. García-Martínez, A. Molina, Spectroelectrochemistry for the study of reversible electrode reactions with complex stoichiometries. Electrochem. Commun. **123**, 106915 (2021)
91. A.C. Gonzalez-Baro, D. Izquierdo, A. Heras, A. Colina, UV/Vis spectroelectrochemistry of o-vanillin: study of the antioxidant properties. J. Electroanal. Chem. **859**, 113844 (2020)

92. S. Benaglia, V.G. Gisbert, A.P. Perrino, C.A. Amo, R. Garcia, Fast and high-resolution mapping of elastic properties of biomolecules and polymers with bimodal AFM. Nat. Protoc. **13**(12), 2890–2907 (2018). https://doi.org/10.1038/s41596-018-0070-1
93. C.S. Goldsbury, S. Scheuring, L. Kreplak, Introduction to atomic force microscopy (AFM) in biology. Curr. Protoc. Protein Sci. **58**, 1–19 (2009). https://doi.org/10.1002/0471140864.ps1707s58
94. B. Contents, *AFM—Atomic Force Microscope Description Lab Manual*, pp. 1–55
95. JPK Instruments AG, *The NanoWizard® AFM Handbook* (2005), p. 40
96. N.E. Leadbeater, J.R. Schmink, Use of Raman spectroscopy as a tool for in situ monitoring of microwave-promoted reactions. Nat. Protoc. **3**(1), 1–7 (2008). https://doi.org/10.1038/nprot.2007.453
97. N. Gierlinger, T. Keplinger, M. Harrington, Imaging of plant cell walls by confocal Raman microscopy. Nat. Protoc. **7**(9), 1694–1708 (2012). https://doi.org/10.1038/nprot.2012.092
98. P. Sunitha, K. Lavanya, A.A. Kumar, L.K. Kanthal, B. Kausik, Int. Res. J. Pharm. **3**(8), 11–13 (2012)
99. K. Jovic, T. Nitsche, C. Lang, J.P. Blinco, K. de Bruycker, C. Barner-Kowollik, Hyphenation of size-exclusion chromatography to mass spectrometry for precision polymer analysis—a tutorial review. Polym. Chem. **10**(24), 3241–3256 (2019). https://doi.org/10.1039/C9PY00370C
100. J.I. Goldstein et al., The SEM and its modes of operation, in *Scanning Electron Microscopy and X-Ray Microanalysis* (Springer US, Boston, MA, 2003), p. 21. https://doi.org/10.1007/978-1-4615-0215-9
101. K. Green, Cyclic voltammetry (CV), in *JoVE Science Education Database* (2021). https://www.jove.com/es/v/5502/cyclic-voltammetry-cv. Accessed 21 Nov 2021
102. D. Rood, B. Bidlingmeyer, Gas chromatography problem solving and troubleshooting. J. Chromatogr. Sci. **42**(1), 54–56 (2004)
103. M. Hesse, Mass spectrometry. A textbook. By Jürgen H. Gross. Angew. Chem. Int. Ed. **43**(35), 4552 (2004). https://doi.org/10.1002/anie.200385205
104. M.A. Ferry, Diagnosing and resolving mass spec problems, in *Environmental Consulting and Supplies* (2017)
105. Waters Corporation, Troubleshooting and diagnostics tips and tricks, in *UPLC Troubleshooting* (2013) [Online]. Available: https://www.waters.com/webassets/cms/library/docs/local_seminar_presentations/sp_sem_uplc_troubleshooting_inst_2014.pdf
106. P. Echlin, Sample stabilization for imagining in the SEM, in *Handbook of Sample Preparation for Scanning Electron Microscopy and X-Ray Microanalysis* (2009), pp. 137–183. https://doi.org/10.1007/978-0-387-85731-2
107. P. Echlin, Sample stabilization to preserve chemical identity, in *Handbook of Sample Preparation for Scanning Electron Microscopy and X-Ray Microanalysis* (Springer Science, 2009), pp. 185–233. https://doi.org/10.1007/978-0-387-85731-2
108. P. Echlin, *Handbook of Sample Preparation for Scanning Electron Microscopy and X-Ray Microanalysis*. Springer US, Boston, MA (2009). https://doi.org/10.1007/978-0-387-85731-2
109. W. Kaim, J. Fiedler, Spectroelectrochemistry: the best of two worlds. Chem. Soc. Rev. **38**(12), 3373–3382 (2009). https://doi.org/10.1039/b504286k
110. N. Elgrishi, K.J. Rountree, B.D. McCarthy, E.S. Rountree, T.T. Eisenhart, J.L. Dempsey, A practical beginner's guide to cyclic voltammetry. J. Chem. Educ. **95**(2), 197–206 (2018)